Networked Sensing Systems

Scrivener Publishing
100 Cummings Center, Suite 541J
Beverly, MA 01915-6106

Publishers at Scrivener
Martin Scrivener (martin@scrivenerpublishing.com)
Phillip Carmical (pcarmical@scrivenerpublishing.com)

Networked Sensing Systems

Edited by

Rajesh Kumar Dhanaraj
Malathy Sathyamoorthy
Balasubramaniam S
and
Seifedine Kadry

Scrivener Publishing

WILEY

This edition first published 2025 by John Wiley & Sons, Inc., 111 River Street, Hoboken, NJ 07030, USA and Scrivener Publishing LLC, 100 Cummings Center, Suite 541J, Beverly, MA 01915, USA
© 2025 Scrivener Publishing LLC
For more information about Scrivener publications please visit www.scrivenerpublishing.com.

All rights reserved. No part of this publication may be reproduced, stored in a retrieval system, or transmitted, in any form or by any means, electronic, mechanical, photocopying, recording, or otherwise, except as permitted by law. Advice on how to obtain permission to reuse material from this title is available at http://www.wiley.com/go/permissions.

Wiley Global Headquarters
111 River Street, Hoboken, NJ 07030, USA

For details of our global editorial offices, customer services, and more information about Wiley products visit us at www.wiley.com.

Limit of Liability/Disclaimer of Warranty
While the publisher and authors have used their best efforts in preparing this work, they make no representations or warranties with respect to the accuracy or completeness of the contents of this work and specifically disclaim all warranties, including without limitation any implied warranties of merchantability or fitness for a particular purpose. No warranty may be created or extended by sales representatives, written sales materials, or promotional statements for this work. The fact that an organization, website, or product is referred to in this work as a citation and/or potential source of further information does not mean that the publisher and authors endorse the information or services the organization, website, or product may provide or recommendations it may make. This work is sold with the understanding that the publisher is not engaged in rendering professional services. The advice and strategies contained herein may not be suitable for your situation. You should consult with a specialist where appropriate. Neither the publisher nor authors shall be liable for any loss of profit or any other commercial damages, including but not limited to special, incidental, consequential, or other damages. Further, readers should be aware that websites listed in this work may have changed or disappeared between when this work was written and when it is read.

Library of Congress Cataloging-in-Publication Data

ISBN 9781394310869

Front cover images supplied by Pixabay.com
Cover design by Russell Richardson

Set in size of 11pt and Minion Pro by Manila Typesetting Company, Makati, Philippines

Printed in the USA

10 9 8 7 6 5 4 3 2 1

Contents

Preface xvii

1 **Introduction to Network Sensing Systems in Society 5.0: Issues and Challenges** 1
Ankit Kumar, Anurag Kumar Kanojiya and Subitha D.
1.1 What is Society 5.0? 2
    1.1.1 Advancements in Society 5.0 Over Society 4.0 3
    1.1.2 Integration and Interconnectivity 3
    1.1.3 Data Utilization and Analysis 4
    1.1.4 Personalization and Customization 4
    1.1.5 Sustainability and Ethical Considerations 4
    1.1.6 Human-Centric Design and Empowerment 4
1.2 Network Sensing Systems in Society 5.0 5
1.3 Issues and Challenges 6
    1.3.1 Data Privacy and Security 6
    1.3.2 Importance of Privacy and Security 7
1.4 Encryption and Decryption Techniques: Safeguarding Data Integrity 8
    1.4.1 Decryption Technology 9
    1.4.2 Challenges and Decisions 10
    1.4.3 Interoperability Challenges in Society 5.0: A Tripping Block on the Road to a Hyperconnected Future 10
1.5 Understanding Interoperability on Society 5.0 10
    1.5.1 The Smart City Dilemma: A Case Study in Interoperability Woes 11
    1.5.2 Ensuring Integration and Data Exchange 11
    1.5.3 Standardization Challenges and Solutions 12
    1.5.4 Heterogeneity Challenges and Solutions 12
1.6 Importance of Robust Communication and Power Plans 12
    1.6.1 The Requirement and Imperative of Flexible and Dependable Infrastructure 13

      1.6.2   Infrastructure Resilience   14
      1.6.3   Key Characteristics of Sturdy Infrastructure   14
      1.6.4   Infrastructure Resilience's Importance in Society 5.0   15
      1.6.5   Infrastructure Resilience Techniques   16
      1.6.6   Reduced Operational Costs   16
      1.6.7   Safeguarding the Digital Age: Security and Privacy in Infrastructure   17
  1.7 Environmental Effects and Energy Efficiency   17
      1.7.1   Supporting Sustainability through Energy-Efficient Approach   18
      1.7.2   Building Automation Systems: Transforming Buildings into Energy-Conscious Entities   19
      1.7.3   Connected Appliances: Transforming Everyday Devices into Energy-Conscious Partners   20
      1.7.4   Benefits of Connected Appliances   20
      1.7.5   Challenges and Considerations   21
      1.7.6   Energy-Efficient Manufacturing: Optimizing Industries for a Sustainable Future in Society 5.0   21
      1.7.7   Benefits of Energy-Efficient Manufacturing   22
  1.8 Utilizing Renewable Energy Sources   22
      1.8.1   Methods of Harnessing Cleaner Energy in Society 5.0   23
  1.9 Conclusion   24
      1.9.1   Future Directions   24
      References   26

2 Remote and Urban Environmental Area Sensing, Connectivity Issues, and Solutions Based on Emerging Technologies   31
Abinaya M., Vadivu G., Balasubramaniam S and Sundaravadivazhagan B.
  2.1 Introduction   32
      2.1.1   Urban Environment Remote Sensing Overview   32
      2.1.2   Smart Factory   32
      2.1.3   Benefits of Remote Sensing in Cities   33
  2.2 Connectivity Challenges in Urban Remote Sensing   33
      2.2.1   Conventional Remote Sensing Systems' Technological Limitations   33
      2.2.2   Logistical Obstacles in the Integration and Transmission of Data   34
      2.2.3   Problems Affect Data Analysis and Quality   34
      2.2.4   Cutting Edge Technologies to Handle Connectivity Issues   35

2.3	Artificial Intelligence for Enhancing Data Processing and Analysis		35
	2.3.1	Artificial Intelligence for Data Insights	35
	2.3.2	Real-Time Monitoring With IoT Sensors	36
	2.3.3	Advanced Imagery and Data Acquisition	36
	2.3.4	Secure Data Management With Blockchain	36
	2.3.5	Immersive Data Visualization With AR/VR	37
2.4	Case Study		37
	2.4.1	Monitoring Urban Air Quality With IoT Sensors	37
	2.4.2	Analyzing Satellite Images for Urban Development and Planning	38
	2.4.3	Drone Aerial Vehicle-Based Monitoring for Environmental Control	40
	2.4.4	Augmented and Virtual Reality Uses in Planning and Monitoring Urban Environments	41
	2.4.5	Combining Emerging Technologies and Remote Sensing	43
2.5	Frameworks for Integrating Multiple Data Sources		45
	2.5.1	Platforms for Collaborative Work and Data Sharing	45
	2.5.2	Regulatory Aspects and Policy Implications	48
	2.5.3	Prospective Pathways and Difficulties	50
2.6	The Possible Effects of Next-Generation Connectivity and 5G		52
	2.6.1	Privacy and Ethical Issues With Urban Remote Sensing	52
	2.6.2	New Technologies' Scalability and Affordability	54
	2.6.3	AR and VR's Place in the Future of Urban Environment Management	56
2.7	Conclusion		57
	2.7.1	Future Gap	58
	References		58
3	**Efficient Network and Communication Technologies for Smart and Sustainable Society 5.0**		**63**
	P. Kanaga Priya, R. Sivaranjani, Malathy Sathyamoorthy and Rajesh Kumar Dhanaraj		
3.1	Introduction		64
	3.1.1	Evolution of Societal Paradigms	64
	3.1.2	Transition from Industry 4.0 to Society 5.0	68
	3.1.3	Definition and Key Characteristics of Society 5.0	70
	3.1.4	Importance of Efficient Network and Communication Technologies	72
	3.1.5	Critical Technologies Shaping Smart and Sustainable Society 5.0	72

3.2	Literature Survey		73
3.3	Internet of Things for Smart Connectivity		76
	3.3.1	IoT Applications in Smart Cities, Agriculture, Healthcare, and Industry	76
	3.3.2	Challenges in IoT Implementation	79
	3.3.3	Opportunities in IoT Implementation	79
3.4	Next-Generation Cutting Edge Communication Technologies: 5G and Beyond		80
	3.4.1	Evolution of Cellular Communication Standards	80
	3.4.2	5G Networks' Features and Capabilities	81
	3.4.3	Emerging Trends and Technologies Beyond—5G (B5G) and 6G	82
3.5	Edge Computing: Decentralized Processing for Low Latency		83
	3.5.1	Understanding Edge Computing Architecture	84
	3.5.2	Edge Computing's Benefits for Analytics and Data Processing	85
	3.5.3	Use Case and Deployment Scenario	86
3.6	Blockchain Technology: Securing Data Integrity and Trust		87
	3.6.1	Fundamentals of Blockchain Technology	87
	3.6.2	Applications in Secure Data Sharing, Supply Chain Management, and Decentralized Finance	89
	3.6.3	Challenges and Potential Solutions	89
3.7	Artificial Intelligence in Network Optimization		90
	3.7.1	Role of AI and Machine Learning in Network Management	91
	3.7.2	AI-Driven Approaches for Resource Allocation and Optimization	91
3.8	Energy-Efficient Networking for Sustainability in Society 5.0		92
	3.8.1	Strategies for Reducing Energy Consumption in Communication Networks	94
	3.8.2	Green Networking Technologies and Practices in Society 5.0	94
3.9	Challenges and Opportunities in Implementing Efficient Network Technologies		95
3.10	Future Directions and Recommendations		96
	3.10.1	Research Priorities for Advancing Network and Communication Technologies	97
	3.10.2	Policy Recommendations for Fostering Sustainable Development in Society 5.0	97
	3.10.3	Collaborative Efforts Toward Achieving a Smart and Sustainable Society 5.0	98

		3.11 Conclusion	98
		References	99
4	**Advanced Techniques for Human-Centric Sensing in Environmental Monitoring**		**101**
	S. Aathilakshmi, Visali C., T. Manikandan and Seifedine Kadry		
	4.1	Introduction	102
	4.2	A Basic Human-Centric Sensing Mechanism	106
	4.3	Types of Advanced HCS Environmental Monitoring System	110
		4.3.1 Multispectral Sensors	110
		4.3.2 Thermal Sensors	111
		4.3.3 LiDAR	111
		4.3.4 Hyperspectral Sensors	111
		4.3.5 Photogrammetry Sensors	112
	4.4	Applications in Environmental Monitoring	113
		4.4.1 Smart Sensor	113
		4.4.2 Wireless Network Technology	114
		4.4.3 Passive Sensing Technology	115
		4.4.4 Activity Recognition Technology	116
		4.4.5 Gesture Recognition Technology	116
	4.5	Conclusion and Future Prospects	117
		References	118
5	**Energy-Aware System for Dynamic Workflow Scheduling in Cloud Data Centers: A Genetic Algorithm with DQN Approach**		**121**
	Hariharan B., Anupama C.G., Ratna Kumari Neerukonda and Rajesh Kumar Dhanaraj		
	5.1	Introduction	122
	5.2	Related Works	124
	5.3	Dynamic Workflow Scheduling System	127
		5.3.1 System Architecture	127
		5.3.2 Genetic Algorithm for Dynamic Workflow Scheduling	128
		5.3.3 Deep Q-Learning for Dynamic Workflow Scheduling	130
		5.3.4 Energy Consumption for Dynamic Workflow Scheduling	132
	5.4	Problem Formulation and Proposed System Architecture	133
		5.4.1 Hybrid Approach	133
		5.4.2 Implementation of GA	135
	5.5	Simulation Set-Up and Experimental Results	136

		5.5.1	Makespan Computation	137
		5.5.2	Energy Consumption Calculation	141
	5.6	Conclusion		142
		References		142

6 Efficient Load Balancing and Resource Allocation in Networked Sensing Systems—An Algorithmic Study   145
Lalitha Krishnasamy, Divya Vetriveeran, Rakoth Kandan Sambandam and Jenefa J.

6.1	Introduction to the Networked Sensing Systems		146
6.2	Understanding the Load Balancing Challenges		147
	6.2.1	Types of Load Balancing	148
	6.2.2	Load Balancing Technologies	149
6.3	Importance of Efficient Resource Allocation		150
6.4	Overview of Existing Approaches		151
	6.4.1	Probabilistic Clustering	151
	6.4.2	Non-Probability Clustering	152
6.5	Artificial Intelligence for Resource Handing		155
	6.5.1	Naïve Bayes	157
	6.5.2	Multi-Class SVM	157
	6.5.3	AdaBoost	158
	6.5.4	Clustering	159
	6.5.5	Learning-Based Resource Allocation (LB-RA)	159
	6.5.6	Neural Networks for Load Balancing and Resource Allocation	160
	6.5.7	Reinforcement Learning	162
6.6	Real-World Applications		163
6.7	Performance Metrics		165
	6.7.1	Throughput	165
	6.7.2	Reaction Time	165
	6.7.3	Latency	165
	6.7.4	Versatility	165
	6.7.5	Asset Use	165
	6.7.6	Optimization	166
	6.7.7	Fairness	166
	6.7.8	MTTF	166
	6.7.9	Cost Effectiveness	166
	6.7.10	Adaptability	166
6.8	Research Directions		166
	6.8.1	Edge Computing	167
	6.8.2	ML and AI	167

		6.8.3 Autonomous Resource Management	167
		6.8.4 Containerization and Orchestration	167
		6.8.5 Hybrid and Multi-Cloud Environments	167
		6.8.6 Energy-Efficient Computing	168
		6.8.7 Quantum Figuring	168
		6.8.8 Asset the Executives	168
		6.8.9 Security and Protection Contemplations	168
		6.8.10 Cross-Domain Resource Allocation	168
	6.9	Conclusion and Future Work	169
		Acknowledgments	169
		References	170
7	**Sustainable Cities and Communities: Role of Network Sensing System in Action**		**173**
	Hitesh Mohapatra, Soumya Ranjan Mishra,		
	Amiya Kumar Rath and Manjur Kolhar		
	7.1	Introduction	174
	7.2	Literature Review	177
	7.3	Proposed Study	181
		7.3.1 Star Topology	181
		7.3.2 Mesh Topology	182
		7.3.3 Tree Topology	183
		7.3.4 Clustered Topology	183
	7.4	Performance Analysis	185
	7.5	Mapping of Topology with Smart City's Applications	189
		7.5.1 Mapping of Star Topology with Smart Parking Application	189
		7.5.2 Mapping of Mesh Topology with Smart Grid Application	191
		7.5.3 Mapping of Tree Topology with Smart Education Model	192
		7.5.4 Mapping of Cluster Topology with Smart Health Care Model	194
	7.6	Conclusion	195
		References	196
8	**Air Pollution Monitoring and Control Via Network Sensing Systems in Smart Cities**		**199**
	S. Sharmila Devi		
	8.1	Introduction	199
	8.2	Related Works	201
	8.3	Air Quality System	203

8.4		Air Quality Monitoring Techniques	204
8.5		Conventional Air Pollution Monitoring	205
	8.5.1	Manual Measurement and Evaluation of Air Quality	205
	8.5.2	Automated Continuous Monitoring Devices	206
	8.5.3	Monitoring Air Quality with Sensing Technology	207
8.6		Wireless Sensor Network for Air Monitoring	209
	8.6.1	Wireless Sensor Networks	209
	8.6.2	WSN Network Topologies	209
	8.6.3	Zigbee Standard	211
8.7		Architecture of Wireless Sensor Networks	212
	8.7.1	Fire and Flood Detection	214
	8.7.2	Biocomplexity	214
	8.7.3	Habitat Monitoring	215
	8.7.4	Factors Influencing the Efficacy of Inexpensive Sensors in the Monitoring of Air Pollution	215
8.8		WSN-Based Air Pollution Monitoring in Smart Cities	216
8.9		Conclusion	221
		References	221

9 Interconnected Healthcare 5.0 Ecosystems: Enhancing Patient Care Using Sensor Networks 225
Ashwini A., Kavitha V. and Balasubramaniam S

9.1		Introduction to Healthcare 5.0	226
	9.1.1	Evolution from Healthcare 4.0 to Healthcare 5.0	227
9.2		Real-Time Monitoring Using Sensor Networks	229
9.3		Advancements in Remote Patient Monitoring	231
	9.3.1	Challenges in Healthcare 4.0	233
9.4		Early Disease Detection Through Sensor Networks	235
9.5		Leveraging Multisensor Data for Comprehensive Health Insights	237
9.6		Security Measures for Protected Health Information	240
9.7		Overcoming Infrastructure and Connectivity Barriers	241
9.8		Improving Treatment Plans Through Sensor-Generated Insights	242
9.9		Conclusion	243
		References	244

10 Farming 4.0: Cultivating the Future with Internet of Things Empowered on Smart Agriculture Solutions 247
Ashwini A., S.R. Sriram, J. Manoj Prabhakar and Seifedine Kadry

- 10.1 Introduction to Smart Agriculture and IoT Integration 248
  - 10.1.1 Evolution of Agriculture: From Traditional to Smart Farming 249
- 10.2 IoT Sensor Networks in Farming 250
  - 10.2.1 IoT Sensors and Their Applications 251
- 10.3 Smart Pest and Disease Control in Crop Production 253
  - 10.3.1 Meticulous Fertilization and Nutrition Control 253
  - 10.3.2 Accurate Irrigation Techniques and Water Administration 255
- 10.4 Automation and Robotics in Agriculture 257
  - 10.4.1 Agricultural Operations Using Automatic Systems 259
  - 10.4.2 AI in Farm Automation 260
- 10.5 Cloud Computing for Agricultural Data Management 262
- 10.6 Big Data Analytics for Predictive Farming 264
- 10.7 Sustainable Practices with IoT in Agriculture 266
- 10.8 The Future Landscape of Farming 4.0 267
- 10.9 Conclusion 268
- References 268

11 Public Safety Management in Smart Society 5.0: A Blockchain-Based Approach 273
P.N. Senthil Prakash, S. Karthic and M. Saravanan

- 11.1 Introduction 274
- 11.2 Security Challenges in Society 5.0 278
- 11.3 Blockchain in Society 5.0 279
  - 11.3.1 Blockchain for Refinery Industry 281
  - 11.3.2 Blockchain in Identity Management 281
  - 11.3.3 Blockchain and Its Impact in Healthcare 283
  - 11.3.4 Blockchain for Supply Chain Management 285
  - 11.3.5 Blockchain in Asset Management 285
  - 11.3.6 Blockchain in Copyright Management 287
- 11.4 Conclusion 289
- References 290

12 Virtualization of Smart Society 5.0 Using Artificial Intelligence and Virtual Reality   297
Sakthivel Sankaran, M. Arun and R. Kottaimalai

- 12.1 Introduction to Smart Society 5.0   298
  - 12.1.1 Smart Society 5.0 and Its Key Characteristics   298
  - 12.1.2 Evolution from Previous Smart Society Models   299
- 12.2 Foundations of Virtual Reality   301
  - 12.2.1 Brief History and Development of Virtual Reality   301
  - 12.2.2 Key Components and Technologies in VR   302
  - 12.2.3 VR Hardware and Software Ecosystems   303
- 12.3 Artificial Intelligence in Smart Societies   304
  - 12.3.1 Overview of AI Technologies Shaping Smart Societies   304
  - 12.3.2 Role of AI in Data Analytics, Automation, and Decision Making   306
  - 12.3.3 AI-Driven Applications in Healthcare, Transportation, and Education   307
- 12.4 Integration of AI and VR   311
  - 12.4.1 How AI and VR Technologies Complement Each Other   311
  - 12.4.2 Examples of AI-Enhanced Virtual Reality Applications   312
  - 12.4.3 Possibilities and Obstacles When Fusing AI with VR   312
- 12.5 AI and VR in Education   314
  - 12.5.1 Virtual Classrooms and Immersive Learning Experiences   314
  - 12.5.2 AI-Enabled Adaptive Learning Systems   314
  - 12.5.3 Skill Development and Training Using VR and AI   315
- 12.6 Smart Society 5.0 Healthcare Innovations   315
  - 12.6.1 Virtual Healthcare Consultations and Simulations   315
  - 12.6.2 AI-Driven Diagnostics and Treatment Planning   316
  - 12.6.3 VR and AI-Based Treatments and Rehab   316
- 12.7 Challenges and Future Directions   316
  - 12.7.1 Current Obstacles to Integrate VR and AI in Smart Communities   316
  - 12.7.2 Prospective Developments and Emerging Patterns in AI and VR in Smart Societies   317
  - 12.7.3 Consider the Role of Emerging Technologies   317
- 12.8 Conclusion   318

	12.8.1	Summary of AI and VR Technologies in Smart Societies	318
	12.8.2	Vision for the Future of Smart Societies with AI and VR	318
	References		319

13 Battery Power Management Schemes Integrated with Industrial IoT for Sustainable Industry Development 323
D. Karthikeyan, A. Geetha, K. Deepa and Malathy Sathyamoorthy

13.1	Introduction		324
13.2	Current Battery Technologies		325
	13.2.1	Metal–Air Battery	326
	13.2.2	Lithium–Sulfur Battery	327
	13.2.3	Batteries Beyond Lithium	328
13.3	Battery Energy Storage and Management		328
13.4	IoT and Cloud Computing Technology in BMS		334
13.5	Sustainable Developments via BMS		337
	13.5.1	SDG8	337
	13.5.2	SDG9	346
	13.5.3	SDG12	346
	13.5.4	SDG13	348
13.6	Conclusion		348
	References		349

14 Trends, Advances, and Applications of Network Sensing Systems 351
Ashwini A., Shamini G.I. and Balasubramaniam S

14.1	Introduction to Network Sensing Systems		352
	14.1.1	Relevance in Different Sectors	353
14.2	Real-Time Trends in Sensor Technology		355
	14.2.1	Advanced Sensing Modalities	355
	14.2.2	Power-Efficient Designs	356
14.3	Advancements in Data Analytics		357
	14.3.1	Big Data Analytics for Sensor-Generated Data	359
14.4	Applications in Healthcare		361
	14.4.1	Remote Patient Monitoring	361
	14.4.2	Smart Healthcare and Medical Establishments	362
	14.4.3	Fall Detection and Old Care	362
14.5	Natural Disaster Detection with Response		363
	14.5.1	Early Detection Systems	363

		14.5.2	Satellite Imagery and Tracking	363
		14.5.3	Resilient Communications Networks	363
		14.5.4	Predictive Analysis and Modeling	364
	14.6	Agricultural Sensing Systems		365
		14.6.1	Crop Monitoring and Management	365
		14.6.2	Soil Sensing and Precision Agriculture	365
		14.6.3	Weather Monitoring and Forecasting	366
		14.6.4	Livestock Monitoring and Management	366
		14.6.5	Data Analytics and Decision Support System	366
		14.6.6	Remote Monitoring and Automation	366
	14.7	Intelligent Transportation Systems		367
	14.8	Smart City Applications		368
	14.9	Challenges		369
	14.10	Conclusion		370
		References		370

About the Editors   375

Index   377

Preface

With today's improvements in wireless and mobile connectivity, Internet of Things (IoT) sensor technologies, and digital innovation, sustainability principles have started to reinforce one another. To switch to more resource-efficient solutions, use resources responsibly, and streamline operations, businesses must embrace digital transformation. Energy management, air pollution monitoring, fleet management, water management, and agriculture are a few examples of potential actuation sectors. Simultaneously, the expansion of IoT deployments and their integration into the contexts of 5G and upcoming 6G mobile networking necessitate that the solutions themselves be green and sustainable, incorporating, for instance, the use of energy- and environmentally-aware technical solutions for communications.

By offering previously unattainable solutions, networking can contribute to a more sustainable society by enabling the collection of data from new and heterogeneous sources in unique ways and from novel sources using novel technology. In addition, the networking-based solution itself needs to be sustainable or environmentally friendly. For instance, changing the network architecture and moving network equipment to key locations can reduce wasteful energy use. These goals drive the search for solutions, which range from "better" and novel sensing objects that need to be energy-efficient using mobile sensing devices.

The goal of "Networked Sensing Systems" is to present and highlight the most recent developments in sustainable networked sensing systems in a variety of contexts with the common goal of enhancing human well-being and halting climate change. Regardless of their area of expertise, the objective is to offer workable solutions that meet the major problems and difficulties in building a sustainable smart society 5.0. This book will serve as a potential platform to discuss networked sensing systems for a sustainable society, namely systems and applications based on mobile computers and wireless networks, while taking into account multidisciplinary approaches that emphasise the human element in resolving these difficulties.

1

Introduction to Network Sensing Systems in Society 5.0: Issues and Challenges

Ankit Kumar, Anurag Kumar Kanojiya and Subitha D.*

School of Computer Science and Engineering, Vellore Institute of Technology, Chennai, India

Abstract

Network Sensing System in Society 5.0 provides unprecedented connectivity and data-driven solutions to numerous societal problems. But they also raise many questions and problems that need to be resolved to ensure they are used effectively and fairly. This content explores the key issues and challenges of the Society 5.0 community. Since sensors collect and transmit a lot of data continuously, data privacy and security become an important issue. Strong protection and encryption are required to prevent data from leakage, unauthorized access, and misuse. The integration and data sharing of many sensors and protocols depend on the interoperability and standardization of the sensor. So, the systems can be made more scalable and efficient using different structures and procedures. Transparency, impartiality, and fairness are among some of the ethical concerns under data analysis and algorithmic decision making. So, it is necessary to take necessary actions to ensure fair results and get rid of biases to maintain trust and prevent bad outcomes. Network Detection System 5.0 focuses on reliable power plans and strong communication methods; therefore, infrastructure flexibility and dependability are very important. Such vulnerabilities in the systems might have the potential to cause disturbances and may interfere with vital activities. Hence, it is necessary to invest in redundancy and resilience in infrastructure. Energy and safety are important issues due to the environmental impact of sensors and data processing. Energy-efficient solutions and renewable energy sources can reduce these impacts and promote sustainability. The development of technology, policy, and business management creates uncertainty about responsibility and control. Guidelines and policies should be developed to address legal, ethical, and social issues while promoting innovation and community service. Public

*Corresponding author: subitha.d@vit.ac.in

Rajesh Kumar Dhanaraj, Malathy Sathyamoorthy, Balasubramaniam S and Seifedine Kadry (eds.) *Networked Sensing Systems*, (1–30) © 2025 Scrivener Publishing LLC trust and acceptance are key to implementing Network Detection System 5.0. Transparent communication, community engagement, and meaningful feedback are critical to solving privacy, security, and consequence issues.

***Keywords*:** Interoperability, standardization, transparency, energy efficiency, renewable smart grid, robust communication

1.1 What is Society 5.0?

Society 5.0 represents an enormous evolution in how data are accumulated, processed, and implemented to enhance various components of human existence [1–3]. While the idea of making use of statistics to improve efficiency and comfort is not always new, Society 5.0 distinguishes itself by means of expanding the scope of application throughout society in an incorporated manner. Unlike previous iterations, wherein computerized structures operated within specific domain names, like temperature control or transportation, Society 5.0 envisions a complete technique to optimizing all aspects of lifestyles.

The key distinction lies in the basic integration of structures to ensure happiness and safety in various sectors as well as strength, transportation, healthcare, buying, training, employment, amusement, and so forth. In Society 5.0, the focus shifts from isolated solutions to interconnected structures that work together seamlessly. This community allows the gathering of numerous certain real-world facts that are then processed by means of AI and other state-of the-art IT structures. What sets Society 5.0 apart is the ability to use these vast amounts of data to directly shape human behavior and actions resulting in a more fulfilling and richer lifestyle Unlike previous models where data guide only individual policies, Society 5.0 influences social behavior at a broad scale. Essentially, Society 5.0 creates an iterative cycle of constantly collecting, analyzing, and transforming data into meaningful insights that improve tangible real-world results This cycle works across society and ensures that the benefits of data-driven decision making extend to all aspects of human life. As we dive deeper into the discussion, it is clear that Citizenship 5.0 represents a significant step forward in social improvement and embodies a future where data-driven insights pave the way for a happier, more comfortable society promising sustainability.

1.1.1 Advancements in Society 5.0 Over Society 4.0

Society 5.0 represents a significant advancement over Society 4.0 addressing various issues and challenges while offering significant improvements in several key areas. Let us see how Society 5.0 is superior to its predecessor and overcomes the limitations faced in Society 4.0 depicted in Figure 1.1.

1.1.2 Integration and Interconnectivity

Using Cyber-Physical Systems and the Internet of Things, Society 4.0 introduced the idea of connection between machines. However, those systems were mostly functioning in isolation, which resulted in dispersed data and ineffective system-to-device connection. Also, it had a vast communication gap between the interconnected devices. Society 5.0, forces integration and relationships between different entities. It made it easier for systems and domains to work jointly and communicate with one another promoting complete optimization and combined interactions.

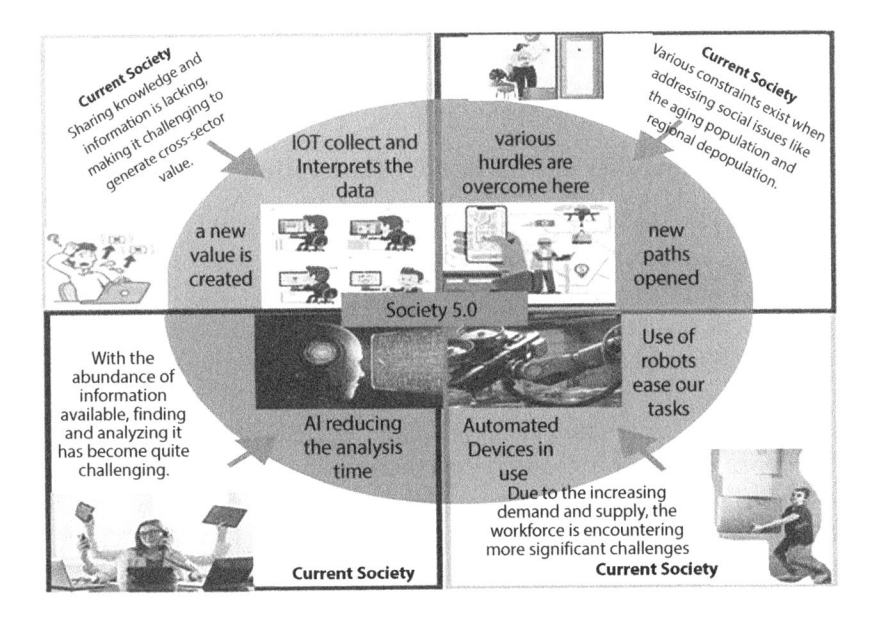

Figure 1.1 Interaction of society 5.0 and the current society.

1.1.3 Data Utilization and Analysis

Although Society 4.0 placed a force or intensity of expression that gives impressiveness or importance to something on connection and data collecting, it had created some trouble with the in-efficient use and analysis of the large volumes of data. The extraction of significant insights was hampered by no longer used or useful analytical methodologies and limited processing capabilities. Society 5.0 leverages advanced technologies like Artificial Intelligence (AI) and Big Data analytics to unlock the full potential of data. It employs sophisticated algorithms to analyze complex datasets in real time extracting actionable insights to drive decision making and innovation.

1.1.4 Personalization and Customization

Based on the user's preferences and actions, Society 4.0 introduced customized experiences and services. However, these efforts have largely failed due to a lack of awareness of unique needs and preferences. By exploiting complicated AI algorithms to analyze a wealth of information about unique interests, behaviors and situations, Society 5.0 elevates personalization to a new level. This can provide highly customized and flexible experiences in various sectors such as retail, healthcare, and education.

1.1.5 Sustainability and Ethical Considerations

Society 4.0 [4] largely ignores the problem of sustainability and ethics by prioritizing economic expansion and technological development. Rapid technological progress has worsened social inequality and destroyed the environment. Society 5.0 involves a balanced and holistic policy approach, with a strong emphasis on social responsibility, ethics, and the environment. It seeks to use technology for the greater good in addition to promoting equity, environmental protection, and ethical decision making at all levels of life with greater integration, improved data management, customized experiences, navigation, ethical considerations, and human-centered design. Society 5.0 has a huge step up from Society 4.0.

1.1.6 Human-Centric Design and Empowerment

Automation and digitization [5, 6] were brought about by Society 4.0 [4], while it was criticized for putting efficiency and productivity above the empowerment and welfare of people. Workers often believed that

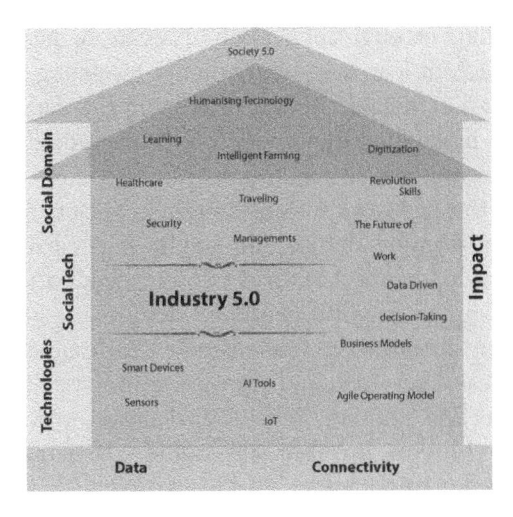

Figure 1.2 Industry 5.0—pyramid with human-centric technique.

methods pushed by technology were forcing them out or excluding them. Humanizing design and empowerment are given the most importance in Society 5.0 in Figure 1.2 ensuring that technology advances, rather than eclipses, mankind. To increase creativity, productivity, and well-being, it promotes the growth of human potential using technology and promotes cooperation between people and machines.

1.2 Network Sensing Systems in Society 5.0

Society 5.0 is characterized by extensive integration, advanced data testing, reputation stability, and the concept of web sentiment processing as the core represents a well-designed basic system, and it is a real analysis of high sensitivity and elevation analysis. Monitoring world events, analytics, trends and in real time, networked sensing systems play a key role in enabling seamless and streamlined integration of various social services from transportation and healthcare to energy management and urban planning through community development.

Data-driven decision-making principle [7] is a foundation of sensing systems based on networks, where data from IoT devices, sensors, and social media platforms are collected in real time from diverse sources. This information stream offers invaluable insights into the complex nature of the dynamics of the society and allows the stakeholders to make smart decisions about how to prevent the emerging problems and take advantage of the opportunities. Through application of advanced data analytics methods, including machine learning and predictive modeling, network sensing systems can obtain actionable information from the data, and this information can then be applied to activities such as predictive maintenance, resource optimization, and risk management.

In addition to that, network sensing system in Society 5.0 also promotes creative and user-friendly design and public participation resulting in technological developments that serve people's needs and goals. Because of increase in openness, involvement, and democratic governing bodies, these systems have encouraged civilians to be the designers of their urban environments, health services, and transportation systems. Citizen feedback mechanisms and mutually beneficial decision-making platforms in network sensing systems increase stakeholder participation promoting citizen contact and giving them chance to participate in decision-making processes. Moreover, network sensing system in Society 5.0 do follow some ethical and sustainable principles by trying to protect privacy and data security to promote social fairness and reduce negative effect on the environment. These systems use smart meters, sensors for monitoring the environment, and renewable energy technologies to make it simpler to use resources more efficiently, reduce carbon production, and promote environment-friendly behavior. Furthermore, moral standards, encryption technologies, and data privacy rules assure private data security while maintaining public trust in the digital economy.

1.3 Issues and Challenges

1.3.1 Data Privacy and Security

Modeled upon a data-driven society [8], Society 5.0 stresses the concept of data-driven decision making and its capacity to impact growth in society. This viewpoint deviates from the traditional comprehension of the media by prioritizing data in the advancement of societal development. In the context of Society 5.0, data are more than just processed information; it also serves as a catalyst for practical findings that inspire transformative change in a wide range of sectors. The core data-driven society in Society 5.0 is enshrined in the Japanese government's "Development Plan 2018," where two terms explicitly reflect the data-driven society, as described in official documents in the field of computer physical systems (CPS) [9] and the Internet of Things (IoT) [10], which harnesses the power to digitize networks in various industrial sectors. This digitization facilitates the collection of vast amounts of data, which is then transformed into actionable intelligence and applied to real-world situations into, thereby increasing the efficiency and effectiveness of decision-making processes.

The evolution of how data affects the real world of Society 5.0 is multifaceted. First, data indirectly influence social outcomes through human decision-making processes. For example, traditional approaches to urban planning have limited data sources which prevents the accuracy of the decision making process, but in a data-driven society [26], real-time data from multiple sources resides that smartphones, CCTV cameras, and transport cards provide a comprehensive understanding of cities dynamics empower stakeholders to make informed decisions, and thus the public fabric. Second, the public use of data in Society 5.0 is characterized by the direct impact of data through automated processes. For example, the complexity of adapting traffic signals to different needs for traffic management is beyond human capacity. In response, AI-controlled systems [11] use continuous input data to iteratively adjust control algorithms independently adapting to changing conditions and traffic. This paradigm shift from human- to AI-controlled systems model data as the primary driver of social work in Society 5.0.

More specifically, Society 5.0 indicates an evolution of paradigms toward a data-driven society in which data go beyond its usual use as information to become an essential part of social progress. By using data, Society 5.0 aims to improve efficiency, encourage innovation across a range of industries, and affect decision-making processes, for example. Data have a crucial role in deciding the future of society, as shown by the revolutionary potential of data-driven techniques, which use innovations in traffic management and urban planning as examples.

1.3.2 Importance of Privacy and Security

Operational approach of Society 5.0 revolves around the use of data to drive out societal progress; a huge responsibility is devolved to data privacy and security [12–14]. These two fundamental elements are very indispensable for sustaining data management integrity and for ethical application of the data. With the data increasingly decisive in the making decisions and also transforming societies, a major concern that should be taken care of is the confidentiality of the data and the privacy of an individuals' information.

Surveillance of the information is the basis for the reliability and credibility of the stakeholders, which should be based on data privacy and also security. Through the application of well-defined regulations and protocols, organizations not only mitigate many legal risks but also create a platform that allows them to publicize the data and innovate. Obedience with the legal frameworks as GDPR and also CCPA is very critical reflecting the accountable control of the ethical data and also responsibility.

Strong security mechanisms [24], including encryption and access control, ensure the checking on the possible threats, such as data breaches and also unauthorized access. This preventive approach reduces the chance of misuse and enhances the secure handling of the data across the whole digital territory. Ethical considerations assume the first place in the directing of the data use in proper ways. Maintaining the privacy rights of the individuals and ensuring equality of the access to data are ethically very important components of the stewardship, which both promote openness and accountability within the data-driven environment.

Furthermore, the safeguard of data confidentiality and privacy will provide a secure platform for the development of innovation in the social structure of Society 5.0. Through such way of confining research and experiments on the data, organizations, in turn, will be able to continue innovating, and the technology will keep on being improved while protecting their intellectual property rights, and also assuring data integrity. In fact, data privacy and also security are not luxury building blocks of humanization in Society 5.0. At every stage of use, there is integration of these living organisms into data-driven decision-making processes, which guarantees just and responsible utilization of data to promote innovation and drive social change.

1.4 Encryption and Decryption Techniques: Safeguarding Data Integrity

Encryption and decryption methods are essential for guaranteeing the security, confidentiality, and integrity of sensitive data in the era of Society 5.0, when data are the essential resource for society. By making data illegible for unauthorized users and reducing the likelihood of unlawful access and exploitation, these techniques serve as the cornerstone of data protection initiatives. Let us explore the complexities of encryption and how it is decrypted in the context of Society 5.0. In Society 5.0 ecosystems, encryption is the first line of defense for data security [12]. Different encryption techniques are used to convert plaintext data into cipher text. Symmetric

Encryption is the method that encrypts and decrypts data using the same key. One of the most straightforward symmetric encryption [15] methods is the Caesar Cipher [16], which is also one of the simplest to break. Many additional symmetric encryption [15] methods have since been developed by cryptologists, including some that are currently in use to encrypt data like passwords. Asymmetric Encryption [17], also referred to as public-key cryptography, is a type of data encryption in which the matching decryption key, also known as the private key, and the encryption key, also known as the public key, are distinct. Only the matching private key can be used to decrypt an exchange that has been encrypted using the public key. In Society 5.0 contexts, key exchange techniques and secure communication channels are made available by algorithms like Elliptic Curve Cryptography [18] (ECC) and Rivest–Shamir–Adleman (RSA) [19]. Homomorphic Encryption [20] is an enhanced technique for encryption that preserves data privacy and encourages data analysis and interaction by enabling computations to be done on encrypted data without the need for decryption.

1.4.1 Decryption Technology

Decryption is the process of restoring encrypted text information to its original text allowing users to securely access encrypted information. Key technical points include the following:

Key Management: Effective management of encryption keys is crucial for a secure decryption process. Key management systems, including key identification, storage, rotation and destruction, ensure the confidentiality and integrity of encryption keys [19] in a community 5.0 environment.

Authentication: Verifying the user's identity and providing appropriate decryption authority are crucial to ensuring data security. Multi-factor authentication (MFA) [21] and biometric authentication [21] methods increase the security of the decryption process and reduce the risk of unauthorized access.

Decryption Algorithm: Using an industry-standard decryption algorithm corresponding to the encryption method used is crucial for successful data decryption. The organization uses advanced decryption algorithms to ensure data integrity and confidentiality throughout the decryption process.

1.4.2 Challenges and Decisions

While encryption and decryption technologies provide significant benefits in data protection in Society 5.0, some challenges and decisions need to be addressed as follows:

Performance: Strong Encryption algorithm will increase performance And affects system latency and response capacity. Security measures along with performance considerations are necessary for the integration of communication and decryption technologies.
Key Management Complexity: Managing encryption keys across multiple systems and platforms can be complex. Implementing a strong key management and encryption key management solution is critical to managing key lifecycles.
Compliance: When using encryption and decryption technologies, data protection laws and regulations (such as GDPR and CCPA) [22] must be followed. For staying within the law and fulfill regulatory requirements, it is really important to make sure that encryption solutions follow the rules set by regulators. Encryption and decryption strategies are vital for retaining data integrity and secrecy in Society 5.0. Organizations can shield sensitive records in information-pushed ecosystems by way of using robust encryption algorithms, effective key management techniques, and solid decryption techniques.

1.4.3 Interoperability Challenges in Society 5.0: A Tripping Block on the Road to a Hyperconnected Future

Society 5.0, which promises us of a hyper-connected and intelligent society, promises a future in which technology smoothly integrates into all aspects of our life. Consider a scenario in which smart cities optimize traffic flow, linked homes anticipate our wants, and intelligent healthcare systems deliver individualized treatment. However, this utopian vision is dependent on a critical factor: compatibility.

1.5 Understanding Interoperability on Society 5.0

Interoperability refers to the capability of various systems and gadgets to communicate and trade facts seamlessly. In the context of Society 5.0, this interprets to make that fact from diverse sources, including clever sensors, wearables, and infrastructure structures, may be simply incorporated and analyzed to allow clever decision making and foster innovation throughout numerous domains. However, reaching this seamless trade gives numerous challenges that threaten to impede the progress of Society 5.0. Let us delve deeper into these demanding situations through a real global state of affairs.

1.5.1 The Smart City Dilemma: A Case Study in Interoperability Woes

Imagine Alice, a resident of a bustling clever metropolis, experiencing a sudden scientific emergency at domestic. Her clever domestic gadget, prepared with several health sensors, detects an anomaly in her vital organs and triggers an emergency reaction about this apparently straightforward situation, but that signal can be disturbed by way of the subsequent interoperability hurdles as follows:

Standardization Issues
May be that statistics transmitted using Alice's clever home machine is probably incompatible with the layout used by the emergency reaction center's gadget. This incompatibility creates a conversation barrier delaying the critical alert and potentially impacting the timeliness of scientific intervention.

1.5.2 Ensuring Integration and Data Exchange

The achievement of Society 5.0 relies on the potential of diverse structures and devices to connect and bypass records correctly. This necessitates overcoming several interoperability hurdles. Diving deeper into Interoperability Issues [25] in Society 5.0 is shown in Figure 1.3.

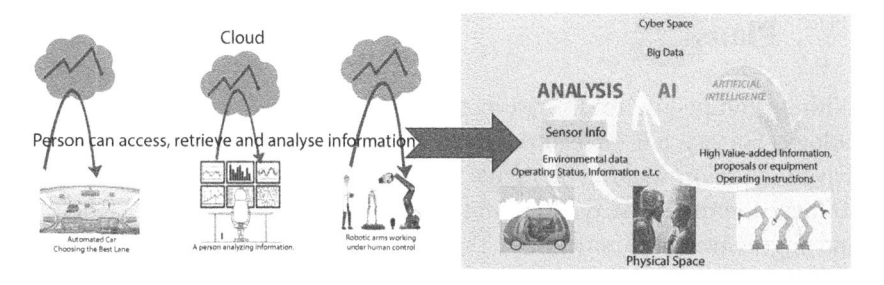

Figure 1.3 Cloud analyses data for AI.

1.5.3 Standardization Challenges and Solutions

The current landscape is riddled with competing standards across various sectors hindering interoperability. For example, the healthcare industry utilizes diverse standards, like HL7 (Health Level Seven) and DICOM (Digital Imaging and Communications in Medicine), while the automotive industry employs CAN (Controller Area Network) and OBU (On-Board Unit) standards.

Collaborative Standardization Efforts: Establishing common ground requires collaboration between industry players, government agencies, and international standardization bodies. Initiatives like the Industrial Internet Consortium (IIC) and the Open Web Alliance (OWA) are fostering collaboration to develop interoperable standards across different domains.

1.5.4 Heterogeneity Challenges and Solutions

Integration is made extremely challenging by the wide variety of devices and systems that are available ranging from cutting-edge AI platforms to antiquated infrastructure. Numerous operating systems, communication protocols [12], and data formats [26] are often used by these systems.

The answer is to build robust frameworks for interoperability that can react to and translate data between different systems. These frameworks use techniques, like data virtualization, format translation, and semantic mediation [27], to make communication between various components easier.

It is also important to invest in processes and technologies to adapt to different products and systems in Society 5.0. From semantic interoperability to API-based integration and intermediary platforms, there are many ways to bridge communication and facilitate data exchange.

1.6 Importance of Robust Communication and Power Plans

With its compelling future vision of highly connected sophisticated technology, Society 5.0 envisions a society of intelligent automation, data-driven decision-making systems, and interconnection. This ambitious strategy is predicated on the vital cornerstones of robust infrastructure and aggressive communication. This essential cornerstone of the contemporary world is crucial in facilitating the following:

Smart Living, which includes encouraging optimal living conditions with an emphasis on quality and extending the lifeline of a process, as well as effective resource management, automation, and real-time data interchange [26, 27]. Improved connectivity allows for continuous connection between people, things, and systems. It also makes it possible to collaborate globally and encourages the unexpectedly rapid transmission of information.

Better Decision Making: Data-driven understanding [26, 27] from interconnected systems can carry out more effective resource allocation and planning techniques, which ultimately results in more knowledgeable information. Thus, it is very critical to guarantee infrastructure flexibility and dependability, since doing so is essential for realizing the goals of Society 5.0. This chapter addresses the need for resilient and adaptable infrastructure, looks at resilience-building techniques, and looks at the key factors to keep in mind as we move toward a Society 5.0 that is future proofed.

1.6.1 The Requirement and Imperative of Flexible and Dependable Infrastructure

In getting used to a changing world in an ever-changing environment with fluctuating needs, infrastructure needs to be flexible and expandable. Because of this, systems must change from being stiff and static to being flexible and adaptable to the ever-changing demands of a technologically advanced civilization.

Adaptability and Scalability
As a result of the rapid development of connected devices and a growing number of data transmissions, communication networks that can handle greater the flow of data and a broader spectrum of standards are necessary. Performance should not be lost in combining new technology like Big Data analytics [26] and the Internet of Things into flexible networks.

Electric Grids: Sustainable power generation systems that can quickly adapt to alternative sources of energy, such as ocean, solar, and wind power, are important for upgrading on to a sustainable future. These grids must be scalable to manage increase in user demand and maintain a constant supply of energy.

1.6.2 Infrastructure Resilience

The idea of Society 5.0, a tremendously greater, hyperconnected future, calls for a robust infrastructure. The seamless operation of this interconnected society is made by means of the framework that serves as the premise for primary facilities like electricity, transportation, and conversation. However, unexpected occurrences, including screw-ups in gadgets, pc hacking, or natural failures, may additionally harm those important components and feature a sequence response. Developing long-lasting systems is therefore important to keep non-stop operation and reducing the results of disruptive occasions.

1.6.3 Key Characteristics of Sturdy Infrastructure

Redundancy: Single-point disasters are lessened through using numerous transmission paths and backup structures. To assure continued flow even in the occasion of a system failure, it is necessary to have backup structures and communiqué routes available in Figure 1.4.

Distributed Systems: The important infrastructure is spread out within systems, which can be decentralized, which reduces its vulnerability to localized disturbances. To save from a single-point failure creating a vast disturbance, assets and services are disbursed across geographically distant places. To ensure uninterrupted access even in case of an electrical failure

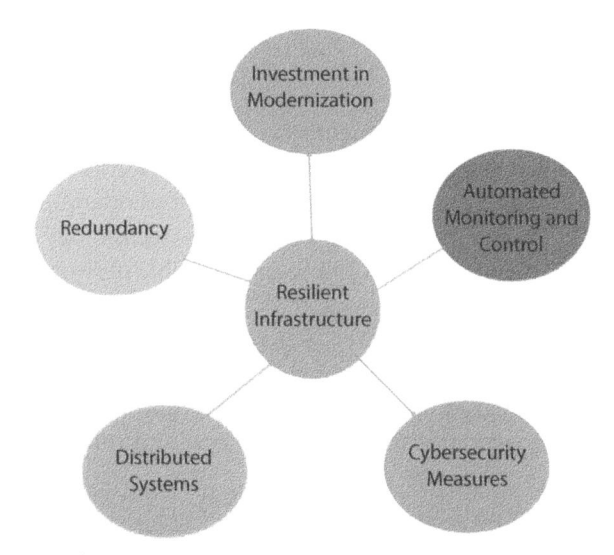

Figure 1.4 Key characteristics of resilient features.

at one area, information might be stored throughout multiple geographically dispersed storage facilities.
Approaches for Privacy and Security: Protection in the path of cyber attacks consists of setting dependable safety suggestions in database regions, which include encryption and everyday vulnerability check-ups. To select out and take away functionality dangers to safety in advance, preventative strategies and persistent surveillance mechanism are implanted.
Automated Monitoring and Control: Real-time tracking and automatic reaction abilities allow quick detection and reducing the severity, seriousness of problems. These systems when are aware of developing troubles takes robotically corrective steps minimizing downtime and ensuring an easy and strong operation. For example, computerized fault detection systems can find and isolate faults in a grid preventing big power outages.
Investing in Modernization: Reliability and efficiency can best be accelerated by constantly changing previous infrastructure with more modern, greater resilient technology. This consists of persevering to fund studies and improvement tasks in addition to enforcing contemporary fixes like smart grids.
Software-Defined Networking (SDN) [30]: The use of this technology allows an extra freedom and flexibility in handling the operations of communique networks by way of isolating the manipulate layer (which manages network congestion) from the statistics layer (which forwards messages).
Self-Recovery Materials: These forms of materials have the capacity to restore small damages or breaches on their own enhancing the sturdiness of infrastructure and reducing preservation expenses.

1.6.4 Infrastructure Resilience's Importance in Society 5.0

Developing a sturdy foundation is readily promoting the foundational thoughts of Society 5.0 and approximately decreasing dangers. The strategies to do it are outlined below:

Preserving Critical Service Regularity: When there are disturbances, resilient engineering makes sure that essentials like power, transportation, and communication keep functioning. By doing this, downtime may be minimized, and essential society services, like scientific research, responses to emergencies, and business operations may additionally flourish.
Improving Recovery Approaches: Durable infrastructure minimizes damage and accelerates the restoration speed at some stage during disruptive events.

Improving Recovery Techniques: Durable infrastructure minimizes damage and quickens up the recuperation technique throughout disruptive occasions.

Supporting Innovation and Boom: By offering a dependable platform for the improvement and sensible use of innovative technologies, a longtime infrastructure base boosts entrepreneurship and boom within the economic system.

1.6.5 Infrastructure Resilience Techniques

Embracing Cost-Effectiveness and Efficiency for a Sustainable Future in Society 5.0: Infrastructure serves as the foundation for clever automation and interconnection; it is imperative to maximize fee-effectiveness and performance. To achieve this, a flexible infrastructure that can adapt to changing demands is essential. It should be capable of quickly reconstructing and recovering each day.

Dynamic Resource Utilization: By adjusting operations according to cutting-edge needs, flexible architecture allows for the fine and feasible use of resources in much less energy being used. For instance, clever grids are capable of decreasing energy waste at some stage in off-top hours with the aid of regulating the drift of electricity based on demand.

1.6.6 Reduced Operational Costs

Flexible infrastructure gives a good-sized benefit in decreasing operational expenses throughout diverse additives of the commercial organization. The following are some key ways of Reducing Operational Costs:

a. Resource Optimization: Flexible infrastructure allows you to scale resources (like servers or storage) up or down based on your actual needs. This gets rid of excess resource uses as well as of sitting idle and the abuse of power.
b. Smarter Energy Management: Flexible infrastructure can integrate with smart technologies to screen and optimize power use. For instance, sensors can detect when a room is unoccupied and automatically adjust the lights or heating/cooling.
c. Reduced Maintenance Needs: Often, bendy infrastructure uses modular components, which can be easier and

inexpensive to maintain or replace in assessment compared to complicated, constant structures. Think about constructing furniture with pre-fabricated pieces instead of custom carpentry.

1.6.7 Safeguarding the Digital Age: Security and Privacy in Infrastructure

Society 5.0 is all about tech making our lives wonderful, but with great energy comes fantastic responsibility (i.e., to maintain our records safe). The following are the ways to be a privacy pro:

Fort Knox Your Logins: Think of complicated passwords as your digital bodyguards. Add some other layer of security with two-component authentication, like a code sent to your telephone. It is like having a double deadbolt on your on-line money owed.
Patch Up With Defenses: Regularly update the devices and software. These updates are like digital patches fixing vulnerabilities that hackers may try and exploit.
Be Data Ninja: Not all organizations need your complete life tale. In Society 5.0, you could regularly pick to proportion anonymized facts. This could help agencies analyze what they need without compromising user privacy. It is like sharing book summaries with friends as opposed to lending them the entire book.
Know Where Your Data Goes: Before delivering your statistics, understand how it is going to be used. Reputed groups can be transparent about statistics practices and come up with manipulation over yours. It is your statistics in the end!

By following those recommendations, we will build a Society 5.0 where each person feels stable and enjoys the blessings of a generation.

1.7 Environmental Effects and Energy Efficiency

In the subsequent phase, 1.7.1 "Supporting sustainable development through electricity-saving solutions," we will be able to have a look at enormous methods for increasing sustainability. This section focuses on reducing energy consumption through products like fluorescent light bulbs and smart product use while outlining their benefits for the environment. In addition, we will also discuss how to lessen our dependence on fossil fuels through the use of wind, ocean, and other renewable energy sources. With an emphasis on the development of Society 5.0, we explore the different ways in which technological advances and energy production from renewable sources might be combined to enhance efficiency and simplify the workflows. The importance of technology in creating a safer and greener future will be addressed during these talks.

1.7.1 Supporting Sustainability through Energy-Efficient Approach

Super Smart Society, or Society 5.0, sets a strong emphasis on intelligent technology and mutual dependence. These developments lay the way for significant improvements in energy efficiency. Within the context of Society 5.0, this section examines several technologies and methods that can lead to a more sustainable energy future.

Smart Grids: Optimizing Energy Delivery in Society 5.0
Conventional power grids guarantee an uninterrupted supply of electricity irrespective of the demand in real time operating on a one-size-fits-all basis. However, these power plants may struggle to satisfy demand during peak periods and produce more electricity than desired during off-peak hours, which frequently result in energy waste.

With its emphasis on intelligent automation and interconnectivity, Society 5.0 paves the door to a more intelligent method of delivering energy through smart grids. By facilitating two-way communication between utilities, generators, and customers, these sophisticated grids allow for dynamic control, real-time monitoring, and increased efficiency.

In Society 5.0, smart grids [31] optimize energy delivery in the following ways:

Reaction to Demand: Demand response programs are made easier by smart grids. These initiatives encourage users to modify their energy-use patterns in response to current electricity rates. This may involve the following:
Duration of Use Pricing: Depending on the time of day, consumers are charged varying prices for power. Generally speaking, prices are higher during times of peak demand and lower during off-peak hours. By scheduling non-essential energy use at off-peak times (such as using laundry machines at night) can help the consumers save their expenses.

Demand-Side Management (DSM) [32]: Utilities can offer programs or rebates to encourage consumers to install energy-efficient appliances and technologies that reduce their overall electricity demand.
Improved Distribution Efficiency: Smart grids utilize smart meters and sensors throughout the distribution network. These devices can identify and pinpoint inefficiencies in energy transmission and distribution. Real-time data allows for the following:
Benefits of Smart Grids: By implementing smart grids, Society 5.0 can gain numerous benefits as follows:

- Reduced Energy Consumption: Demand response programs and improved distribution efficiency lead to a decrease in overall energy consumption minimizing wasted energy and promoting sustainability.
- Lower Costs: Smart grids can help utilities optimize generation and distribution potentially leading to lower electricity prices for consumers.
- Increased Reliability: Real-time monitoring and control capabilities enhance grid stability and reliability minimizing power outages and disruptions.

1.7.2 Building Automation Systems: Transforming Buildings into Energy-Conscious Entities

The attention of Society 5.0 on smart automation and connection is also present within the bodily environment, as building automation structures (BAS) are revolutionizing the usage of electricity in the large buildings. These state-of-the-art structures integrate the functioning building structures, like lighting fixtures, ventilation, heating/cooling (HVAC), and appliances. By the use of sensors, data analytics, and instrumentation mechanically, constructing automation structures (BAS) can effectively lower the strength usage and improve occupant comfort. In Society 5.0, BAS enhances electricity performance through the following methods:

Benefits of Building Automation Systems

- Reduced Energy Consumption: BAS can lead to energy savings of up to 30% by optimizing operation of building systems and minimizing energy waste.

- Lower Operating Costs: Reduced energy consumption translates to lower utility bills and operating costs for building owners.
- Improved Comfort: By automatically adjusting building systems based on occupancy and weather conditions, BAS maintains a comfortable environment for occupants.
- Enhanced Sustainability: Reduced energy consumption contributes to a cleaner environment and reduces greenhouse gas emissions.

1.7.3 Connected Appliances: Transforming Everyday Devices into Energy-Conscious Partners

By focusing on connectivity and intelligent automation, Society 5.0 goes beyond buildings and infrastructure to replace everyday equipment. Connected devices equipped with sensors and Internet connectivity are becoming increasingly common, and they offer a great potential for improving energy efficiency. These smart devices communicate with users and central systems enabling the following:

- Remote Control and Monitoring: Users can now control their devices remotely using smartphones or voice assistants.
- Demand Response: Appliances connected in the system can directly or indirectly participate in programs that respond to demand and supply or automatically modify their working based on current energy prices.

Intelligent Operation: By adjusting cooking cycles largely based on the kind and quantity of food, connected ovens reduce the amount of energy used for preheating or overcooking. Intelligent refrigerators have the ability to aggressively control cooling by monitoring internal temperature and setting frequency.

1.7.4 Benefits of Connected Appliances

The integration of linked home equipment offers several benefits in Society 5.0 as follows:

- Enhanced Convenience: Remote control and scheduling offer greater convenience and flexibility for users allowing them to integrate appliance operation into their daily routines.

- Enhanced Efficiency: By optimizing their function for optimal performance, connected appliances can save energy consumption and achieve desired outcomes precisely prepared food.
- Data-Driven Insights: Real-time energy consumption data provided by connected appliances empower users to make informed decisions about their energy consumption habits.

1.7.5 Challenges and Considerations

Despite their advantages, linked appliances have many limitations as follows:

- Security Issues: To protect users from breaches of privacy, appliances that are connected have to ensure the privacy of the data they collect.
- Standardization: Conflicts between different models and brands might result because of the absence of international norms for data formats and protocols for communication.
- Price: Compared to conventional models, connected appliances may have a greater initial cost. Nevertheless, over time, these expenses may be balanced by long-term energy savings and greater convenience.

1.7.6 Energy-Efficient Manufacturing [23]: Optimizing Industries for a Sustainable Future in Society 5.0

Since the world is advancing toward environmental sustainability, the need for produced goods is continuously increasing, which is posing a hurdle in environmental sustainability. So, along with a focus on intelligent automation and connectivity, Society 5.0 is also providing a unique opportunity for shifting the manufacturing sector toward environmentally friendly methods.

Businesses can significantly reduce their energy usage per unit of manufacturing using methods such as automation, data analysis, and cutting-edge technology paving the way for a more sustainable future.

Society 5.0 increases energy efficiency in manufacturing:

- Traditional manufacturing often relies on manual monitoring and control of processes. In Society 5.0, this is enhanced by the use of sensors and advanced monitoring systems.
- Automated Adjustments: Based on real-time data, intelligent systems can automatically adjust process parameters, like temperature, pressure, and machine settings, to optimize energy use.
- Predictive Maintenance: Sensor data can be analyzed to predict potential equipment failures. Proactive maintenance prevents unnecessary energy waste caused by malfunctioning machinery.
- Precision Operations: Robots carry out activities more consistently and precisely, which reduces the amount of waste and energy needed for rework or error correction.
- Optimized Machine Usage: Robots can operate 24/7 with minimal downtime allowing for efficient machine utilization and maximizing production output without additional energy consumption.

1.7.7 Benefits of Energy-Efficient Manufacturing

The adoption of energy-efficient practices in manufacturing offers numerous benefits as follows:

- Increased Productivity: Automated systems and optimized processes can lead to increased production output while maintaining or even improving product quality.
- Enhanced Resource Efficiency: Reduced material waste and optimized energy use contribute to a more sustainable manufacturing process minimizing environmental impact.
- Improved Worker Safety: Replacing human workers with robots for hazardous or repetitive tasks can enhance worker safety and reduce the risk of accidents.

1.8 Utilizing Renewable Energy Sources

Powering a Sustainable Future [33] (with a focus on Society 5.0 improvements): The fight against weather change and environmental degradation brings our attention to renewable electricity resources. Installing technology, like solar, wind, hydro, and geothermal electricity, plays an important role in Society 5.0. With its emphasis on interconnectivity and automation, it unlocks interesting opportunities for the subsequent generation of renewable power solutions. This segment also dives into how Society 5.0 advancements can revolutionize how we harness more cleaner electricity and construct a sustainable future.

1.8.1 Methods of Harnessing Cleaner Energy in Society 5.0

- **Ocean Energy [33]:** Going beyond conventional wave and tidal electricity, Society 5.0 technologies can free up the substantial capacity of the ocean waves. Underwater mills can use kinetic energy from ocean currents, and at the same time, progressive structures can convert the rise and fall of ocean tides into energy. Additionally, studies into salinity gradient strength (harnessing the difference in salt awareness between freshwater and seawater) hold promise for producing clean power in coastal areas. Bio-energy With Advanced Techniques: Biomass, organic matter used as fuel, offers a renewable energy source. However, traditional methods of bioenergy production can raise sustainability concerns. Society 5.0 facilitates advancements in techniques, like anaerobic digestion, which breaks down organic waste materials to produce biogas, a methane-rich fuel source usable in vehicles and power generation.
- **Microalgae Cultivation [34, 35]:** It is giving new hope for advanced biofuels. These tiny microorganisms may be farmed successfully, with the usage of little land and water, and generating oils that may be converted into biofuels. Society 5.0 era can optimize and automate algae farming procedures resulting in accelerated biofuel production.
- **Piezoelectric Energy Harvesting [36]:** This approach converts mechanical energy (such as pressure or vibration) into electrical energy. For example, imagine streets embedded with piezoelectric materials that generate power from the weight of passing automobiles. In Society 5.0, integrating piezoelectric materials into building structures, sidewalks, or even clothing can provide small amounts of energy. This energy could be used to power low-energy devices or contribute to the power grid.

1.9 Conclusion

In summary, Society 5.0 is a major breakthrough in the use of information technology to support all aspects of human life in a coherent and integrated manner. Unlike previous versions, the development of Citizen 4.0 represents a paradigm shift in how we collect, process, and use data to improve all aspects of life. It involves integrating systems across sectors, such as energy, transport, health, education, and recreation, to ensure excellent social, safety, and productivity. One of the defining characteristics of Society 5.0 is that large amounts of data can be used to directly influence people's actions and behaviors and ultimately fill and support people's lives. The data-driven approach goes beyond individual policies to create social policy on a broader scale through a continuous cycle of data collection, analysis, and interpretation. By using artificial intelligence and other advanced IT systems, Society 5.0 can use the world's big data to improve decision-making processes and improve overall health.

The progress of Society 5.0 through its leaders has emerged to overcome the challenges and limitations of the past. Although the exchange of information is not without difficulty, the financial resources that Society 5.0 provides in changing the way we use clean energy, supporting the future and improving the overall quality of life, are enormous. Organization 5.0's emphasis on communication and the use of smart technology lays the foundation for public safety and the environment giving great hopes for the development of renewable energy solutions. As we delve deeper into the field of Society 5.0, it is clear that Citizen 5.0 marks a major leap forward in social life. Information guided by insight here paves the way for a happier, safer life. The integration and cooperation of various departments to ensure that the results of decision-making information disseminate to all areas of human life are created to provide people with a useful, safe, and pleasant environment.

In essence, Society 5.0 represents an era of change where information, technology, and human life come together to create the future feature of innovation, sustainable development, and health. By embracing the principles of Society 5.0 and harnessing the power of connected systems, we can create a path to a more prosperous and harmonious life using intelligent information for the benefit of everyone.

1.9.1 Future Directions

Improved connectivity solutions: Imagine a future where information flows between systems enabling better communication and collaboration across regions. By encouraging collaboration and investing in business solutions, we will triumph over modern-day demanding situations and pave the manner for a greater connected Society 5.0.

Advanced Communication and Power Systems: Imagine a global future where conversation has an awful lot of strength as power and is supported by constant and related digital gadgets. In addition, efficient distribution of energy and the wireless connectivity supports the growing infrastructure. By utilizing renewable energy sources such as solar and wind power, we not only minimize our environmental impact but also pave the way for a stronger, more sustainable future for the upcoming generations.

Data Evaluation and New Intelligence: Data are the new foreign money of Society 5.0. Technology and intelligence is the key to unlocking its complete capability. Future traits will focus on superior structures that can examine massive amounts of information and provide immediate guidelines. From predictive models to personalized hints, AI generation will remodel decision making throughout all industries and force innovation and productivity anywhere.

Focus on Sustainability Desires: As we flow to Society 5.0, it is important to be conscious of our desires: creating sustainable and sustainable equality of life for all. This aligns our efforts with those of the United Nations Sustainable Development Goals (SDGs) using generation to resolve worldwide troubles. Whether tackling climate alternate, lowering inequality, or growing monetary growth, Society 5.0 has the capacity to drive progress toward a better world for each person.

Encourage Citizen Engagement: Society 5.0 is more than just an era project. It is also a technological step forward. It is about empowering humans to create their own destiny. In this destiny, citizens are companions in decision-making techniques using virtual generation and participatory approaches to specify their opinions and promote wonderful alternates. We can make certain that everyone is accountable via transparency, responsibility, and inclusion. A seat at the table promotes and integrates virtual literacy [28, 29]: In an international community in which technology is anywhere, digital literacy is not a luxury; this is the manner of existence, if important. The destiny direction of Society 5.0 is the importance of the virtual divide and is making sure that everybody has admission to the know-hows and abilities they want to be triumphant in inside the digital age. We need to paint collectively to create shared and equal possibilities for all with the aid of providing schooling and training programs to aid marginalized communities.

Ethical Considerations and Rights: The vision of Society 5.0 is interesting, though it raises important questions and issues. From privacy laws to algorithmic bias, there is absolutely no need to address those challenges and enact stronger laws to protect individuals and the public. By emphasizing ethics and responsibility in terms of age and vigilance, we are able to ensure that people remain strong at all times in a world of Society 5.0.

Public Destiny: Society 5.0 promises incredible integration and the creation of a more sustainable, authentic, and human-centered globalization. By embracing technological innovation, encouraging entrepreneurship, and respecting the importance of ethics, we can create a future in which the generation becomes a mechanical force that complements everyone's strengths and beautiful personalities.

References

1. Narvaez Rojas, C., Alomia Peñafiel, G.A., Loaiza Buitrago, D.F., Tavera Romero, C.A., Society 5.0: A Japanese concept for a superintelligent society. *Sustainability*, *13*, 12, 6567, 2021.
2. Carayannis, E.G., Dezi, L., Gregori, G. *et al.*, Smart Environments and Techno-centric and Human-Centric Innovations for Industry and Society 5.0: A Quintuple Helix Innovation System View Towards Smart, Sustainable, and Inclusive Solutions. *J. Knowl. Econ.*, 13, 926–955, 2022. https://doi.org/10.1007/s13132-021-00763-4.
3. Özdemir, V. and Hekim, N., Birth of industry 5.0: Making sense of big data with artificial intelligence,"the internet of things" and next-generation technology policy. *OMICS: J. Integr. Biol.*, *22*, 1, 65–76, 2018.
4. Nair, M.M., Tyagi, A.K., Sreenath, N., The Future with Industry 4.0 at the Core of Society 5.0: Open Issues, Future Opportunities and Challenges, in: *2021 International Conference on Computer Communication and Informatics (ICCCI)*, Coimbatore, India, pp. 1–7, 2021, 10.1109/ICCCI50826.2021.9402498.
5. Horvat, D., Kroll, H., Jäger, A., Researching the effects of automation and digitalization on manufacturing companies' productivity in the early stage of industry 4.0. *Procedia Manuf.*, *39*, 886–893, 2019.
6. Schumacher, A., Sihn, W., Erol, S., Automation, digitization and digitalization and their implications for manufacturing processes, in: *Innovation and Sustainability Conference Bukarest*, pp. 1–5, Elsevier, Amsterdam, The Netherlands, 2016, October.
7. Smyrnaiou, Z., Liapakis, A., Bougia, A., Ethical use of artificial intelligence and new technologies in Education 5.0. *J. Artif. Intell. Mach. Learn Data Sci.*, 1, 4, 119–124, 2022.

8. Mishra, P., Thakur, P., Singh, G., Sustainable smart city to society 5.0: State-of-the-art and research challenges. *SAIEE Afr. Res. J.*, *113*, 4, 152–164, 2022.
9. Humayed, A., Lin, J., Li, F., Luo, B., Cyber-physical systems security—A survey. *IEEE Internet Things J.*, *4*, 6, 1802–1831, 2017.
10. Tran-Dang, H., Krommenacker, N., Charpentier, P., Kim, D.S., Toward the internet of things for physical internet: Perspectives and challenges. *IEEE Internet Things J.*, *7*, 6, 4711–4736, 2020.
11. Cugurullo, F., Caprotti, F., Cook, M., Karvonen, A., McGuirk, P., Marvin, S. (Eds.). *Artificial Intelligence and the City: Urbanistic Perspectives on AI*. Taylor & Francis, New York, NY, 2023.
12. Al Ameen, M., Liu, J., Kwak, K., Security and privacy issues in wireless sensor networks for healthcare applications. *J. Med. Syst.*, *36*, 93–101, 2012.
13. Pramanik, S., Pandey, D., Joardar, S., Niranjanamurthy, M., Pandey, B.K., Kaur, J., An overview of IoT privacy and security in smart cities, in: *AIP Conference Proceedings*, vol. 2495, AIP Publishing, 2023, October.
14. Rao, P.M. and Deebak, B.D., Security and privacy issues in smart cities/industries: technologies, applications, and challenges. *J. Ambient Intell. Hum. Comput.*, *14*, 8, 10517–10553, 2023.
15. Gui, Z., Paterson, K.G., Patranabis, S., Rethinking searchable symmetric encryption, in: *2023 IEEE Symposium on Security and Privacy (SP)*, IEEE, pp. 1401–1418, 2023, May.
16. Verma, R., Kumari, A., Anand, A., Yadavalli, V.S.S., Revisiting shift cipher technique for amplified data security. *J. Comput. Cognit. Eng.*, *3*, 1, 8–14, 2024.
17. Meraouche, I., Dutta, S., Tan, H., Sakurai, K., Learning asymmetric encryption using adversarial neural networks. *Eng. Appl. Artif. Intell.*, *123*, 106220, 2023.
18. Tidrea, A., Korodi, A., Silea, I., Elliptic curve cryptography considerations for securing automation and SCADA systems. *Sensors*, *23*, 5, 2686, 2023.
19. Du, S. and Ye, G., IWT and RSA based asymmetric image encryption algorithm. *Alexandria Eng. J.*, *66*, 979–991, 2023.
20. Yang, W., Wang, S., Cui, H., Tang, Z., Li, Y., A review of homomorphic encryption for privacy-preserving biometrics. *Sensors*, *23*, 7, 3566, 2023.
21. Rodrigues, A.R.L., *Enhanced Multi-Factor Authentication for Mobile Applications*, 2023, (Master's thesis).
22. Zhang, M., Meng, W., Zhou, Y., Ren, K., CSChecker: Revisiting GDPR and CCPA Compliance of Cookie Banners on the Web, in: *2024 IEEE/ACM 46th International Conference on Software Engineering (ICSE)*, IEEE Computer Society, pp. 958–958, 2024, March.
23. Mansour, M., Gamal, A., Ahmed, A.I., Said, L.A., Elbaz, A., Herencsar, N., Soltan, A., Internet of things: a comprehensive overview on protocols, architectures, technologies, simulation tools, and future directions. *Energies*, *16*, 8, p. 3465, 2023.

24. Dorel, B.A.D.E.A., Elena, R.D., Cristina, B.O.M., Elida, TODARITA., *Urban resilience and security in today's society: interoperability of two concepts old but yet new, different but however together.*
25. Afzal, M., Widding, K., Hjelseth, E., Hamdy, M., Systematic investigation of interoperability issues between BIM and BEM, in: *ECPPM 2022-eWork and eBusiness in Architecture, Engineering and Construction 2022*, pp. 713–720, CRC Press, 2023.
26. Yang, Y., Zeng, H., Chen, T., Lv, M., Design of a distributed offloading and real-time data unified access platform for IoT within command and control communication networks. *Cluster Comput.*, 26, 3, 1–15, 2023.
27. Sandeep, M., Khatri, S., Chandavarkar, B.R., Data Format Heterogeneity in IoT-Based Ambient Assisted Living: A Survey, in: *Proceedings of Second International Conference on Computational Electronics for Wireless Communications: ICCWC 2022*, Springer Nature Singapore, Singapore, pp. 505–515, 2023, January.
28. Rujiani, C.L., Syahputra, E.R., Andriana, S.D., Implementation Of Application Programming Interface (API) Using Representational State Transfer (REST) Architecture For Development E-Learning Unhar Medan. *Int. J. Data Sci. Visualization (IJDSV)*, 1, 1, 2023.
29. Ahmad, S., Ali, S., Waqar, N., Naz, N.S., Mehmood, M.H., Comparative evaluation of the maintainability of RESTful and SOAP-WSDL web services, in: *2023 International Conference on Business Analytics for Technology and Security (ICBATS)*, IEEE, pp. 1–9, 2023, March.
30. Dora, J.R. and Hluchy, L., Detection of Attacks in Software-Defined Networks (SDN), How to conduct attacks in SDN environments, in: *2023 IEEE 17th International Symposium on Applied Computational Intelligence and Informatics (SACI)*, IEEE, pp. 000623–000630, 2023, May.
31. Mazhar, T., Irfan, H.M., Haq, I., Ullah, I., Ashraf, M., Shloul, T.A., Ghadi, Y.Y., Imran, Elkamchouchi, D.H., Analysis of challenges and solutions of IoT in smart grids using AI and machine learning techniques: A review. *Electronics*, 12, 1, 242, 2023.
32. Williams, B., Bishop, D., Gallardo, P., Chase, J.G., Demand side management in industrial, commercial, and residential sectors: a review of constraints and considerations. *Energies*, 16, 13, 5155, 2023.
33. Farghali, M., Osman, A.I., Chen, Z., Abdelhaleem, A., Ihara, I., Mohamed, I.M., Yap, P.S., Rooney, D.W., Social, environmental, and economic consequences of integrating renewable energies in the electricity sector: a review. *Environ. Chem. Lett.*, 21, 3, 1381–1418, 2023.
34. Satya, A.D.M., Cheah, W.Y., Yazdi, S.K., Cheng, Y.S., Khoo, K.S., Vo, D.V.N., Bui, X.D., Vithanage, M., Show, P.L., Progress on microalgae cultivation in wastewater for bioremediation and circular bioeconomy. *Environ. Res.*, 218, 114948, 2023.

35. Lacroux, J., Llamas, M., Dauptain, K., Avila, R., Steyer, J.P., van Lis, R., Trably, E., Dark fermentation and microalgae cultivation coupled systems: outlook and challenges. *Sci. Total Environ.*, *865*, 161136, 2023.
36. Zheng, X., He, L., Wang, S., Liu, X., Liu, R., Cheng, G., A review of piezoelectric energy harvesters for harvesting wind energy. *Sens. Actuators, A*, *352*, 114190, 2023.

2

Remote and Urban Environmental Area Sensing, Connectivity Issues, and Solutions Based on Emerging Technologies

Abinaya M.[1]*, Vadivu G.[1], Balasubramaniam S[2] and Sundaravadivazhagan B.[3]

[1]Department of Data Science and Business Systems, SRM Institute of Science and Technology, Chennai, India
[2]School of Computer Science and Engineering, Kerala University of Digital Sciences, Innovation and Technology, Thiruvananthapuram, Kerala, India
[3]Department of Information Technology, University of Technology and Applied Science—Al Mussanah, Oman

Abstract

This chapter states the interlink between urban environmental problems and the remote sensing technology involved, to avoid connectivity potential challenges, and a new development area. Due to the world's fast urbanization, the need for effective urban-based environment management and monitoring has increased. Understanding of large urban areas and the techniques of remote sensing provide useful instruments for acquiring information. Integrating remote sensing technologies in urban environments faces some hurdles due to the logistical and technical problems with connectivity. This chapter explores and addresses problems and gives a novel solution with new technologies like Augmented Reality (AR), Artificial Intelligence (AI), Virtual Reality (VR), Data Analytics, and the Internet of Things (IoT). By using the new and advanced technology, legislators, environmental scientists, and urban planners can use the remote sensing method of environmental urban development and management.

Keywords: Augmented reality, virtual reality, remote sensing technology, artificial intelligence, Internet of Things (IoT), advanced data analytics, sustainable urban development, environmental management

---

Corresponding author: am0150@srmist.edu.in

2.1 Introduction

2.1.1 Urban Environment Remote Sensing Overview

Remote sensing is used for the collection of data with the latest technology about the Earth without the use of any physical contact. It involves collecting data remotely with the help of sensors placed on spacecraft, satellites, and in other platforms [1]. These sensors capture the electromagnetic waves that the Earth's surface generates. Such radiation can then be generated and processed to gain the most useful information [2].

2.1.2 Smart Factory

Urban environments have begun to employ remote sensing methods, such as satellite images, LiDAR, also known as Light Detection and Ranging, and drone photography, to observe and regulate the different aspects of the urban settings [3].

Some examples where remote sensing used in cities are listed as follows:

1. Classification of Land Use: Within the metropolitan areas, there are different land cover types, like residential, commercial, industrial, park, and agricultural land, which can be mapped and then classified with the help of remote sensing data [4].
2. Monitoring the Environment: Remote sensing would also help to observe the factors like vegetation health of the plant, changes in land usage over time, and the quality of air and water. The data help to determine how urbanization affects the environment and informs sustainable development strategies [5].
3. Infrastructure Planning: Comprehensive data about the networks of roads, buildings, bridges, and uses are currently in place; remote sensing data can do help with infrastructure planning such as planning new construction projects, evaluating the infrastructure vulnerabilities, and then streamlining transportation networks with the gathered data from the satellite [6].
4. Disaster Management: By providing correct information and precise data about natural disasters like floods, earthquakes, wildfires, and landslides, remote sensing plays an important role in disaster management. The surveys were taken

aerially, and satellite images can be used in the support of emergency response efforts and to determine the areas that need assistance for the evaluation of any damage [7].

2.1.3 Benefits of Remote Sensing in Cities

- Broad Coverage: The usage of remote sensing technologies facilitates the acquisition of data from different geographical regions, and it enables the exhaustive observation and examination of urban environments [8].
- Regular Monitoring: Data will be acquired by satellites and remote sensing platforms regularly giving current information about the changes that are taking place in urban areas over time [9].
- Cost Effectiveness: The need for costly field surveys makes it possible to collect data from dangerous or inaccessible locations. It is frequently less expensive than the traditional ground-based methods of data collection [10].
- Urban planners, legislators, and environmental scientists support sustainable urban development and environmental management using remote sensing methods to collect valuable insights into urban dynamics and make well-informed decisions [11].

2.2 Connectivity Challenges in Urban Remote Sensing

2.2.1 Conventional Remote Sensing Systems' Technological Limitations

In urban environments, traditional remote sensing systems face technological challenges because of things like the following:

- **Limited Spatial Resolution:** It may be difficult for traditional satellite sensors to distinguish between various land cover types or features in urban areas due to their inability to capture fine-scale details [12].
- Urban regions are more frequent and humid than the rural ones. This will result in the inaccuracy of data that we get from remote sensing, and the features get disrupted [13].

- **Effects of Shadowing:** In urban areas, buildings that are tall create shadows that hinder the underlying features making it quite difficult to accurately predict an image [14].
- The more dynamic range of remote sensing sensors will produce high-density data in urban areas with more complex geometries that lead to data and saturation loss.

2.2.2 Logistical Obstacles in the Integration and Transmission of Data

Logistical obstacles in data integration and transmission in urban remote sensing are more significant.

- **Limited Access to Urban Environments:** Deploying ground-based sensors or gaining access to specific areas to gather data is difficult in urban areas due to the high population [15].
- **Limitations of Infrastructure:** Due to inadequate infrastructure, it is difficult to transmit data in real time and to integrate remote sensing data with the current urban area. Examples of this are the limited internet connection and poor electricity supply [16].
- **Interoperability of Data:** Analyzing and integrating the different datasets are challenging because of data in different formats and the varying resolutions [17].

2.2.3 Problems Affect Data Analysis and Quality

The quality and analysis of remote sensing data are affected by issues based on connectivity in urban areas as follows:

- **Data Loss and Gaps:** Data loss or gaps can be created by network connectivity outages and signal inference, so the accuracy and the completeness of remote sensing datasets may be different [18].
- **Data Transmission Delays:** can limit the timeliness and usefulness of remote sensing data for decision making. These delays can be caused by slow or unreliable data transmission [19].
- **Reduced Temporal and Spatial Coverage:** Problems with connectivity may limit the temporal and spatial coverage of

remote sensing data making it more difficult to monitor and analyze urban dynamics thoroughly.

2.2.4 Cutting Edge Technologies to Handle Connectivity Issues

New technologies provide creative ways to address connectivity issues in urban remote sensing [20] as follows:

- Satellite technology advancements have resulted in the creation of high-resolution satellite constellations that can capture detailed imagery of urban areas with enhanced temporal and spatial resolution.
- Unmanned Aerial Vehicles (UAVs): Unlike traditional satellite sensors, UAVs, also known as drones, offer versatile and affordable platforms for gathering high-resolution aerial imagery and LiDAR data in urban settings.
- Edge Computing and Cloud-Based Solutions: By enabling real-time processing and analysis of remote sensing data at the point of acquisition, edge computing and cloud-based platforms lessen the need for centralized infrastructure and increase data accessibility.
- Integration of sensor networks and Internet of Things (IoT) devices with remote sensing systems enables smooth data transmission and integration allowing for extensive monitoring of urban environments in almost real time.
- Urban remote sensing practitioners can overcome connectivity issues and fully utilize remote sensing data for environmentally conscious and sustainable urban development by utilizing these cutting-edge technologies.

2.3 Artificial Intelligence for Enhancing Data Processing and Analysis

2.3.1 Artificial Intelligence for Data Insights

- **Machine Learning algorithms:** Automate tedious tasks like image classification and change detection enabling faster and more accurate analysis.

- **Deep Learning Models:** To a deeper understanding of urban areas, one can predict complex patterns and relationships in large datasets.
- **Predictive Analytics:** For proactive planning and resource allocation, it will help in forecasting future trends and potential challenges in assisting [21].

2.3.2 Real-Time Monitoring With IoT Sensors

- **Environmental Monitoring:** To deploy efficient environmental management and track the quality of air, temperature of the surroundings, noise levels in the environment, and other vital parameters in real time.
- **Infrastructure Monitoring:** To prevent maintenance costs and to reduce downtime, monitoring the health and performance of complex building structures, like buildings and bridges, and the network of transporting is enabled.
- **Public Safety Monitoring:** To improve the emergency response times and track real-time traffic flow and identify the threads [22].

2.3.3 Advanced Imagery and Data Acquisition

- **High-Resolution Satellite Constellations:** For accurate land use mapping and urban growth monitoring entire city images are necessary.
- **Agile Drone Technology:** Flexibility in data collection and capturing of data in difficult areas or providing close-up views for corresponding features.
- **LiDAR (Light Detection and Ranging) Technology:** Precise volume and elevation of measurements for infrastructure planning and disaster management are facilitated by creating a 3D model [23].

2.3.4 Secure Data Management With Blockchain

- **Decentralized Ledger System:** Preventing unauthorized data and their manipulation, fostering trust, and their transparency in data sharing create the hope of data immutability.
- **Secure access Control:** Giving permission to authorized users and restricting access to private and secure data [24].

- **Improved Collaboration:** Integration of data sharing between the different stakeholders for decision making and resource allocation [25].

2.3.5 Immersive Data Visualization With AR/VR

- **Augmented Reality (AR):** A deeper understanding of urban environments can overlay the data visualizations in the real world making the user interact with spatial data.
- **Virtual Reality (VR):** Various scenarios and informed decision making make the user analyze the urban models for enhanced and immersed simulations.
- **Enhanced Stakeholder Engagement:** To understand a very difficult urban planning proposal and participate in collaborative decision-making processes, interactive technique and method are provided by immersive technology. Sustainable urban development and management of the urban areas are enhanced using innovative technology and their integration with the following technologies: The stakeholders with availability of data and the analysis for decision process, resource optimization algorithm, and building urban communities.

2.4 Case Study

2.4.1 Monitoring Urban Air Quality With IoT Sensors

Overview

The major reason for pollution is air, which affects the health of humans and the environmental system. To reliably evaluate the air state and the cost in metropolitan areas, conventional techniques for monitoring frequently the lack of geographic coverage and granularity are needed. The Internet of Things (IoT) plays an important role in economics and reliability, scalability monitoring of the system, and the quality of air in urban areas [26].

A majority of the cities with air pollution problems set up a new technique for the network of Internet of Things (IoT) sensors in different areas to track important air quality substances like nitrogen dioxide (NO_2), particulate matter (PM), sulfur dioxide (SO_2), carbon monoxide (CO), and ozone (O_3) in residential areas and their surroundings, industrial areas and the nearby zones, and high-traffic areas. To record the spatial variations in the quality of air, different sensors were placed.

Gathering and Analyzing Data

A cloud platform or centralized database is used to gather data on air quality parameters, such as the air quality index (AQI), for sending real-time data. The gathered data are processed and verified using new techniques in data analytics techniques, such as machine learning algorithms, for finding the hotspots and the pattern of air pollution [27].

Communication and Visualization

To access and understand data, user-friendly platforms, such as interactive maps, dashboards of the air level noted, and mobile apps for measuring the air quality, are used to visualize and process quality, which makes it easiest for stakeholders, like the government, policymakers, general public, and agencies. When the quality of air is not healthy or above the threshold, residential people and the corresponding authorities may receive real-time alarm signals and notifications [28].

Effects and Results

The quality of air in the affected area is estimated by the temporal and the spatial variations of air pollution and the number of hotspots to improve, based on that the targets and the decision-making information are chosen. Behavioral changes and awareness are the objectives of sustainable urban development and public health protection.

Conclusion

New technologies, like IOT, have potential with their sensors for monitoring of air quality in the urban community, and the value of data in real time, analysis of data, and communication for the data are addressed in the case study for urban areas and showcases their potential. Cities can take steps to protect public health and improve their environmental quality for both the current and future generations.

2.4.2 Analyzing Satellite Images for Urban Development and Planning

Urban planners and legislators make decision making in a good manner for the promotion of sustainability and standard enhancement of the urban environment [28]. Satellite imagery and its analysis are excellent tools for urban development and its planning for providing information about the environmental dynamics, infrastructure of the buildings, and the usage of land.

Case Study: Using Satellite Data to Manage Urban Growth
Overview: A case study on urban development and the growth in a rapid city are monitored and tracked with satellite images. High-growth potential areas and urban planning guide policies to manage the growth and sustainably are discussed in this study [29].

Techniques
Data Acquisition: Sentinel and Landsat satellite data with high-resolution satellite images were obtained as a source.
Image Processing: Software was used and processed for processing the satellite imagery to create maps of the land cover and identify urban features.
Change Detection: To identify changes in land cover over time and quantify urban expansion, change detection techniques were used. Spatial analysis techniques, like urban morphology and changes in land use, are well examined.
Visualization: By using software like GIS (Geographic Information System), the findings and analysis were made into charts, maps, and other visual modes of communication and interpretation.
Results: Due to deforestation and the demolition of agricultural land into urban areas, increase in the study areas of urbanization are viewed through satellite imagery analysis. High-density population, development of infrastructure, and degradation of the environment are identified with the help of spatial analysis. The results show the significance of different perspectives for urban planning and the policymakers for giving priorities for development safeguarding the entire natural resources and balancing the ecological threat of urban expansion.

Uses

1. **Urban Planning:** To support urban population expansion and enhance urban livelihood, satellite image analysis is helpful for land plan creation amending the laws and investment in infrastructure.
2. **Environmental Management:** Prioritize the areas at risk of environmental degradation and identify the changes in vegetation and land cover, and the efforts for conservation.
3. **Disaster Management:** To identify the areas of vulnerability, the areas that are affected and damaged are identified with the help of satellite imagery analysis to prevent response in the planning, management of disaster risk assessment, and for recovery from post-disaster.

Conclusion: For making the decisions and management of sustainability in urban growth, analysis is used with the help of satellite images for the planning of urbanization and their development to protect natural resources and improve the quality of life for the citizens due to lifestyle changes and because of urbanization.

2.4.3 Drone Aerial Vehicle-Based Monitoring for Environmental Control

Providing high-resolution images, real-time monitoring, evaluation of environmental conditions, and the possibility of threats are based on drone-based surveillance for environmental management. Proactive decisions for safeguarding natural resources and for reducing environmental risks are made with the help of drones for management [29].

Case Study: Using Drones for Wetland Conservation Through Surveillance
Overview: Using drone-based surveillance, a wetland ecosystem risk for the degradation of the environment and human activity was monitored and managed. The study focuses on the evaluation of the wetland ecosystem's health for detecting problems and developing conservation plans for the protection of biodiversity.

Techniques
Drone Deployment: A commercial-grade drone is fitted with high-resolution cameras and sensors to survey the wetland area. Drone Aerial Survey: Taking in-depth pictures of terrain from perspectives and different angles.
Environmental Monitoring: Gathering of information on anthropogenic disturbances, vegetation health, water quality, and habitat diversity is crucial for the environmental factor data that are gathered by the drone.
Data Analysis: For environmental assessment and to develop the spatial indicators, maps and statistical metrics are collected, examined, and processed using the geographic information system (GIS) methods and the software.
Decision Support: For pollutant control in the environment, habitat restoration in the land, and the planning of land use. Gathered information from the drone-based surveillance for the management.
Results: The drone-based surveillance did not include the essential hints about the wetland ecosystem and pointed out only the several issues to be addressed, such as the endangerment of species, loss of habitat, and

the pollution of water. Environmental managers are able to track changes, target interventions, conservation, and the balancing of ecosystems with the help of high-resolution imagery and data after analysis. Community involvement in the wetland is facilitated by stakeholder engagement and collaboration for the new initiatives.

Uses

1. **Environmental Monitoring:** Early findings of changes in the ecosystems and the intervention for the degradation are possible with the help of real-time environmental condition monitoring using drone-based surveillance.
2. **Habitat Mapping:** A detailed map of habitat types, distribution of species, and connectivity of ecosystems is possible with drone images and its high resolution; with this, we can preserve biodiversity and restore natural habitats.
3. **Illegal Activity Detection:** Detecting and identifying illegal activities, like poaching, illegal dumping, and logging, in remote and not reachable areas can be identified with high precision and efficiency using drones.
4. **Drone-Based Surveillance** provides useful information and data timely to stakeholders for the management of the environment so they can safeguard the natural resources and preserve the ecosystems. Organizations are actively involved in identifying environmental challenges and sustainable development accomplishments for environmental monitoring and management strategies with the help of drones.

2.4.4 Augmented and Virtual Reality Uses in Planning and Monitoring Urban Environments

The fields of urban monitoring and planning of the environment can benefit from the use of new immersive technologies like augmented reality (AR) and virtual reality (VR). These technologies give the immersive feel of visualization for the analysis of spatial data, and also for engaging stakeholders, in making important decisions. Stakeholders are involved in working collaboratively for the designing of environment sustainability, scenes with alternate scenarios, and for the understanding of more complex spatial relationships using AR and VR in urban planning processes [30].

Case Study: Simulating Urban Green Spaces With Virtual Reality
Overview: This case study deals with urban green space initiatives, and the planning in a widely populated city was visualized, and similar content is simulated using Virtual Reality (VR) technology. The study's main goals were to involve the parties who are interested in the planning and assessment of green space interventions for measuring the projects that would stop the rules and regulations of cities, and for providing information regarding sustainable urban development.

Techniques
Urban planners created new three-dimensional (3D) models for the particular area of study, including the green space project planning, pattern of land usage, and the particular infrastructure.
Virtual Reality Simulation: By using the 3D models in virtual reality (VR), users can engage and explore the virtual simulated and the brain-mimicking world in real time.
Scenario Analysis: Different scenes and examples of interventions of diverse green space, which include the extensions of the park, developments of greenways, and gardens on the rooftop, are shown to the stakeholders.
Stakeholder Engagement: Advantages, usability, and the visual appeal of the participants are seen through the VR simulation, and the participants comment on the projects they see through a window.
Decision Support: Stakeholder feedback was collected and processed to refine, for prioritizing the initiatives, to ensure the planning of green space, and to be in line with the goals of the community [31].
Results: A realistic and immersive experience of the new suggested green space projects and for the VR simulation of advantages and disadvantages of various interventions of stakeholders. Encouraging dialogue, fostering consensus, and collaborating on the development of fresh approaches to urban planning participants in the sessions, with a feeling of responsibility and dedication to sustainable urban development projects.

Uses

1. **Urban Design and Visualization:** Urban planners are allowed to visualize and simulate the concepts for urban design, for the streets, public spaces, and for the green infrastructure, AR and VR immersive technologies for the promotion of informed decisions and the engagement of stockholder.

2. **Public Involvement and Engagement:** Augmented Reality (AR) and Virtual Reality (VR) provide immersive and interactive experiences by involving the public in urban planning, gathering feedback, and promoting cooperation among various stakeholders, such as developers, policymakers, and residents.
3. **Design Iteration and Evaluation:** To test and refine the planning proposals in a virtual environment for the designers, and the design in conflicts, and to alter the solutions of investigation before using them, and the immersive simulations for the development.

The use of immersive technologies in the urban area is well monitored and planned, which shows great promise. These technologies have better capabilities to visualize, analyze, and communicate spatial data and scenarios. Urban planners are key stakeholders in the effort to co-create sustainable urban environments for current and future generations by utilizing immersive technologies such as AR and VR.

2.4.5 Combining Emerging Technologies and Remote Sensing

To improve the urban environment, monitoring and planning are used for the integration of remote sensing and emerging technologies, for the collection of data, analysis of data, and the decision-making processes in an effective way. Practitioners in urban environments can gain new insights for comprehending and regulating diversity in real time and for integrating with new revolutionized technologies like Artificial Intelligence (AI), machine learning (ML), Internet of Things (IoT), Virtual Reality (VR), and Augmented Reality (AR) [32].

Case Study: Monitoring of Urban Green Infrastructure Assisted by AI
Overview: A case study on urban green infrastructure in a fast-revolutionizing era was managed and monitored with the integration of remote sensing and the fast-evolving Artificial Intelligence (AI).

The study's focus was to develop an automated system with the help of satellite imagery and Artificial Intelligence (AI) algorithms for the effectiveness and analysis of new urban green infrastructure like roofs, parks, and urban forests.

Techniques

Data Acquisition: Optical and multispectral sensors are used with the high-resolution satellite images of the satellite.

1. **Image Analysis:** Spatial features and spectral signatures are extracted, and the urban green buildings and satellite images are processed and analyzed with the use of remote sensing techniques.
2. **AI Algorithm Development:** Convolutional neural networks (CNNs) were trained on labeled remote sensing data to classify and detect urban green infrastructure like machine learning.
3. **Automated Monitoring System:** To detect alterations in urban green infrastructure using data in real time, it predicts the warning signs of natural calamities in prior and give the corresponding message to the respective departments.
4. **Integration With IoT Sensors:** To improve the nature and the precision of the monitoring process with the help of IoT sensors and to place in urban green spaces for the gathering of all the information related to soil moisture, environmental temperature, and healthy vegetation.

Results: The condition and functionality of urban green infrastructure, allow for a high level of maintenance and management with the integration of these automated systems and remote sensing. Urban practitioners have problems like tasks prioritizing and the allocation of images based on satellite imagery and IOT data for making decisions and supporting the management of data with the spaces using an automated system.

Uses

1. **Management of Urban Green Infrastructure:** For automated management and monitoring of urban green infrastructure, the integration of AI and the new remote sensing method promotes the ecosystem through the enhancement of services, conservation of biodiversity, and the rules based on the urban environment.
2. **Disaster Risk Reduction:** By identifying environmental changes and evaluating risks to urban populations and infrastructure, remote sensing and Artificial Intelligence (AI) technologies offer early warning systems for natural disasters like floods, wildfires, and landslides.

3. **The Development of Smart Cities**, where data-driven decision making, effective resource management, and sustainable development practices are prioritized, is facilitated by the integration of remote sensing with IoT, AI, and other emerging technologies [33].
4. **The Amalgamation of Remote Sensing** revolutionary prospects for augmenting urban environmental surveillance and strategizing, thereby permitting a more efficient and eco-friendly administration of urban environments. Cities can address difficult urban issues, foster resilience, and enhance the quality of life for citizens by utilizing AI, IoT, remote sensing, and other cutting-edge technologies. Sustainability and equality are achieved by this.

2.5 Frameworks for Integrating Multiple Data Sources

2.5.1 Platforms for Collaborative Work and Data Sharing

The usage of collaborative platforms for data sharing and the possibility of collaboration by combining the different data sources in urban planning of the environment and the management of these are possible with these techniques. This gives collaboration, and facilitates transparency and specific well-informed decision making with the stakeholders by providing them a central location to access the data, share the huge variety of datasets, and process them [34].

Overview of the Framework
Urban environmental planning and maintenance stakeholders will easily share data, knowledge about the data, and the reason for collaboration and its expertise through a virtual ecosystem.

This is the integration of different techniques for sharing huge datasets, with remote sensing, spatial data, Internet of Things, and indicators of social–economic factors using different technologies such as cloud computing, applications based on the web, and the interoperability of standards.

To gain comprehensive insights from the platform that is collaborative based on the dynamics of the urban environment, trends in identifying and finding the solution for complex problems are employed using access to analytical tools and data sources.

Important Collaborative Platform Characteristics

1. **Data Integration:** The datasets are obtained from different organizations, like governmental organizations specifically academic institutions, businessmen and their partnerships, and non-governmental organizations, for the integration of collaborative platforms. Stakeholders have enormous access to the data related to the urban environment for processing and for making decisions with the analysis.
2. **Interoperability:** To combine the exchange of data and the integration of different datasets, the platforms are collaborative in providing data interoperability protocols and their standards. This is very useful for the formats to be interoperable and to collaborate with the different standards.
3. **Tools for Visualization and Analysis:** The visualization interactivity of different features is possible with this collaborative system and the exploration, analysis of data, and urban data interpretation. With the help of the tools, stakeholders would develop charts, dashboards, graphs, and maps for analyzing the spatial patterns, relationships, and trends in the data.
4. **Stakeholder Participation:** The stakeholders are actively involved in urban planning and monitoring. Stakeholders can share their ideas in an open forum and get feedback in a place where they work together on projects through online discussion boards, reputed forums, and in a collaborative workspace.
5. **Decision Support:** To create evidence-based decision making and a policy-based formulation, the collaborative platforms give the decision support functionalities and the tools for analyzing the risks for predictive analytics, and for modeling the scenario. By this, the potential solutions are identified.

Case Study: Platform for Urban Data Exchange

Overview: A case study of the Urban Data Exchange Platform is used by the stakeholders to be involved in the planning of the urban environment, project monitoring, for sharing data, and working easily. The platform's focus was to develop a centralized hub for accessing, sharing, and analyzing the data in an urban dataset, such as in IoT sensors, remote sensing of data, socioeconomic indicators of the data, and the spatial data layers.

Qualities
Data Repository: The centralized location for storing and managing the data in an urban dataset, with air quality measurements, maps for infrastructures, and demographic information and information related to satellite images.
Data Visualization: Examining the spatial patterns, data trends, and relationships within urban data through interactive maps, graphs, and charts.
Collaborative Tools: Discussion boards, project workspaces, and shared documents are used to promote communication and cooperation methods in the stakeholders and the platform.
Analytics and Modeling: To help in decision making and the creation of policies, with the new sophisticated analytics and for modeling tasks like planning the scenario, model prediction, and spatial analysis.
Open Data Access: By allowing data access in openness of data, creating the data, and the involvement of the public in urban planning management.

Advantages
Enhanced Data Accessibility: The platform is accessed by a centralized stakeholder to improve the urban environment data and for a user-friendly interface in finding the data, accessing the data, and downloading the relevant datasets.
Enhanced Cooperation: A different set of stakeholders, like governmental organizations, private educational institutions, non-profit organizations, and community organizations, work together and share knowledge.
Informed Decision Making: The platform gives evidence-based policy formulation and information for decision making and addresses the urban challenges in a user's access to timely and trustworthy urban data.
Building Capacity: The best practice manuals and case studies for urban data management and the analysis to gain the knowledge in an exchange of capacity building.

The combination of multiple data sources in urban areas using collaborative platforms for data sharing and collaboration is monitored and planned. These platforms provide a hub for data access, sharing, and urban data analysis among stakeholders, facilitating collaboration on new developments, sustainability, and urban areas. Integration of platforms to enable transparency and innovation, in public participation, in the making of decisions and the processes of open access, the tools for collaboration, and the tools produce developmental urban development.

2.5.2 Regulatory Aspects and Policy Implications

There are different policies and implications of regulations for the combination of multiple data in the planning and monitoring in urban areas. The development of new frameworks, regular standards, and new guidelines governing data collection, sharing the data, privacy in data, data security, and ethical use is important for policymakers and new regulators to develop initiatives for data-driven good urban development.

- **Data Governance:** For the data gathering and control, storing the data, exchanging the data, and usage of urban area data, policy creation must generate thorough frameworks for the governance of data. These frameworks might give the protocols and the potential concerns regarding data creation, access to the data, privacy of data, security, and accountability. They give balance between the innovation of data and individual safeguarding of liberties and interests of the public.
- **Interoperability Standards:** To enable a good combination and exchange of various urban datasets, policymakers can adopt interoperability standards and protocols. Compatibility and interoperability across various applications, methods, and formats of data and enable the stakeholder's standardization to integrate across boundaries and fully organizational usage of data.
- **Open Data Policies:** To increase the openness, creativity of the data, and public involvement in urban environmental monitoring and planning, open access policies are provided by the stakeholders. Public data, academic data, and developer access to the datasets are possible with the initiatives of data, for cooperation, innovation data, and decision making.
- **Data Security and Privacy:** To stop unauthorized access, misuse of data, and the exhibit of sensitive data, issues about data privacy and security are addressed by policymakers. Data protection laws, cybersecurity, and guidelines for encryption are examples of this regulation for the protection and support of individual rights for privacy and data integrity, and less data are breached, and there are less cyberattacks. Policymakers will consider the changes in using urban

data, such as accountability questions, equity of data, transparency, and justice. The accountability for policies related to urban development, rules, and code violations, as well as the social, inclusivity, and human dignity.
- **Regulatory Aspects to Consider:** Regulatory Frameworks: To use the new cutting-edge technologies in an urban setting, like Augmented Reality (AR), Artificial Intelligence (AI), and Internet of Things (IoT), the regulators use the frameworks. To give the stakeholders legal certainty regarding the compliance of regulations, the frameworks will have concerns, like data privacy, data security, data liability, and intellectual property rights, for the impact of the environment.
- **Compliance Requirements:** To guarantee a data-driven project in an urban area for planning and monitoring, standard complaints requirements, and program certification are required. Users' data and data providers' data are used to analyze system complaints and behavior for the purpose of risk reduction.
- **Public Engagement:** To gather information, immediate feedback, and the effects of regulations for the stakeholder's interaction with the public, partners in industry, agencies of government, and the society of civil organizations. The consultation procedures ensure that the frameworks regularly represent different priorities and points of view for assisting in the development of consensus and trust.
- **Monitoring and Enforcement:** Regulators are needed to set up systems, like the audits of data, data inspections, sanctions of data, and enforcement of actions, for monitoring and enforcing the adherence to regulatory requirements. Good regulations make the guidance of supervision in data-driven projects legal and moral safeguarding the general public interests and advancing.
- **Adaptive Regulation:** To keep quick technological development and the need for changing the urban area, regulators may implement regulatory strategies. By implementing and overseeing adaptable frameworks, adjusting standards and regulations, and formulating guidelines to address evolving conditions and risks among stakeholders. To facilitate the interaction efficiently.

Conclusion

The regulators are given individual rights, and public interest in upholding and advancing sustainability at the same time, by developing strong frameworks, standards to follow, and regulations to facilitate innovation to be data driven. Regulators and the creators of policies can give transformative potential of urban area data to develop smarter cities for upcoming generations. This can be done with the active involvement of stakeholders, ethical principles, and regulations.

2.5.3 Prospective Pathways and Difficulties

Enormous challenges and directions are developing due to the change in environmental monitoring and planning of urban areas. This improves the incorporation of data-driven initiatives and procedures for making decisions in cities worldwide.

Integration of Emerging Technologies: Upcoming new projects will focus on incorporating the new cutting-edge technologies into the programs of urban environmental planning and monitoring, including blockchain technology, Augmented Reality (AR), Virtual Reality Artificial Intelligence (AI), and the Internet of Things (IoT). These new tools for data analysis, visualization of data, and decision making are involved in the development of the urban environment.

Improved Data Interoperability: To facilitate the combination and interchange of all urban data, efforts are required for compatibility across the datasets and also for interoperability. Protocol standardization, formatting of the data, and metadata are very easy tools for the stakeholders to use in the examination of urban dynamics.

Community engagement and science initiatives for the citizens will be an upcoming direction to be involved in the urban environmental planning and monitoring projects. Transparency of data, inclusivity of data, and cohesion of social data are gained through the new knowledge that is available locally, and it produces the driven initiatives for the decision-making process.

Resilience and Adaptation Strategies: The new effects of climate change and environmental risks in urban planning and development are used for resilience and adaptation strategies. Extreme environmental weather, natural calamity and disasters, and other environmental hazards can be rectified with the cities with the help of assessment of risks, mapping of vulnerability, and other strategies.

Policy Innovation and Governance Reform: To note and address the new challenges and the evolving opportunities for the integration of data in the evolving environment, innovation policies and governance are upcoming researches. The collaborative governance models and their implementations, decision-making process participation, and regulatory framework adaption in urban development initiatives foster data transparency, accountability of data, and stakeholder engagement for the data.

Problems

Data Accuracy, Reliability, and Quality: Data accuracy, reliability, and data quality in urban areas remain significant challenges in this field. The primary obstacle is the lack of data availability. Strong validation of data, verification, and assurance of data quality and the processes are required to concentrate on the bias and misuse of data. Individuals are required to safeguard the policies and privacy rights to address ethical problems in data collection, data sharing, and data usage. For privacy, data confidentiality, and data autonomy, the advantages of data-driven innovation necessitate the consideration of ethical principles and safeguard regulations.

Digital Divide and Equity: The important problem is the integration of data and the initiatives in urban settings for bridging the data and the advances in the equity for the data and technology. Targeted strategies and innovations are necessary for all resident people, including gated communities, to give them equal access to data resources and opportunities.

Building Institutional Capacity and Promoting Collaboration: Among the different stakeholder collaborations, the promotion of stakeholders' diversity and the building of institutions are the obstacles to data integration and efforts in urban environmental planning. To overcome this problem, we need technical skills, cross-sectoral partnership development, and knowledge exchange encouragement.

Financial Sustainability: Allocating funds and resources effectively is the primary obstacle in integrating data and developing cities with limited data in the present day. The long-term use of data from urban and infrastructure and the initiatives of capacity building need sustainable mechanisms for finance, partnerships for public and private data, and funding model creation.

Conclusion

Navigating future research and tackling challenges require teamwork, thinking creatively, and flexible methods. The objective is to establish sustainable urban data and environments for future generations, develop the transformative potential of data integration in emerging technologies, enhance data interoperability, engage communities, promote innovation policies, and address the associated challenges. In a new era of data-driven world, cities adapt to the challenges of evolving. This needs the constant learning of data, collaboration of data, and data adaptation for upcoming development.

2.6 The Possible Effects of Next-Generation Connectivity and 5G

2.6.1 Privacy and Ethical Issues With Urban Remote Sensing

Ethical and privacy concerns in urban remote sensing are relevant to embracing the potential of 5G connectivity and next-generation urban environmental planning and monitoring. These technologies give earlier possibilities for data collection, data analysis, and the correct decision making for significant privacy and ethical issues to address the rights of people and the public interests [34].

Data Security and Privacy Risk
Data Collection Practices: The extent and magnitude of surveillance and the concerns regarding data collection in an urban environment for connectivity improvement and sensor proliferation. To guarantee the stakeholders the information about the control and the collection of data, in the data collection, transparency procedures and consent for the information are important sensing projects related to remote sensing and should have priority for the anonymization of data and reverse de-identification techniques for reducing and safeguarding the risks in individual privacy for data sharing and analysis of data. Sensitive data as location data or biometric identifiers can be used to identify people with these new innovative anonymization techniques [35].

Accountability and Transparency
Algorithmic Transparency: The accountability of algorithms and the transparency of data are in question for the application of machine learning algorithms and the combination of new artificial intelligence in urban

remote sensing. The decision-making processes are highly unbiased, ensure data accountability and impartiality, employ transparent algorithms, and prioritize the ability to explain data. Transparency, implementation, and mechanisms for auditing are recommended for the policymakers and the regulators.

Accountability Mechanisms: In order to ensure that data-driven projects in urban remote sensing meet the legal and ethical requirements, accountability mechanisms are implemented. To address the privacy gaps or ethical policy changes, regulators in these policies, stakeholders of the industry, and organizations of civil sectors are important for compliance and accountability enforcement [36].

Justice and Equality
Digital Divide: In urban areas, the usage of cutting-edge technologies has the potential to widen the research gaps already among the people gaps. To guarantee that the citizens, including community margin, have equal access to the urban remote sensing advantages for next-generation connectivity, it is essential to connect the digital divide and digital inclusion promotion.

Community Engagement: To address the concerns of equity and the data-driven interventions of guarantee that reflect the need of priorities and the communities, for the development of design implementation, and the data governance initiatives. Residents have more concerns about showing their interests and for holding responsible decision makers for results and the usage of approaches in participation, partnership in community, and decision-making inclusivity [37].

Legal Safeguards and Regulatory Frameworks
Data Protection Laws and Regulations: To protect each person's privacy rights and to control the data gathering, data utility, and data dissemination, for the protection of data and the regulations. A data protection framework gives the protection of data and the mechanisms of enforcement data for personal data protection and privacy rights in an urban environment. Example: General Data Protection Regulation (GDPR) in the European Union.

Ethical Guidelines: The best practice creation and ethical privacy guidelines for urban remote sensing data can help in directing the stakeholders' decision making and behavior responsibility. As a way to promote the ethical use of data for data-driven projects that contribute to the overall welfare benefits in society, rules [38] address issues such as data consent management, accountability, and risk management.

Conclusion

Urban remote sensing raises ethical and privacy issues that should be taken very seriously as cities embrace 5G and other next-generation connectivity for data-driven urban development. To ensure that data-driven initiatives uphold individual rights, foster public trust, and contribute to sustainable and inclusive urban development, policymakers, regulators, and stakeholders must proactively address privacy risks, promote transparency and accountability, advance equity and social justice, and strengthen regulatory frameworks and legal protections.

2.6.2 New Technologies' Scalability and Affordability

The integration of new and developing emerging technologies, such as remote sensing, in urban environmental planning is very crucial. To ensure the adaption and sustainable implementation for the world, scalability of data and affordability of data are important things to consider [39].

1. **Infrastructure Extension:** The ability to scale up the infrastructure support in emerging technologies is known as scalability. The measurement of this encompasses the usage of supplementary data sensors, enhancement of data communication networks, and data allocation of computational resources for the substantial data sets in remote sensing and IOT.
2. **Interoperability:** Scalable solutions are required to communicate and integrate with different technologies and systems easily. The solutions are interoperable for scalability across different cities and regions in the data exchange and diversity of platforms.
3. **Modular Design:** To enable adaptability and the different flexibility for shifting needs and the level-up scaling, the technologies may give the modular design approach. Without the infrastructure need, solutions are needed for the modularity, customization, and expandability of data to meet the challenges and the needs in urban settings.
4. **Cloud-Based Solutions:** Large datasets generated by IoT devices and remote sensing are processed and analyzed cheaply with the help of cloud-based platforms. Infrastructure-as-a-service (IaaS) solutions are scalable and

use this for the on-demand computing power provisioning and capacity of storage in cloud computing.
5. **Cost-Effective Solutions:** When using the new emerging technologies in urban setting areas, specifically in the cities and the remote areas, affordability of data is important to consider. Cost-effective solutions require aligning the operating costs by sharing infrastructure, hardware, and software, as well as leveraging the resources and expertise of the private sector in implementation and financing of new technologies in urban areas through public-private partnerships (PPPs) to ensure affordability. PPPs divide the costs and risks between the data from private and public data parties to adapt financially for the new technologies.
6. **Lifecycle Cost Analysis:** The entire usage of maintaining and deploying the cutting-edge technologies in an urban setting is considered in the evaluations of affordability. The lifecycle cost and technology affordability measures are used to analyse the upfront costs, maintenance costs, expected returns on investment (ROI), and ongoing costs of solutions throughout their usage.
7. **Inclusive Pricing Models:** Inclusive pricing models assure equitable access to technology for the residents, income level, and socioeconomic status, irrespectively. These people are given priority when considering affordability. The communities of the underdeveloped are given less privilege and increased accessibility for technological solutions in financial mechanisms, pricing structures, and pricing subsidies.

Conclusion

The usage of emerging technologies requires careful focus on two key factors: scalability and affordability in urban planning. Cities may produce scalable solutions that give the advantage of cloud-based infrastructure usage, interoperability of data, and data modularity to enhance and modify the data and their technological capacities in all responses to making changes in the demands and the challenging obstacles. Moreover, by this, the cities can guarantee good access to the solutions and for sustainable urban development for all citizens in considerable solutions, partnerships, and pricing models.

2.6.3 AR and VR's Place in the Future of Urban Environment Management

Immersive technologies, like virtual reality (VR) and augmented reality (AR), play an important role to change urban environmental management techniques in upcoming projects. One innovative approach to visualising, analysing, and communicating urban data is to involve stakeholders in proactive measures to identify environmental issues and make well-informed decisions. Mixed reality (AR) and virtual reality (VR) are used in several significant domains of urban environmental management. Urban environmental management uses AR and VR in various important areas [40].

Analysis and Visualization of Data
Stakeholders are able to visualize the urban environmental data in more immersive 3D environments, like AR and VR, to improve the comprehension and the relationships of spatial data, data trends, and data patterns [41] using interactive data analysis tools to investigate the urban environments, scenarios, and the possibility of data.

Collaboration and Engagement of Stakeholders
By producing immersive experiences that promote involvement and cooperation in urban environmental management projects, augmented reality and virtual reality (AR/VR) help to increase stakeholder engagement. The stakeholders can collaboratively exchange data viewpoints and co-create solutions in an environmental challenge for virtual meetings, workshops, and design collaboration sessions in virtual reality (VR) environments.

Planning and Decision Support
Legislators, environmental managers, and urban planners can view alternative scenarios, weigh trade-offs, and prioritize interventions through a head-mounted display using AR and VR as decision support tools. In the use of scenario modeling, assessment of impact, and the analysis of risks, the technologies help the stakeholders in developing strategies, which are flexible and for sustainable development based on the facts from urban data [42].

Education and Public Awareness [43, 44]
Civic duty and environmental stewardship are through the applications of AR and VR in terms of public awareness and local teaching about environmental issues and are encouraged. Engaging the participants or the audiences with the new immersive storytelling and the virtual tours in

interactive educational experiences using AR and VR environments practices them to behave in a more proper sustainable development [45].

Readiness and Preparedness for Disasters
To help emergency responders and the members of the community to build the skills and develop the mechanism that is necessary to practice along with the environmental hazards, simulations of augmented reality and virtual reality to produce more realistic scenarios for training, and the preparation for disasters with virtual simulations, stakeholders use the routes for evaluation, emergency coordinate responses, and potential understanding of impacts like wildfires, storms, and floods [46].

Observation and Investigation
The surveillance and real-time monitoring of environment data related to urban settings are possible with the help of AR and VR technologies for the stakeholders to spot the changes in hotspots and new threats. AR and VR technologies have the importance of revolutionizing urban environmental management for providing future cutting-edge environmental monitoring, disaster resilience, management of stakeholders, visualization of data, and decision support. Cities can address difficult issues in the environment, sustainability encouragement, and quality of life for the citizens of AR and VR. These technologies have significant potential for improving urban management of environmental data [47, 48].

2.7 Conclusion

This chapter discusses remote sensing technology for dealing with issues in the urban environment, efficient connectivity, and capability. The importance of combining cutting-edge technologies, like Artificial Intelligence, the Internet of Things, Augmented Reality, and Virtual Reality, to improve urban environmental management and get around connectivity issues is highlighted by key findings. The creation of different cooperative platforms for data sharing, the adoption of all moral and privacy standards, and the expenditure of money on all the scalable and reasonable-cost technological solutions are recommended. Urban environmental remote sensing and the development in all 5G connectivity, Artificial Intelligence (AI)-powered analytics, and technological democratization have a future scope. Urban environmental management techniques are in revolution, to make easy for the decision process by the stakeholders, encourage participation, and also creating stronger, viable, and enjoyable cities for the coming generations.

2.7.1 Future Gap

Further studies might concentrate on handling the remaining obstacles and constraints in the fusion of remote sensing technology and developing urban environmental management methodologies. More specifically, there are creative ways to improve the cost effectiveness, and scalability, and the accessibility of these technologies must be examined especially in environments with limited resources and populations. To ensure responsibility and implementation, more research into the area of ethical, social, and environmental management of remote sensing is essential. Future research also concentrates on the policy interventions and the capacity-building programs that work to motivate the stakeholders to adopt and use remote sensing technology.

References

1. Toth, C. and Jóźków, G., Remote sensing platforms and sensors: A survey. *ISPRS J. Photogramm. Remote Sens.*, 115, 22–36, May 2016.
2. Atkinson, P.M. and Lewis, P., Geostatistical classification for remote sensing: an introduction. *Comput. Geosci.*, 26, 4, 361–371, 2000.
3. Shen, H., Li, H., Qian, Y., Zhang, L., Yuan, Q., An effective thin cloud removal procedure for visible remote sensing images. *ISPRS J. Photogramm. Remote Sens.*, 96, 224–235, 2014.
4. Cheng, G., Han, J., Zhou, P., & Guo, L., Multi-class geospatial object detection and geographic image classification based on collection of part detectors. *ISPRS J. Photogramm. Remote Sens.*, 98, 119–132, 2014.
5. Zhang, B., Wu, D., Zhang, L., Jiao, Q., Li, Q., Application of hyper spectral remote sensing for environment monitoring in mining areas. *Environ. Earth Sci.*, 65, 649–658, 2012.
6. Ozden, A., Faghri, A., Li, M., Tabrizi, K., Evaluation of Synthetic Aperture Radar satellite remote sensing for pavement and infrastructure monitoring. *Procedia Eng.*, 145, 752–759, 2016.
7. Bello, O.M. and Aina, Y.A., Satellite remote sensing as a tool in disaster management and sustainable development: towards a synergistic approach. *Procedia-Social Behav. Sci.*, 120, 365–373, 2014.
8. da Cunha, L.K., *Exploring the benefits of satellite remote sensing for flood prediction across scales*, The University of Iowa, 2012, Doctoral dissertation, United States.

9. Hedley, J.D., Roelfsema, C.M., Chollett, I., Harborne, A.R., Heron, S.F., J. Weeks, S., Mumby, P.J., Remote sensing of coral reefs for monitoring and management: a review. *Remote Sens.*, 8, 2, 118, 2016.
10. Mumby, P.J., Green, E.P., Edwards, A.J., Clark, C.D., The cost-effectiveness of remote sensing for tropical coastal resources assessment and management. *J. Environ. Manage.*, 55, 3, 157–166, 1999.
11. Smith, D.Y., Participatory Planning and Procedural Protections: The Case for Deeper Public Participation in Urban Redevelopment. *Louis U. Pub. L. Rev.*, 29, 243, 2009.
12. Amani, M., Ghorbanian, A., Ahmadi, S.A., Kakooei, M., Moghimi, A., Mirmazloumi, S.M., Brisco, B., Google earth engine cloud computing platform for remote sensing big data applications: A comprehensive review. *IEEE J. Sel. Top. Appl. Earth Obs. Remote Sens.*, 13, 5326–5350, 2020.
13. Xu, Y. and Zhu, Y., When remote sensing data meet ubiquitous urban data: Fine-grained air quality inference, in: *2016 IEEE International Conference on Big Data*, pp. 1252–1261, 2016.
14. Pearce, W.A., Cloud shadow effects on remote sensing. *IEEE Trans. Geosci. Remote Sens.*, 5, 634–639, 1985.
15. Spencer, M. and Ulaby, F., Spectrum issues faced by active remote sensing: Radio frequency interference and operational restrictions technical committees. *IEEE Geosci. Remote Sens. Mag.*, 4, 1, 40–45, 2016.
16. Gagliardi, V., Tosti, F., Bianchini Ciampoli, L., Battagliere, M.L., D'Amato, L., Alani, A.M., Benedetto, A., Satellite remote sensing and non-destructive testing methods for transport infrastructure monitoring: Advances, challenges and perspectives. *Remote Sens.*, 15, 2, 418, 2023.
17. Prados, A., II, Leptoukh, G., Lynnes, C., Johnson, J., Rui, H., Chen, A., Husar, R.B., Access, visualization, and interoperability of air quality remote sensing data sets via the Giovanni online tool. *IEEE J. Sel. Top. Appl. Earth Obs. Remote Sens.*, 3, 3, 359–370, 2010.
18. Wehner, D. and Frey, T., Offshore unexploded ordnance detection and data quality control—a guideline. *IEEE J. Sel. Top. Appl. Earth Obs. Remote Sens.*, 15, 7483–7498, 2022.
19. Hellegers, P.J.G.J., Soppe, R., Perry, C.J., Bastiaanssen, W.G.M., Combining remote sensing and economic analysis to support decisions that affect water productivity. *Irrig. Sci.*, 27, 243–251, 2009.
20. Ma, E., Lai, J., Wang, L., Wang, K., Xu, S., Li, C., Guo, C., Review of cutting-edge sensing technologies for urban underground construction. *Measurement*, 167, 108289, 2021.
21. Zhang, L. and Zhang, L., Artificial intelligence for remote sensing data analysis: A review of challenges and opportunities. *IEEE Geosci. Remote Sens. Mag.*, 10, 2, 270–294, 2022.
22. Beeri, O. and Peled, A., Geographical model for precise agriculture monitoring with real-time remote sensing. *ISPRS J. Photogramm. Remote Sens.*, 64, 1, 47–54, 2009.

23. Lillesand, T., Kiefer, R.W., Chipman, J., Remote sensing and image interpretation, in: *John Wiley & Sons*, 2019.
24. Akhtarkavan, E., Majidi, B., Manzuri, M.T., Secure communication and archiving of low altitude remote sensing data using high capacity fragile data hiding. *Multimed. Tools Appl.*, 78, 10325–10351, 2019.
25. Hu, Z., Bai, Z., Yang, Y., Zheng, Z., Bian, K., Song, L., UAV aided aerial-ground IoT for air quality sensing in smart city: Architecture, technologies, and implementation. *IEEE Netw.*, 33, 2, 14–22, 2019.
26. Barthwal, A. and Acharya, D., An IoT based sensing system for modellingand forecasting urban air quality. *Wirel. Pers. Commun.*, 116, 3503–3526, 2021.
27. Pettinger, L.R., Field Data Collection–An Essential Element in Remote Sensing Applications, in: *Proceedings of the international workshop on Earth resources survey systems*, Washington, DC, pp. 49–64, May 1971.
28. Jude, S., Mokrech, M., Walkden, M., Thomas, J., Koukoulas, S., Visualising potential coastal change: communicating results using visualisation techniques, in: *Broad Scale Coastal Simulation: New Techniques to Understand and Manage Shorelines in the Third Millennium*, pp. 255–272, 2015.
29. Valdez-Delgado, K.M., Garcia-Salazar, O., Moo-Llanes, D.A., Izcapa-Treviño, C., Cruz-Pliego, M.A., Domínguez-Posadas, G.Y., Danis-Lozano, R., Mapping the Urban Environments of Aedes Aegypti Using Drone Technology. *Drones*, 7, 9, 581, 2023.
30. Lu, W., Zhao, L., Xu, R., Remote sensing image processing technology based on mobile augmented reality technology in surveying and mapping engineering. *Soft Comput.*, 27, 1, 423–433, 2023.
31. Leebmann, J., Application of an augmented reality system for disaster relief. *Int. Arch. Photogrammetry, Remote Sens. Spatial Inf. Sci.*, 34, 5, W10, 2006.
32. Malathy, S., Sangeetha, K., Vanitha, C.N., Dhanaraj, R.K., Integrated architecture for IoTSG: internet of things (IoT) and smart grid (SG), in: *Smart Grids and Internet of Things: An Energy Perspective*, pp. 127–155, 2023.
33. Wu, L., Jiang, X., Yin, Y., Cheng, T.C.E., Sima, X., Multi-band remote sensing image fusion based on collaborative representation. *Inf. Fusion*, 90, 23–35, 2023.
34. Alkhelaiwi, M., Boulila, W., Ahmad, J., Koubaa, A., Driss, M., An efficient approach based on privacy-preserving deep learning for satellite image classification. *Remote Sens.*, 13, 11, 2221, 2021.
35. Sarti, F., Inglada, J., Landry, R., Pultz, T., Risk management using remote sensing data: Moving from scientific to operational applications. *Present. X SBSR, April*, 23, 27, 2001.
36. Pincheira, M., Donini, E., Giaffreda, R., Vecchio, M., A blockchain-based approach to enable remote sensing trusted data, in: *2020 IEEE Latin American GRSS & ISPRS Remote Sensing Conference (LAGIRS)*, pp. 652–657, March 2020.

37. York, N.D., Pritchard, R., Sauls, L.A., Enns, C., Foster, T., Justice and ethics in conservation remote sensing: Current discourses and research needs. *Biol. Conserv.*, 287, 110319, 2023.
38. Davis, D.S. and Sanger, M.C., Ethical challenges in the practice of remote sensing and geophysical archaeology. *Archaeol. Prospect.*, 28, 271–278, 2021.
39. Leckie, D.G., Advances in remote sensing technologies for forest surveys and management. *Can. J. For. Res.*, 20, 4, 464–483, 1990.
40. Toth, C.K., The future of remote sensing: Harnessing the data revolution. *Geoacta*, 42, 1–6, 2017.
41. Templin, T., Popielarczyk, D., Gryszko, M., Using augmented and virtual reality (AR/VR) to support safe navigation on inland and coastal water zones. *Remote Sens.*, 14, 6, 1520, 2022.
42. Rai, A., Jain, K., Khoshelham, K., Augmented Reality in Education and Remote Sensing, in: *IGARSS 2022-2022 IEEE International Geoscience and Remote Sensing Symposium*, pp. 6856–6859, 2022.
43. Balasubramaniam, S., Kadry, S., Kumar, K.S., Osprey Gannet optimization enabled CNN based Transfer learning for optic disc detection and cardiovascular risk prediction using retinal fundus images. *Biomed. Signal Process. Control*, 93, 106177, 2024.
44. Balasubramaniam, S., Arishma, M., Dhanaraj, R.K., A Comprehensive Exploration of Artificial Intelligence Methods for COVID-19 Diagnosis. *EAI Endorsed Trans. Pervasive Health Technol.*, 10, 1–10, 2024.
45. Buyukdemircioglu, M. and Kocaman, S., Development of a Smart City Concept in Virtual Reality Environment. *Int. Arch. Photogrammetry, Remote Sens. Spatial Inf. Sciences-ISPRS Arch.*, 43, B5-2022, 51–58, 2022.
46. Balasubramaniam, S., Kadry, S., Dhanaraj, R.K. *et al.*, Res-Unet based blood vessel segmentation and cardio vascular disease prediction using chronological chef-based optimization algorithm based deep residual network from retinal fundus images. *Multimed. Tools Appl.*, 83, 1–30, 2024.
47. Bhattacharya, P., Saraswat, D., Dave, A., Acharya, M., Tanwar, S., Sharma, G., Davidson, I.E., Coalition of 6G and blockchain in AR/VR space: Challenges and future directions. *IEEE Access*, 9, 168455–168484, 2021.
48. Banfi, F. and Mandelli, A., Interactive virtual objects (ivos) for next generation of virtual museums: from static textured photogrammetric and hbim models to xr objects for vr-ar enabled gaming experiences. *Int. Arch. Photogrammetry, Remote Sens. Spatial Inf. Sci.*, 46, 47–54, 2021.

3

Efficient Network and Communication Technologies for Smart and Sustainable Society 5.0

P. Kanaga Priya[1]*, R. Sivaranjani[2], Malathy Sathyamoorthy[2] and Rajesh Kumar Dhanaraj[3]

[1]*Department of Computer Science and Engineering, Sri Eshwar College of Engineering, Coimbatore, Tamil Nadu, India*
[2]*Department of Computer Science and Engineering, KPR Institute of Engineering and Technology, Coimbatore, Tamil Nadu, India*
[3]*Symbiosis Institute of Computer Studies and Research (SICSR), Symbiosis International (Deemed University), Pune, India*

Abstract

This book chapter discusses the "Efficient Network and Communication Technologies for Smart and Sustainable Society 5.0," which requires modern network and communication technologies in a rapidly changing environment. In this chapter, network infrastructure efficiency and communication methods for sustainability are discussed. Applications of edge computing, blockchain, Internet of Things (IoT), 5G, and Artificial Intelligence (AI) for network optimization are explored. This chapter includes case studies and research on how these technologies might minimize environmental impact, enhance efficiency, cut energy consumption, and boost socioeconomic growth. The pros and cons of these technologies are discussed, and it suggests research and policy changes to make society more intelligent and sustainable.

Keywords: Network, smart, society 5.0, communication technologies, edge computing, blockchain, Internet of Things (IoT), sustainable

---

*Corresponding author: kanagapriyacse@gmail.com

Rajesh Kumar Dhanaraj, Malathy Sathyamoorthy, Balasubramaniam S and Seifedine Kadry (eds.) Networked Sensing Systems, (63–100) © 2025 Scrivener Publishing LLC

3.1 Introduction

This part covers the concept of Society 5.0, which is considered to be crucial for the fast-growing modern world. Summarizing the starting point for a full discussion of the technology-sustainability nexus, the chapter explores how Society 5.0 can significantly change world politics. The commencement of the long voyage of human society into the impending Smart and Sustainable Society 5.0 marks a pivotal era where different advanced technologies converge to create a harmonious and equitable future for all. The transition of this paradigm not only extends the growth of technology but also goes beyond and incorporates sustainability and innovation. A novel concept of Smart and Sustainable Society 5.0 emerges as a solution to major global problems and a balancing tool for human–technology–nature interaction through the application of advanced technologies. This introduction provides the background to a comprehensive analysis of ways in which artificial intelligence, the IoT, renewable energy, and sustainable development get involved in society's development. The objective is to build a society that is socially and environmentally conscious besides technologically advanced. It is almost important to understand a future where ecological conservation and economic prosperity coexist, and technology is a tool that enables humanity and does not hinder it. Achieving Smart Sustainable Society 5.0 involves transforming enterprises, communities, and economies, which can only be accomplished through collective effort. Therefore, mastering the basics is the first step to adopting this approach.

3.1.1 Evolution of Societal Paradigms

Industrialization and social transformations are deeply interrelated, as the latter can bring about the growth of the economy, progress in technology, migration and building of urban centers, the reorganization of society, and cultural revolutions. Among the many consequences brought by industrialization is the growth of the economy, which, in turn, modifies several aspects including the distribution of wealth job trends and social structure. Technological evolutions, like steam engines, electricity, and digital technology, have been the reason for the ability to form communication, transportation, and healthcare that have been bringing change in the entire society. Urbanization led people to population flow from the rural periphery to urban centers to achieve comfortable living and favorable working conditions. Through the introduction of industrial capitalism, new social classes like the bourgeoisie and workers appeared amid the class stratification and power relations transformations. Cultural change has been possibly brought about by factors, like globalization, mass media, and the development of technology, and it has been impacting the world views, aesthetics, and people's cultural identities. Environmental influence is also essential, as pollution, resource depletion, and habitat destruction become immense. Developing a good link between industry and society is an essential factor for the promotion of equality between the classes and ultimately toward developing a sustainable way of life as well as being able to deal successfully with the beneficial and bad effects of technology.

Industry 1.0: Mechanization With Water and Steam Power
Through steam and water power in equipment and transportation, England's industrial revolution of the 18th century converted agrarian societies into industrial ones. Manufacturing equipment driven by steam developed production zones, extended rail systems, and allowed mass-produced goods to travel to nations with low production levels. Transportation and industry were revolutionized by steam machinery, including the telegraph, fossil fuel cars, and steamships.

Industry 2.0: Mass Production and Electrification
Cheap steel manufacturing and sophisticated chemical and electrical processes marked the start of the Second Industrial Revolution. In 1882, it carried on with the introduction of electricity to cities and enterprises. Electric power, automated procedures, and production lines were important components. Synthetic chemicals, artificial fertilizers, and textile items were manufactured together with the emergence of total quality and scientific management principles.

Industry 3.0: Automation and Digitalization
In Industry 3.0, the rapid growth of computers and cyberspace is referred to as the informatics revolution, which led to the creation of programmable devices and the commercialization of digital technology. This revolution focused on issues, like overcapacity, and enabled more efficient production while lowering labor costs. Artificial intelligence-based intelligent machines also emerged throughout the revolution completely changing the sector.

Industry 4.0: Smart Manufacturing and Cyber-Physical Systems (CPS)
Smart factories, IoT, and Cyber-Physical Systems (CPS) are all part of the revolution known as Industry 4.0, which was first presented in Germany. Big data, cloud computing, machine-to-machine connectivity, and smart

products are all included. Faster and more flexible production is made possible by innovation. Products can be produced more quickly, more intelligently, more competitively, and more smartly thanks to Industry 4.0 technologies like data management, cyber-physical systems, AI, robotics, augmented reality, machine learning, smart factories, IoT, cloud computing, and big data.

Industry 5.0: Human–Machine Collaboration and Sustainable Production

Building on Industry 4.0, Industry 5.0 emphasizes the mutually beneficial association between people and machinery. It imagines a day when human abilities and creativity are complemented by cutting-edge technologies, like robotics and AI, creating more collaborative and human-centered work settings. The evolution of industry is shown in Figure 3.1.

Society 1.0: Hunter–Gatherer Communities

The early human communities throughout the Palaeolithic epoch were characterized by nomadic hunter–gatherer lifestyles. Individuals depend on fishing, hunting, and harvesting natural plants for sustenance. Its key features include small bands of people living in temporary shelters, using primitive tools and weapons made from natural materials, and sharing resources within kinship-based social structures.

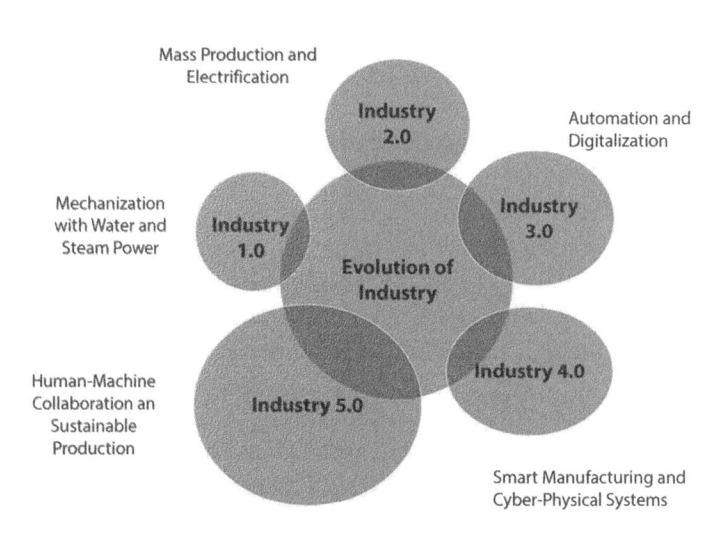

Figure 3.1 Evolution of industry.

Society 2.0: Agricultural Societies

The switch from hunting and gathering to firm farming marked the beginning of agricultural societies. People began cultivating crops, domesticating animals, and establishing permanent settlements. The key features include agricultural villages and early civilizations that emerged supported by irrigation systems, surplus food production, and specialized labor roles. Writing systems, organized religion, and social hierarchies developed.

Society 3.0: Industrialized Societies

The Industrial Revolution in Society 3.0 is the era of industrialization that occurred in the 18th and 19th centuries and brought about dramatic changes in society with the advent of mechanization, mass production, and urbanization. Industrialized societies shifted from agrarian economies to industrial economies. The key features include factory-based manufacturing, urban growth, migration from rural areas to cities, the rise of capitalism and industrial capitalism, and the emergence of labor movements and urban poverty.

Society 4.0: Digital Societies

The Digital Revolution, starting in the late 20th century, ushered in the era of digital societies characterized by widespread adoption of digital technologies, globalization, and interconnectedness. Its key features are as follows: Rapid advancements in information technology, the internet, telecommunications, and digital media revolutionized communication, commerce, and culture. Digitalization has transformed economies, industries, and social interactions.

Society 5.0: Smart and Sustainable Societies

Society 5.0 represents a vision for the future where cutting-edge technologies, such as AI, robotics, and IoT, are seamlessly unified into every aspect of society to boost human well-being, promote sustainability, and discourse complex challenges. Its key features include a human-centric approach to technology, a symbiotic relationship between humans and machines, and emphasis on sustainability, innovation, and inclusive development. Society 5.0 envisions a smarter, more sustainable, and inclusive society empowered by advanced technologies. The evolution of industry is shown in Figure 3.2.

These descriptions provide an overview of the evolutionary process of human societies from their earliest origins to their current state and future aspirations as shown in Table 3.1 highlighting key milestones, transformations, and societal trends across different eras. The evolution of societal paradigms reflects the resilience and adaptability of human societies in the

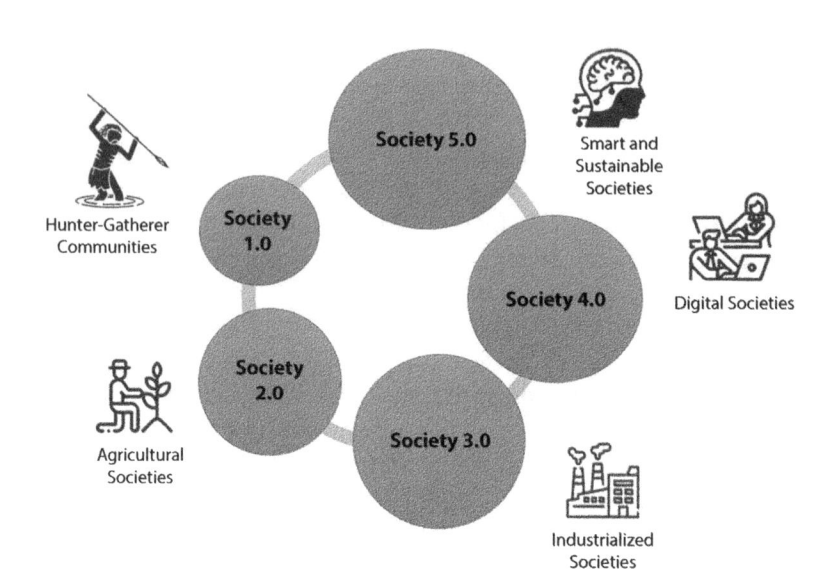

Figure 3.2 Evolution of Society 1.0 to Society 5.0.

face of change [10]. It underscores the importance of embracing innovation while upholding timeless values of sustainability, equity, and community. A better understanding of history can help us navigate the complexities of the modern world toward a more prosperous and inclusive future.

3.1.2 Transition from Industry 4.0 to Society 5.0

The transition to Industry 4.0 has given rise to a general decline in manual workplaces and a rise in automation and the generation of information through the means of digitalization. A smart society is created by integrating technology with personal life. Such an approach helps to improve efficiency of the society as presented in Table 3.2. The idea of AI provided the term "Society 5.0," which marks the development of technology in the life support and monitoring, control, and assessment mechanisms. It assists economic development and quality of life through its support of sustainability, technology, and social responsibility. Limited resource utilization, quick changes, effective distribution of power, and practical implementation should all be required for the success of Society 5.0. This approach envisions creating a community that is well aware of advancements in technology and easily puts technology to use in all components of its daily life.

Table 3.1 Evolutionary process of human societies.

Name of the society	Process description	Period	Explanation
Society 1.0	Hunter society	13,000 BC	Natural life—Living in harmony with the environment and engaging in hunting
Society 2.0	Farming activities	Before 18th century	Performing agricultural activities like irrigation, etc.
Society 3.0	Industrial activities	18th–20th century	Steam engine inventions and industrial production
Society 4.0	Intelligent society	Late 20th–beginning of 21st century	Using information and communication technologies
Society 5.0	Super intelligent society	Future	Providing sustainable solutions to the environment with the help of Industry 4.0

Table 3.2 Transition from Industry 4.0 to Society 5.0.

The problem of Industry 4.0	The solution to the problem (Society 5.0)
Eliminating uniqueness	Difference
Getting rid of dissimilarity	Localization
Release from anxiety	Elasticity
Resources and environmental constraints are not preferred	Sustainability and environmental compatibility

3.1.3 Definition and Key Characteristics of Society 5.0

Society 5.0 symbolizes a visionary model change that exceeds past social models by blending cutting-edge technology with human-centric principles to produce a more sustainable, inclusive, and affluent society. At its core, Society 5.0 embodies the seamless conjunction of the physical and digital worlds, where advanced technologies, such as AI, the IoT, robotics, and big data analytics, are leveraged to address pressing global challenges while enhancing the quality of life for all individuals. The social problems that can be solved with the help of Information and Communication Technologies (ICT), such as big data, Robot, AI, the IoT, and 5G networks are called Society 5.0 as shown in Figure 3.3.

The characteristics of Society 5.0 include the following:

Human-Centered Society
Through Society 5.0, people's requirements as well as their welfare become the priority, and technology is utilized to enhance people and contribute to society.

Integration of Cyber-Physical Systems (CPS)
Society 5.0 refers to the idea of complete incorporation of CPS with the end goal of being more productive and having better living standards through smart infrastructures and services.

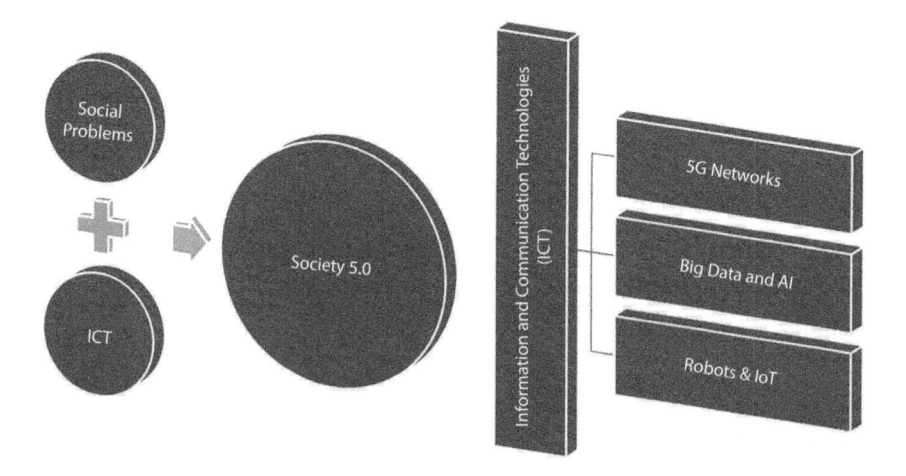

Figure 3.3 The outline of Society 5.0.

Inclusive Growth
The intervention of technology to close the digital gap and offer equitable chances to everyone is one of the ways Society 5.0 aims for inclusive growth.

Sustainability and Resilience
One of the main tasks of Society 5.0 is to eliminate climate change and to go for balance in economic development, environmental protection, and social welfare. It is, therefore, fostering renewable energy, green practices, and sturdy infrastructures.

Innovation Ecosystem
Society 5.0 can create platforms for the development of more professionals in the fields of continuous learning, studying, as well as finding solutions to complex problems. Such platforms can be through the sharpening of innovation and entrepreneurship having partnerships between the state, industry, academia, and civil society.

Data-Based Decision Making
Different sectors, including health, education, transport, and urban planning, apply data as part of the new thinking scheme—decision-making based on evidence. Dip Analysis and AI algorithms are being used to achieve that result in policy making and service delivery.

Ethical and Human-Centric AI
AI will be a central element of a developed number 5 society, and thus, the importance of ethical and responsible use and application is revealed starting a process that is a result of putting human beings on top of priority, dignity, privacy, and autonomy.

International Collaborations
To make mutual goals, Society 5.0 supports pooling and information sharing outside national boundaries. It depicts that space is a common realm and consolidates that global cooperation is a critical measure when looking into universal matters.

3.1.4 Importance of Efficient Network and Communication Technologies

Worldwide Communication and Expanded Data Transfer

The seamless and interconnected nature of the modern world allows people, governments, and businesses to transcend their geographical boundaries and create a space to strengthen economic ties and share information.

Business Operations

The modern age of business communication combines email services, conferencing platforms, file storage systems, and live collaboration technologies. This cluster tunes the decision making, actions, workflows, and productivity of the business, which leads to the efficacy and competitiveness of the business in the market.

Information Sharing and Access

Libraries, clinics, research, relief operations, and other sectors in which networks are important for sharing information rapidly can be critical in the decision-making method.

The Role of Innovation and Development

Processes and interactions between the virtual and physical worlds now lean strongly on advanced communication technologies, including cloud computing, the IoT, and big data analytics, that are supported by an adequate network.

Enhanced Customer Experience

Fast and perfect online transactions will be made available, and this will elevate the expectations of the organization's consumers and will give them quality customer service.

3.1.5 Critical Technologies Shaping Smart and Sustainable Society 5.0

To accomplish the creation of an environment that will support the Society 5.0 concept, a wide range of networks, devices, as well as information and communication technologies, have to be deployed in turn. They encompass cloud computing technology, which provides scalable computing power and storage space to handle big data, edge computing technology, which reduces response time and improves the speed of critical applications, big data analytics technology, which facilitates analysis and innovation through

Table 3.3 Critical technologies shaping smart and sustainable Society 5.0.

Technology name	Scope
IOT	Allows various data gathering in cyberspace
AI	The data collected from IoT are transferred into new types of knowledge
Edge computing	Innovation enables increasing speed and improving real-time processing at the actual system level, which is essential for increasing the IoT's usefulness
Blockchain	A technology to ensure the integrity of data
Network innovation	Technology that quickly and at a high limit can appropriate vast amounts of data

large datasets, IoT technology, which allows real-time data collection and improvement in various sectors using real-time mechanisms, and Artificial Intelligence (AI) technology, which implements intelligent algorithms to automate processes and derive insights. In addition to that, the robotic and drone equipment make both manufacturing and logistics more performance oriented. Also, blockchain delivers transparency and safety in data transmission. Cyber-sensing networks offer real-time optimization and tracking, while cybersecurity technology defends our key data and infrastructure. On the one hand, networking innovation technology provides more connectivity and scalability. On the other hand, device technology makes the Internet of Things neat and efficient. Quantum computing technology has the potential to improve data security, and wearables and virtual reality technologies allow for remote help and personalized healthcare. In the end, it is unbearable to overvalue the worth of networks, IoT, AI, edge computing, blockchain, and these other technologies since they together constitute the framework of a Smart and Sustainable Society 5.0 as shown in Table 3.3 facilitating creative problem solving, effective resource allocation, and safe, interconnected ecosystems for the good of society at large.

3.2 Literature Survey

To address societal issues and bring wealth to all, Society 5.0 is a super-intelligent society that combines cutting-edge technologies, social life, and industry. In this utopian society, economic expansion and technological advancement are not driven by the interests of a small group of people, but rather by the desire for everyone to live life to the fullest. The expectations of Society 5.0 are being met by investigating emerging technologies including fog computing, cloud computing, 5G/5G IoT, edge computing, Internet of Everything, blockchain, and beyond networks. The work of the author sheds light on several open research opportunities and difficulties for Society 5.0, such as the function of 5G–IoT in waste pooling, resource identification, and dispersed resources. It also emphasizes empowering pertinent linked health societal sectors and remote education.

The concept "Society 5.0" was first established at the Japan Business Federation, in the "5th Science and Technology Basic Plan." Then, it later came into practice in January 2016 by the Cabinet of Japan [1]. To increase future investment, the cabinet of Japan implemented a strategy that was aimed at realizing Society 5.0. In contrast to Germany and the US, Fukuda's research examines how Japan's Science, Technology, and Innovation (STI) ecosystem is evolving into Society 5.0. The analysis demonstrates system resilience while identifying the main socioeconomic hazards and categorizing them as capital, labor, and spatial risks [2]. According to this strategy, creating Society 5.0 would be essential to achieving both mid- and long-term growth. This would allow for the introduction of I4.0 technologies, such as robotics, big data, AI, and the Internet of Things, to the social and industrial spheres and solve a variety of societal problems [3–5].

Numerous studies focus on Society 5.0. The authors discussed the various stages that evolved from Industry 1.0 leading up to Society 5.0 and also discussed Society 5.0's objectives and contributions. Specifically, they showcased the growth of people, organizations, and communities capable of setting up, running, and utilizing Society 5.0 systems in the face of COVID-19, which is a global pandemic. Furthermore, they stressed that governmental policies on digital technologies are significantly influenced by institutional compliance with Society 5.0. Consequently, societies capable of setting up and running e-based systems are needed for Society 5.0 [6].

Industry 4.0 and Society 5.0 were defined as the humanization of industrial production. To learn about the difficulties and opportunities faced by the Russian business community, the authors conducted a survey. According to the report, establishing mechanisms to promote investment in socially conscious technologies and giving priority to these principles for national technological growth are important. As Russia's economy moves toward becoming a digital society, it faces comparable challenges to those of other countries [7].

The authors evaluated the concept of Society 5.0 and discussed the integration of Society 5.0 technologies into people's daily lives and the community. The authors found that Japan, Turkey, and Indonesia had the highest levels of research engagement in this field after examining previous studies on Society 5.0 [8]. The challenges of digitalization in education were explored during the shift to Society 5.0 focusing on Turkey's studies and activities. The authors emphasized the importance of prioritizing digitalization in evaluating tools, gathering relevant practices, and achieving objectives in education highlighting the shift from Education 1.0 to Education 4.0 [9].

In a 2020 study, the ideas of Society 5.0 were explored, and the ICT effects were analyzed on older caregivers and the elderly who require care and also analyzed the effects of sociocultural and ethical implications on elderly care. In particular, there was concern expressed about the likelihood that AI robots would offer spiritual social services and care in addition to physical cleaning and cleanliness. This work explains how communication technologies have possible effects on the healthcare sector [11].

The author addresses cyber-physical systems, which comprise hardware, software, and biological elements. The study also covers the key ideas of cyber-physical system design, including big data, cloud computing, artificial intelligence (AI), big data, and digital twins. The author noted that, in addition to software and hardware, developing technologies also take biological elements into account. Systems that not only allow machines to communicate with one another but also keep people informed are in development. One of the industries most impacted by Society 5.0 is healthcare, and the author provided instances of cyber-physical systems in this field [12]. Socioeconomic robotic systems' impact was studied, which is proliferating in the context of globalization. The application of robotic and intelligent systems in the industrial, economic, and sociocultural domains was studied by the authors about digital transformation and change in the economy. The study emphasized that a significant economic revolution is provided by the systems around [13].

To build extremely intelligent societies, the authors have researched the framework of Japanese Society 5.0. Japan is facing several issues, similar to those facing other nations, such as lack of competition, falling birth rates, and aging populations. Society 5.0, which takes a community-centered approach, seeks to address these issues by utilizing technology advancements. Another of its goals is to create a better global order where everyone has access to the technological advancements of modern civilization and to support the nation's growth and development in all spheres. The shift from Industry 4.0 to Society 5.0 was discussed, which places a strong emphasis on human-centric, sustainable resource management and integrates digital efficiency with resilient design to create a resilient society [14]. The significance of Society 5.0 lies in its visionary approach to addressing societal challenges and fostering inclusive prosperity through the integration of cutting-edge technologies and social innovation. By leveraging emerging technologies, such as fog computing, cloud computing, 5G/5G–IoT, edge computing, and blockchain, Society 5.0 aims to enhance efficiency, connectivity, and resilience across diverse sectors. Furthermore, Society 5.0 can be a driver where nations and citizens can cope with crises like the COVID-19 global pandemic and demographic aging. Society 5.0 aims at a model of the future global environment where the use of technology boosts sustainable growth and development at the same time it targets the welfare of humanity through selective decision making and legislative initiatives.

3.3 Internet of Things for Smart Connectivity

The Internet of Things (IoT), a network of objects that can talk to one another, provides both smart services and data sharing. Interconnection, sensors and actuators, data collecting and analyzing, real-time controlling and monitoring, scalability and interoperability, privacy and security, energy efficiency, and user experience are some of the main themes. To achieve efficient data communication and sharing, the devices of the Internet of Things make use of a variety of communication protocols such as Wi-Fi, Bluetooth, Zigbee, and cellular networks. The actuators will respond based on the information that the sensors have analyzed from their surroundings. Then, data are delivered to the cloud systems for analysis and storage. In-time monitoring and control enabled by the IoT renders preventive maintenance and resource management possible. For successful data exchange in various domains, interoperability and scalability matter a lot. For user-friendly interfaces and effective interaction, the user experience counts the most. IoT technology could help businesses increase their productivity, quality of life for people, and efficiency of processes by deploying those ideas.

3.3.1 IoT Applications in Smart Cities, Agriculture, Healthcare, and Industry

In many areas, including smart cities, agriculture, healthcare, and industry, the Internet of Things (IoT) application manifests vast, formative, and very realizing features that are changing the way processes are operated and lives are improved. Here are some key examples of IoT applications in each sector:

Smart Cities
The Internet of Objects revolution triggers the emergence of intelligent cities through new means of resolving resource issues, innovative delivery of public services, and campaigns on living standards. Smart accounting systems, environmental observation, security devices, smart energy management, smart buildings, and urban planning, smart monitoring of traffic, smart infrastructure management, public service requests, and public involvement are the main applications. The IoT sensor allows for tracking the proper functioning of urban infrastructure systems, thus discovering and solving any occurring problems soon. For public road safety, a computer system that analyzes traffic conditions in real time for collecting data on traffic flow, congestion, and vehicle movement is implemented. IoT systems include sensors that measure environmental factors, analyze trash levels, and offer improved public safety and emergency services. IoT-connected meters and grids, which are part of smart energy management, serve as an avenue for cutting energy waste and promoting the adoption of new sustainable energy. The Internet of Things is used in the creation of smart buildings, as well as in the development of urban planning, to improve the efficiency of operations.

Agriculture
Greater levels of automation, data analytics, and sensor technologies allow devices associated with the Internet of Things (IoT) in agriculture to revolutionize how farming operations function. One of the most important fields of application is precision farming, which is based on data processing, crop and livestock monitoring, and climate monitoring. Furthermore, it covers remote monitoring and control of the processes. IoT sensor technology gives sensors the capability to measure things like temperature, humidity, nutrient content, and moisture levels. These measurements help farmers control pests, explosives, and fertilizers more efficiently. Amid options like RFID tags, GPS trackers, and health sensors, all the units are used to monitor livestock as it gives the farmers a primary way to see whether the animals are receiving good treatment, the feeding and breeding procedures, as well as the pasturing practices are being optimized. The current data and temperature data are crucial in agriculture planning, as they help farmers plan their planting, harvesting, and crop protection activities accordingly. The costs of human labor for the operation of such equipment are reduced,

and production is increased due to the Internet of Things (IoT) awakened automation. Farmers being capable of obtaining the best out of resources, producing bigger yields and doing it environmentally can be achieved through using the Internet of Things applications.

Healthcare
With the facilitation of patient care, raising operational efficiency, and providing unreached healthcare delivery, the Internet of Things (IoT) wholly transforms the healthcare industry. The scope of IoT applications for healthcare is very broad, and it includes telemedicine, wearable health devices for monitoring, management of chronic disease, hospital asset tracking, remote patient monitoring, infection control, disaster management, and emergency response. The presented technologies gradually allow doctors to see their patients around, patients to follow the prescribed regime, monitor the environment, and nurses to check the essential signs of patients from a distance. In addition, pharmacists even help a great deal by decreasing inventory waste, optimizing processes, and managing chronic diseases. The main point of IoT systems in patient and environmental safety monitoring is for preventing the spread of infections, infrastructure durability, and managing conditions. It is patients' vital statistics, sleep routines, and their level of physical activity that are tracked by bio surveillance devices. They are stored in the electronic health records. Emergency response systems, which are supported by Internet of Things technologies, enable medical interventions to be effective even in several natural disasters.

Industry
The application of the Internet of Things (IoT) technology in industrial areas is creating a new revolution that transforms traditional processes and activities, therefore, generating opportunities for the industrial sectors to have a significant change in various domains. Predictive Maintenance, one of the essential applications, is emerging in industrial equipment and machines by sensor IoT collecting their real-time data on the functioning, operating, and environmental temperature. The data thus provides a basis for pre-emptive maintenance, which allows organizations to detect potential trouble in the system before it happens and, hence, minimizes unplanned downtime, repair costs, and inconveniences to employees. Furthermore, IoT yields great utility with tracking devices that follow the movement and condition of materials and goods all along the supply chain. Immediate visual feedback allows efficient control of the inventory process, as well as the demand forecast, and better logistics resulting in the

diminishing of expenses and an overall increase in efficiency. Also, at its core Smart Manufacturing lies the power of IoT sensors, robots, and automation systems, which are used in the optimal control of production process quality as well as monitoring. Timely information about the processes enables managers to accomplish faster decision-making processes, thus boosting productivity, flexibility, and competitiveness in a manufacturing environment. To sum up, the deployment of the Internet of Things (IoT) in industry demonstrates the adoption of new standards that enhance various industrial processes, making them smarter, more efficient, and more resilient.

3.3.2 Challenges in IoT Implementation

The IoT systems go through challenges related to cybersecurity, data management, scalability, interoperability, regulatory compliance, interaction with legacy systems, and breakdown of reliability. Cybersecurity problems cover data breaches, hacking, and invading privacy. With numerous technology relays and protocols, there will be interoperability challenges, which also call for tilting the scales toward less complex and broader solutions. Data Operations and analytics specialists are required with advanced infrastructure and technologies for data management and analysis. The observance of the security, interoperability, and data privacy rules is dependent on compliance with the regulations and the regulatory compliance itself. Connecting Internet devices directly to the older systems could be very tedious and, in some cases, time consuming. Examples, like industrial automation, transportation, or healthcare, sharing the same aim are uptime and reliability. Thus, resilience in those cases is a must.

3.3.3 Opportunities in IoT Implementation

Organizations may achieve some IoT benefits like customer-centric exposure, data-driven insights, predictive maintenance, smart cities, innovations, and digital transformation. With the help of IoT, jobs are going to be automated, while data are one of the most sought-after products for decision making. IoT generates a market with personalized experiences because of which loyalty and customer interaction are being increased. It eases predictive maintenance operations; these occurrences result in less downtime and enhance the lifespan of assets. The IoT facilitates telemedicine, as well as individual care and remote monitoring, among other features that are now inherent to the healthcare sector. The organizations can then capitalize on the capability of the Internet of Things by formulating a smart implementation of their strategic plans, which address the roadblocks over time, till they get to the stage of sustainable growth and full potential harnessing of the new technologies.

3.4 Next-Generation Cutting Edge Communication Technologies: 5G and Beyond

Telecommunications are transforming because of next-generation communication technologies like 5G and beyond, which provide faster speeds, reduced latency, and broader connectivity. These technologies have the power to transform whole industries, open the door to brand-new services and applications, and spur creativity in a variety of fields. 5G allows more connected devices per square kilometer, lower latency, and higher upload and download rates. Network slicing is also introduced, which enables network operators to divide their network into virtual networks for particular use cases. Investigate holographic communication, bio-communication, terahertz communication, quantum communication, and artificial intelligence integration beyond 5G, 6G, and beyond. These technologies will improve connection paving the way for the development of cutting-edge technologies, like the Internet of Things, augmented reality, AI, and driverless cars, as well as spur economic expansion, innovation, and increased competitiveness internationally.

3.4.1 Evolution of Cellular Communication Standards

From early analog systems to 5G technologies, the development of cellular communication standards has been fueled by technical breakthroughs and standardization initiatives to improve data speeds, coverage, dependability, and efficiency.

First Generation (1G)
The first commercial mobile, 1G cellular networks, were presented in the 1980s and had basic voice calling with low data rates for limited-service areas.

Second Generation (2G)
Since the appearance of the 2G networks at the very beginning of the 1990s, voice quality as well as security and spectrum efficiency improved

when digital communication was implemented instead of analog transmission. The data networks along these networks were also made available to the users of this.

Third Generation (3G)
The era of 3G networks began in the early 2000s, with faster data speeds and multimedia services, it was a significant step forward. Such innovations helped to spur a boom in mobile multimedia streaming and internet data services.

Fourth Generation (4G LTE)
The appearance of 4G LTE networks in the late 2000s served as a means for widespread mobile broadband, streaming movies, and online gaming due to the data speed rise, decreased latency, and improved spectral efficiency.

Fifth Generation(5G)
5G networks allow for new apps, such as augmented reality and the Internet of Things, by offering lightning-speed data rates, low latency, as well as an enormously increased number of connections. These technologies comprise things like millimeter-wave frequencies, massive MIMO, beamforming, and network slicing.

3.4.2   5G Networks' Features and Capabilities

Unlike the preceding 4G networks, 5G networks are proficient in providing a large number of connections, very low latency, very fast data rates, improved coverage, seamless network slicing, cloud computing, and additional security features. They do an excellent job of downloading and uploading quickly, playing games in real time, and streaming high-definition videos. They are the vehicle for IoT networks and smart city initiatives that require large connectivity coverage. High availability and reliability are other attributes of 5G that are also dedicated to ensuring smooth operation even in a high-density area. Besides, they allow for the transmission on low- and high-band frequencies generating not only in-space but also out-of-station communication. For low-latency applications, like augmented reality and driverless cars, in particular, edge computing lowers latency by permitting real-time data analysis processing.

3.4.3 Emerging Trends and Technologies Beyond—5G (B5G) and 6G

Currently, as Beyond-5G (B5G) and 6G technologies, the next level in wireless communication is under development. They hold promise in delivering speeds that are even faster, with latency that has been reduced, and an enhanced network connection than the current 5G connectivity. However, the next few years will see these emerging trends that change the way businesses work, and start new industries based on these applications. Given below are some of the key emerging technologies:

Terahertz Communication
Compared to the used millimeter-wave and microwave frequencies, B5G and 6G show impressive performance improvements that include much better data speed and bandwidth by experimenting with the terahertz (THz) frequency implementation for wireless communication.

Massive MIMO and Beamforming
Massive MIMO (Multiple Input Multiple Outputs) and Beamforming comprise two techniques crucial in 5G communication. B5G and 6G technologies leverage beamforming as a tool to enhance the capability of spectrum efficiency, coverage, capacity, and so on.

A huge MIMO system is accomplished by the use of a maximal number of antennas in the service of multiple users simultaneously, and thus, the system capacity and throughput are increased. The purpose of beamforming is to accurately focus radio waves in certain directions. Coverage is thus enhanced as interference is significantly reduced especially in densely populated metropole areas.

Intelligent Reflecting Surfaces (IRS)
The radio operators target the passive reflecting surfaces, which are manipulated by the high-tech machinery to face the radio waves. To know about this technology better, it is necessary to learn about reconfigurable intelligent surfaces and metasurfaces, which are the two types of surfaces.

By utilizing the phase changing and amplification, the IRS reaches wider areas, with further range, and higher signal quality be it indoor or outdoor. It results from efficient use of time processing and moving base stations to the next level.

Quantum Communication
One of the most important tools of the B5G and 6G technologies is applying the principles of quantum communication for the encryption of communication channels and making them safe and impenetrable. Quitting the classical keys round for quantum-encoded ones to transmit the cryptographic keys is done by quantum key distribution (QKD) and high privacy is achieved for both parties.

Intelligent Network Management
Artificial intelligence (AI) and machine learning (ML) techniques are widely adopted by B5G and 6G networks to provide intelligent network management and optimization.

Speeding up responses, increasing efficiency, and making networks more dependable is not only possible with versatile orchestration and advanced analytics but also autonomous operations powered by artificial intelligence.

Bio-Communication
Concurrently, research has been carried out, which explores the possibility of employing biological systems in terms of DNA and protein. Utilizing the in-built information-processing and transmitting agency features of biological molecules, bio-based communication technologies are being considered the safest and energy-saving choices for communication.

3.5 Edge Computing: Decentralized Processing for Low Latency

Through onsite data processing, edge computing decreases latency and bandwidth consumption in distributed computing. This is perfect for real-time sensitive applications like augmented reality, driverless cars, and the Internet of Things. Quicker response time and better performance of data processing close to the edge or network are achieved. Through filters and analysis locally, edge computing facilitates applications that take a short time to travel with a small bandwidth. Being able to store sensitive data near its source and with a lower likelihood of data leakage or unlawful access also frees up some privacy and security rights. It is anticipated that the networked systems and services are going to be relying heavily on edge computing technology in the future.

3.5.1 Understanding Edge Computing Architecture

Human-driven society, which aims to make use of modern technologies, such as the Internet of Things (IoT), artificial intelligence (AI), and big data analytics to optimize a relationship between social progress and economic growth, relies on edge computing architecture significantly. Edge computing in the Society 5.0 context means processing data in a manner that involves both the points close to the sources of the generated data and the centralized cloud servers instead of relying solely on the centralized servers. This method includes a few benefits like reduction of delay, improved data security, and better productivity. Some of the key components and layers of edge computing architecture for Society 5.0 are discussed below:

Edge Devices
The principle of edge computing architecture is high-end devices that are capable of gathering large amounts of data from multiple sources, which include sensors, actuators, wearables, cell phones, and IoT endpoints.

Edge Gateways
Edge gateways will lower the latency and bandwidth utilization by removing, reordering, compressing, and encrypting data as it is in transit between edge devices and the cloud.

Edge Servers
Edge servers are placed close to edge devices and deliver the computational ability that is useful for processing data, executing AI algorithms, and running apps in real time. The advantage of this is that it allows being proactive and receiving immediate responses.

Fog Computing
Due to classical distributed architecture in fog computing, the computing power spreads over network layers closer to edge devices than classical cloud data centers.

Distributed Data Processing
Through the architecture of edge computing, distributed data processing methods are used to analyze data streams from edge devices, which then help to mine the learned insights. At the network edges, this is the technology that enables the conduction of real-time analytics and decision-making processes and, at the same time, minimizes data flow as well as latency.

Security and Privacy
In applications where edge computing also serves as a means for processing highly delicate data, security and privacy issues are major aspects of the design process. Security mechanisms, such as encryption, access controls, and secure booting, are implemented by edge devices and gateway devices to protect sensitive and critical data.

Interoperability and Standards
It is necessary to provide certainty of compatibility and seamless integration among different edge computing systems and devices by implementing interoperability and standardization activities. Channeling data among edge devices, gateways, and cloud platforms is streamlined due to the availability of industry-established protocols, APIs, and data format standards.

Collaborative Ecosystem
The collaboration of edge devices, gateways, servers, and cloud services is made possible by the edge computing architecture with the help of which intelligent, context-aware applications and services can be delivered. In Society 5.0, all these innovative applications, such as smart cities, driverless cars, medical monitoring, and industrial automation, are enabled by this collaborative approach, which, in turn, is also the economic driver.

3.5.2 Edge Computing's Benefits for Analytics and Data Processing

Within the concept of Society 5.0 in which the approaches of technology's most advanced part hope to provide a balanced society, through which people's comfort and economic power will simultaneously work, edge computing offers different benefits in this area. The following are some of the crucial returns of edge computing in data processing and analytics for Society 5.0:

Reduced Latency
Another element that is supported by edge computing is lower reaction times and reduced data processing latency through the process of data processing closer to the points of generation. The processing of data needs to be even faster for applications in the real-time environment of Society 5.0. Such applications include industrial automation, healthcare monitoring, and driverless cars, where the decisions need to be made on the prompt.

Improved Data Privacy and Security
The issues of data protection and privacy have become increasingly important for raising global concerns. Processing of usual data can be left on local edge devices or gateways by edge computing that omits sending raw data to centralized cloud servers, which results in saving bandwidth. A decrease in the data breach probability, illegal access, and dangerous exposure of data is highly vital for the privacy protection and security of data.

Bandwidth Optimization
The architecture of edge computing can do this in the sense that it calculates and minimizes its bandwidth consumption by locally refining and filtering data before sending it to the cloud where it will be analyzed extensively. This prevents data transfer expenses from outspending the budget and reduces the consumption of network bandwidth where the connectivity is either irregular or cannot be guaranteed to be there.

Real-Time Insights and Decision- Making
Edge computing brings about real-time analytics and decision making as it operates at the ends of a network processing data through the network edge. Thus, the ability of Society 5.0 apps to customize the services is realized by utilizing the immediate provision of answers and data on predictions, optimizations, and deviations, which are happening all the time.

3.5.3 Use Case and Deployment Scenario

Edge computing, as a technology in Society 5.0, expands the possibilities for achieving revolutionary use cases and deployment scenarios. It allows driving data to be analyzed in real time where it makes the roads safer, creates perfect traffic flows, and eliminates congestion. In addition, it also provides for environmental monitoring, for implementation of preventive measures, acts as a prompt in the case of an emergency arising in a crisis, and improves public health. Thus, thanks to the growth of the IoT, telemedicine, point-of-care diagnostics, and remote monitoring of patients are technically possible in the healthcare sector. It allows supply chain streamlining, quality control, and predictive maintenance in the industrial area. Real-time production and energy distribution or consumption networking are done by smart grid integration, and also through the renewable energy sources and energy storage devices in support of energy management. It is in a position to read the energy consumption data from smart meters and IoT devices to the end of supporting the demand response strategies.

It accesses the existing capacities rather than the expanding by installing new ones. In addition, it can extract maximum energy from the buildings, thus minimizing expenses. This technology has also made precision farming possible.

3.6 Blockchain Technology: Securing Data Integrity and Trust

Blockchain technologies could bring a great deal of trust and data integrity into Society 5.0. It provides a decentralized, immutable ledger for storing information and transactions, which makes these transactions ensure their integrity. Blockchain transactions furthermore offer the reliability of auditing and transparency of stakeholders. The smart contracts along with the cryptographic keys allow people to retain possession and rule over their data. Blockchain-based identity management systems provide users with increased privacy and better security against identity thefts. The ability of blockchain to provide supply chains with more transparency and traceability allows stakeholders to view the origin, authenticity, and quality of products. Smart contracts aim to eliminate intermediaries and achieve affable transactions through computerized automation and execution. Blockchain-based platforms for data sharing, which are secure, allow for the exchange of meaningful information between various parties. It also allows for tokenization of assets and, therefore, places value generation and ownership transfers between peers on the agenda.

3.6.1 Fundamentals of Blockchain Technology

The basis of Society 5.0 is blockchain technology, which makes it a simpler process to incorporate sophisticated technologies into a human-oriented society that combines social and economic progression. The following are the fundamentals of blockchain technology in Society 5.0:

Decentralization
Data can be stored between several computers instead of centralizing the data in one spot; with blockchain, data can be stored on a network of decentralized nodes. Thus, this scalability leads the network to the state of the system, which meets the objectives of empowering of an individual and at a community level by Society 5.0.

Immutable Ledger

Blockchain technology is a transaction and data-logging device that never gets tampered with and is incapable of any data change. In terms of recording data into the blockchain, its integrity and authenticity are guaranteed since no data deletion or modification is possible after it is loaded.

Transparency and Traceability

Transactions between blocks are traceable and transparent, which means that a user can check for the source data and prove when the information was created. Such transparency can be translated into the open trail of transactions, which makes stakeholders feel more confident by increasing the level of accountability of the participants and boosting the necessity for regulatory compliance.

Cryptographic Security

The blockchain secures data and transactions on the network using so-called cryptographic measures like hashing and digital signatures. These cryptographic methods provide the basis of protecting the information from any attempts to modify data or illegal access during data verification enabling the users to secure their information.

Smart Contracts

Smart contracts are programmable agreements, which are recorded and enforced on the blockchain simultaneously with specific rule preconditions. They allow for trustless transactions and provide an alternative for the functions of intermediaries, which facilitate and enforce agreements. Parties to an agreement no longer depend on those intermediaries. In the Society 5.0 era, smart contracts are being used in sectors such as supply management, decentralized finance, and DAOs (decentralized autonomous organizations).

Tokenization

Blockchain creates a digital record that represents ownership of assets, and the technology makes it possible to tokenize such assets as digital currencies, stocks, and physical assets. They can use the tokenized representation of assets on the blockchain to do safe transfer of ownership, and peer-to-peer asset exchange, and also to discover new avenues for wealth generation and exchange in Society 5.0.

3.6.2 Applications in Secure Data Sharing, Supply Chain Management, and Decentralized Finance

Blockchain technology is a prerequisite for safe data sharing in a Society 5.0 world, where digitalization initiatives strive to create a social system that balances economic superiority with community enhancement. The following are some applications of blockchain in these areas:

Secure Data Sharing
With the help of blockchain, it is possible for data sharing across borders, supply chain traceability, and interchange of health data in a safe environment. It can be performed through cryptographic keys that allow the patients to use their medical records while maintaining privacy. Moreover, it yields accountability and a sense of integrity in the operations of the supply chain. Data compliance is guaranteed by its decentralized encryption technology.

Supply Chain Management
Blockchain technology reduces the risks of fraud and forgeries through the ability to monitor and audit supply chain operations in real time. Supplier verification can be made easier with the help of credential records and certifications. As such, the blockchain serves as a transparent ledger that tracks stock movements and replenishment requests, which simplifies the process of having an efficient inventory management strategy.

Decentralized Finance
Through transparency, blockchain technology enables one to form decentralized exchanges (DEX) for cryptocurrency trading. The implementation of smart contracts brings out lending and borrowing platforms, which are decentralized, hence, giving an opportunity to underserved individuals to get to the financial services they need. The algorithms play a role in automated market making, which, in turn, facilitates market trading and provision of liquidity.

3.6.3 Challenges and Potential Solutions

Society 5.0 has great potential where blockchain technology can be applied for the better, but it does have obstacles. One of the issues is scalability where the number of transactions is normally slow. In addition, the costs of these transactions are higher. Therefore, to fix the crowding problem, sharding, layer 2 protocols, and consensus algorithm will be integrated. If interoperability is a concern in the case of the inability of the blockchain networks and platform, it needs to be addressed immediately to provide efficient data transmission and cooperation. Cross-chain communication techniques, protocols, and interoperability standards also help to bridge the gap assisting in bringing integration and connectedness in Society 5.0.

The issues of privacy and confidentiality are quite significant and are especially common in sectors such as banking and healthcare. Privacy-enabling techniques, such as secure multi-party computing, zero-knowledge proofs, and homomorphic encryption, will be of great importance to maintain transparency and auditability while data privacy is protected. Adopting permissioned blockchains, while including access limitations, will create a more safeguarded network for conduct of the sensitive information and transactions. Moreover, user experience and adoption are also a problem in Society 5.0. The broader deployment of blockchain technology is impeded by so many critical issues such as complexity and usability. To expand the acceptability, apps, which are accessible on mobile devices and simple interfaces, can be created. In addition, linkage with current systems can also be made seamlessly. The adoption and involvement of users and stakeholders in Society 5.0 can be stimulated if there is proper education about the benefits and applications of blockchain technology to users as well as stakeholders.

3.7 Artificial Intelligence in Network Optimization

Artificial intelligence (AI) in Society 5.0 is capable of transforming networks and infrastructure into a single system using the integration of new technologies. The AI-powered network management systems use machine learning algorithms to observe, enhance, and optimize network security, reliability, and performance of the network. Intelligent vehicle networks offer predictive maintenance, allocation of dynamic resources, intelligent traffic routing, detection of security risks and their consequences, and quality of service optimization as the extra advantages of these systems. AI-based cybersecurity solutions allow the detection of risks in real time and mitigation where the software examines and finds out if there is any deviant activity. Edge computing capabilities with integrated networks and AI-driven ones enable applications to run better and faster and produce data for making better decisions.

3.7.1 Role of AI and Machine Learning in Network Management

Society 5.0 integrates AI and ML into network management for troubleshooting network performance, identifying traffic patterns, as well as security problems, which becomes possible without any human involvement. To identify failures and malfunctions in the future, they use past data for analyzing and recognizing patterns. Based on actual demand and workload patterns, AI and ML replace human resources by dynamically increasing, decreasing, or modifying bandwidth, capacity, and computing resources, which results in better utilization of the resources. AI-driven network routing algorithms and traffic control models implemented for routing and traffic engineering reduces data processing such as latency and congestion. Through transmission network analysis and identification of malicious work, they can encounter security risks in real-time mode. Delivering standard user demands, various preferences, and service level agreements takes place through adjusting networks based on QoS (Quality of Service). It guarantees stable operation, believably, and highly positive rates for the members of Society 5.0.

3.7.2 AI-Driven Approaches for Resource Allocation and Optimization

AI-inspired optimization and resource allocation techniques rely on machine learning algorithms and data-driven foresight to maximize resource usage in areas that include manufacturing, cloud computing, and network technologies where resources need to be allocated quickly and in a dynamic manner. The following are some key AI-driven approaches for optimization and allocation of resources:

Reinforcement Learning
An agent needs to interact with its environment by introducing the kind of machine learning known as reinforcement learning (RL) to effectively make decisions that will maximize the reward. RL algorithms are unique in that they can react by adjusting the allocation of resources according to environmental cues, such as the usage of the resources in an optimized manner and the satisfaction of the users. RL applications, such as load balancing in cloud computing, adjusting traffic engineering in networks, and smart grid energy management, all have been fielded with RL approaches.

Genetic Algorithms
Genetic algorithms are optimization tools that are based on the concepts of natural selection and evolution. GAs can be employed in resource allocation while solving problems to find solutions with the highest performance values so that these solutions require the least amount of resources to achieve the goals and constraints. GAs are greatly useful in solving complicated tasks that can be decomposed into smaller problems, such as production schedule optimization in production management or job scheduling in distributed computing.

Convolutional Neural Networks
The primary use of convolutional neural networks (CNNs) is image recognition, categorization, and other such tasks. CNNs can be applied in resource management to process large amounts of data, like reading through the traffic patterns or network readings, to detect any abnormality. In this way, the employed CNNs examine the anomaly, the predictive maintenance, as well as any existing faults from a network infrastructure perspective. It allows the allocation of resources to ensure that preventable losses or poor performance are mitigated.

Swarm Intelligence
The collective behavior of animals and social insects serves as the model for swarm intelligence algorithms such as Particle Swarm Optimization (PSO) and Ant Colony Optimization (ACO). The mimicking of swarm cooperativity using swarm intelligence algorithms can be used to search for optimal answers by focusing on resource distribution.

To increase the performance, swarm intelligence techniques have been taken into account for different optimization problems such as vehicle routing, task scheduling, and network routing.

3.8 Energy-Efficient Networking for Sustainability in Society 5.0

In the context of Society 5.0, a society where advanced technologies aimed at creating a community-oriented society that tries to find a balance between societal growth and economic progress while ensuring a minimum negative impact on the environment, data-efficient networking becomes a fundamental part of sustainability. The following are key strategies and technologies for achieving energy-efficient networking in Society 5.0:

- Green Networking Technologies—Significantly less energy is used in network infrastructure when low-power, dynamic power management, and energy-efficient hardware is used. By using these strategies, devices can adjust their power usage to meet the needs of their job and use less energy when there is little activity.
- Optimized Network Architecture—Energy-efficient resource consumption is made possible by network virtualization technologies, such as SDN and NFV. By reducing the distance over which data must be transmitted, edge computing uses less energy. Data journey distance is decreased by Content Delivery Networks (CDNs) with distributed caching and nodes improving performance.
- Efficient Protocol Design—In wireless sensor networks, energy-aware protocols give priority to energy saving, and traffic optimization minimizes idle times and duplicate data transmissions. For effective resource allocation, network traffic is prioritized using Quality of Service (QoS)-based traffic management, taking into account user preferences and application needs.
- Renewable Energy Integration—Carbon emissions and dependency on fossil fuels are decreased via energy-collecting systems and green data centers. By enabling sustainable operation in isolated or off-grid areas, these techniques lower carbon emissions.
- Data-Driven Optimization—Big Data Analytics: Proactively identifying inefficiencies and opportunities for optimization through the analysis of network traffic patterns, resource utilization, and information on energy use permits resource allocation and strategic decision making for energy-efficient networking. ML for Optimization: Energy efficiency and sustainability in network operations are improved using machine learning algorithms to optimize network topologies, forecast future patterns of energy use, and dynamically modify network parameters based on real-time data analysis.

3.8.1 Strategies for Reducing Energy Consumption in Communication Networks

Reducing energy consumption in communication networks is crucial for sustainability in Society 5.0. The following are several strategies for achieving this goal:

Optimized Hardware and Infrastructure
Implement energy-efficient networking hardware, low-power components, and cooling systems to reduce power consumption in network infrastructure.

Network Virtualization and Consolidation
Adopt network virtualization technologies to reduce energy usage by consolidating network services and reducing the number of physical devices such as software-defined networking (SDN) and network function virtualization (NFV). To maximize resource allocation and minimize costs related to distributed network topologies, centralize network administration and control.

Traffic Optimization and Management
Reduce energy usage by streamlining network traffic patterns and transmission protocols to reduce redundant data transfers, packet retransmissions, and idle times. To prioritize network traffic according to application requirements, implement Quality of Service (QoS) techniques. This will ensure effective resource use and minimize energy waste.

Dynamic Power Management
To minimize energy consumption during idle periods and alter power usage based on workload needs, take advantage of network equipment's power-saving capabilities and dynamic power management methodologies.

3.8.2 Green Networking Technologies and Practices in Society 5.0

Green networking methods and solutions constitute major elements that would permit the evolution of environmentally friendly communication systems in a socially smart community concurrent with the adoption of high-end technologies. The following are some key green networking technologies and practices for Society 5.0:

Energy Efficient Hardware
Take advantage of the networking equipment that is suited to the needs of a business while at the same time keeping the power usage low, for instance, switches, routers, and network interface cards (NICs) that utilize energy minimally when they are operating. Implement energy-efficient devices, including storage drives, memory cards, and processors, among the basic options of the equipment that communicates within a network.

Green Data Center Design
To decrease energy consumption and eliminate emissions, plan energy-efficient cooling, lighting, and airflow control systems for data centers. Enhance hot/cold aisle containment methods, free cooling, and green energy generation techniques further so that your data center operations can constantly be more sustainable and energy saving.

Energy-Aware Protocol Design
Employing energy-saving networking protocols and procedures, which are inclusive of power-efficient transmission protocols to be used in wireless communication networks and power-saving routing algorithms for wireless sensor networks. However, it measures the impact of various power-saving technologies on energy consumption. For example, IEEE 802.3az minimizes energy consumption when attachable devices are not in use for EEE (Energy-Efficient Ethernet).

3.9 Challenges and Opportunities in Implementing Efficient Network Technologies

While implementing network technologies, it faces both substantial barriers and opportunities. The following are some key considerations of the challenges. Below are some primary aforementioned issues in Society 5.0.

- Legacy Infrastructure—The materials and equipment of the current network infrastructure may not be conducive to the integration of high-efficiency technologies. Particularly, in locations that have aged, it is difficult to replace, and hence, substituting all units may turn out to be costly.
- Regulatory Barriers—The whole process of novel network technology implementation is impeded by delays in creating access to regulations as well as complying with requirements,

especially for highly regulated sectors such as utilities and telecommunications.
- Cyber Security Risks—Network security and data protection may become more complex, and it requires developing more resourceful security strategies and procedures; when the network technology adoption is acquired, it may lead to new threats and cybersecurity risks.
- Skill Shortages—When an organization tries to use smart networks that incorporate Network Function Virtualization (NFV) and Software-Defined Networking (SDN), they run into trouble since these people are hard to come by.

The opportunities include the following:

- Enhanced Efficiency—Deploying network technologies that use less energy, including NFV (Network Function Virtualization), SDN (Software Defined Networking), and energy-efficient hardware, guarantees not only the network's performance, reliability, and scalability but also cuts down energy consumption and operating costs as well.
- Innovation and Differentiation—The organization may develop innovative products and services, which can enable the company to establish its brand and gain a market edge through the delivery of value-added solutions that cater to changing societal needs in the society of the future.
- Sustainability—Enhanced efficacy network technologies save energy, thus lowering carbon emissions and environmental impact. Hence, the sustainability objectives of different societies as depicted by Society 5.0 are strengthened.
- Improved Connectivity—By implementing network technologies in Society 5.0, economic growth has been promoted by increased connectivity in distant places [15].

3.10 Future Directions and Recommendations

Legislative frameworks, policies of organizations, and technological changes will ultimately shape the determination of where to use the network technology effectively. The future ventures of edge computing system infrastructure, eco-friendly networking, and continued research in the upcoming new technologies, such as quantum networking, 6G, and photonic computing, are the future potential key areas. Policies must coincide with data privacy, security, and environmental regulations, and frequency management ought to be sealed for fifth-generation ones. It is desirable to make sure that the less energy-consuming network infrastructure must be promoted through policy incentives. They should emphasize stakeholder engagement and be strong on the training policy and risk management. Identifying user needs and concern satisfaction involves both community participation and user education. To evaluate the effectiveness of network technologies, stakeholders must monitor the network's performance.

3.10.1 Research Priorities for Advancing Network and Communication Technologies

Social 5.0 is a futuristic civilization that is transforming how we live and developing the platforms that enable us to engage with each other. To attain this ambitious aim, we have to integrate technology improvements with the existing environmental concerns. 5G and beyond, the Internet of Things, edge and fog computing, green networking technologies, privacy and security, AI network management, interoperability standards, 5G applications, and human-friendly design are the key research priorities. Therefore, the fundamental purpose of these research priorities is to create low-latency, ultra-fast, and reliable networks that may be potentially applied to applications such as networking, autonomous cars, and media immersion. They likewise explore AI and learning strategies in intelligence and networking that permit technical, societal, and environmental sustainability. The expanded potential for collaboration among educational institutions, corporations, and state agencies in such sectors may contribute to developments in the technology and communication field eventually moving our society's structure to the age of 5.0, toward a more open, safe, and green world.

3.10.2 Policy Recommendations for Fostering Sustainable Development in Society 5.0

The measures for the encouragement of sustainable advancement in Society 5.0 were shortlisted. It discusses the importance of the following agenda: digital inclusion, sustainable consumption and production, innovation and entrepreneurship, green infrastructure, social inequalities and international cooperation, environmental laws, and promoting resilience to natural disasters and climate change. These recommendations aim to supply barriers to digital separation, reduce carbon emissions, and support sustainable consumption. The policies strive toward the investment in environment-friendly energy sources, transportation, and green infrastructure areas by sustainable motives. The policies need to eliminate social inequality and promote social safety in medical insurance, education, and other programs. The above recommendations stated here would be intended to better the condition of the economy, society, and the environment in the present and future.

3.10.3 Collaborative Efforts Toward Achieving a Smart and Sustainable Society 5.0

A society that is as smart and as healthy as it could be 5.0 is a large-scale idea that includes the use of advanced technologies, a human-centered approach, and environmentally friendly practices to deal with the current societal problems. Partnerships across sectors, national cooperation, foreign cooperation, community participation, and ethical governance are some of the major kinds of cooperative features. Partnering across communities, businesses and schools, civil society, and the government leads to collective actions that allow to co-create complete answers for sustainable development. Sectoral research and development, idea creation and incubation, and co-creation are all part of inter-sectoral cooperation among several fields. The twofold plan is being backed by encouraging foreign cooperation both on creativity and technology transfer in favor of sustainability. The involvement of the community and grassroot groups plays an important role in the initiation, execution, and evaluation of sustainable projects. Education improves people's digital skills and lifetime growth. Information on openness, accountability, and equality is gained through participation and ethical governance in the process of the creation and application of technology.

3.11 Conclusion

To achieve a smart and sustainable society 5.0, contributions from numerous stakeholders across various industries and nations will be required. This chapter explored the core principles, technological innovation aspects, and the policy-related recommendations necessary for achieving Society 5.0 vision of a society where cutting-edge technologies are applied as a tool aimed to enhance human needs and the environment, and to generally tackle societal problems. We can bring the creative power and innovation of advanced communication and network technologies to sustain economic growth and progress, and improve our life standards with the help of multi-stakeholders and extending international cooperation. Through the promotion of community engagement and participatory governance, communities and individuals are allowed to constructively and productively play a part in the shaping of their societies and also contribute to the fostering of self-sufficient communities geared toward the attainment of sustainability goals. The chapter presents the central themes, which are the emergence of paradigm shifts in society from the Industrial Age to the digital era; people witnessed the conversion of upheavals that started with the Industrial Revolution and were later followed by shifts in social and economic dynamics. Society 5.0 brings forth an approach to technology integration based on the human-centric perspective, encompassing harmonization between the three pillars representing humans, technology, and nature. The extensive use of edge computing, IoT, 5G, and other network and communication technologies plays a vital role in making smart and sustainable solutions applicable across various industries ranging from the healthcare, transportation, agriculture, and planning of urban spaces. Stakeholders of multi-sectoral, cross-sectoral, and international collaboration are important in designing comprehensive programs that eradicate these constraints with alternative approaches that are fair for many. From time to time, it is, of course, necessary to look at ethical questions, sustainability of the environment, and social equality in technology design, deployment, and control over networks and communications.

References

1. Fujii, T., Guo, T., Kamoshida, A., A consideration of service strategy of Japanese electric manufacturers to realize super smart society (Society 5.0), in: *Knowledge Management in Organizations: 13th International Conference, KMO 2018*, Žilina, Slovakia, August 6–10, 2018, Springer International Publishing, pp. 634–645, 2018, Proceedings 13.
2. Fukuda, K., Science, technology and innovation ecosystem transformation toward society 5.0. *Int. J. Prod. Econ.*, 220, 107460, 2020.
3. Nieuważny, J., Masui, F., Ptaszynski, M., Rzepka, R., Nowakowski, K., How religion and morality correlate in age of society 5.0: Statistical analysis of emotional and moral associations with Buddhist religious terms appearing on Japanese blogs. *Cognit. Syst. Res.*, 59, 329–344, 2020.

4. Narvaez Rojas, C., Alomia Peñafiel, G.A., Loaiza Buitrago, D.F., Tavera Romero, C.A., Society 5.0: A Japanese concept for a superintelligent society. *Sustainability*, 13, 12, 6567, 2021.
5. Salimova, T., Guskova, N., Krakovskaya, I., Sirota, E., From industry 4.0 to Society 5.0: Challenges for sustainable competitiveness of Russian industry, in: *IOP Conference Series: Materials Science and Engineering*, vol. 497, IOP Publishing, p. 012090, April 2019.
6. Saracel, N. and Aksoy, I., Society 5.0: Super Smart Society. *Soc. Sci. Res. J.*, 9, 2, 26–34, 2020.
7. Salimova, T., Vukovic, N., Guskova, N., Krakovskaya, I., Industry 4.0 and Society 5.0: challenges and opportunities, the case study of Russia. *Smart Green City*, 17, 4, 1–7, 2021.
8. Akin, N.E.M. and Akyol, O.D., An evaluation of the concept of society 5.0 in the light of Academic Publications. *Trends Bus. Econ.*, 35, 577–593, 2021.
9. Mukul, E. and Büyüközkan, G., Digital transformation in education: A systematic review of education 4.0., *Technol. Forecasting Social Change*, 194, 122664, 2023.
10. Nahavandi, S., Industry 5.0—a human-centric solution. *Sustainability*, 11, 16, 4371, 2019.
11. Calp, M.H. and Bütüner, R., Society 5.0: Effective technology for a smart society, in: *Artificial Intelligence and Industry 4.0*, pp. 175–194, Academic Press, 2020.
12. Yildirim, Cyber-physical systems: from the information society to the superintelligent society, in: *International Conference on Electrical and Electronics Engineering (ELECO)*, IEEE, pp. 266–269, 2020.
13. Aktug, S. and Sevinc, M., Society 5.0: the economic revolution of intelligent systems and robots, in: *International Academic Studies Congress Spring, Roting Academy*, Mersin, pp. 139–146, 2021.
14. Mourtzis, D., Angelopoulos, J., Panopoulos, N., A Literature Review of the Challenges and Opportunities of the Transition from Industry 4.0 to Society 5.0. *Energies*, 15, 17, 6276, 2022.
15. Pereira, A.G., Lima, T.M., Santos, F.C., Industry 4.0 and Society 5.0: opportunities and threats. *Int. J. Recent Technol. Eng.*, 8, 5, 3305–3308, 2020.

4

Advanced Techniques for Human-Centric Sensing in Environmental Monitoring

S. Aathilakshmi[1]*, Visali C.[2], T. Manikandan[2] and Seifedine Kadry[3]

[1]Department of Electronics and Communication Engineering, Chennai Institute of Technology, Chennai, India
[2]Department of ECE, Vivekanandha College of Engineering for Women (Autonomous), Thiruchengode, Tamil Nadu, India
[3]Department of Applied Data Science, Noroff University College, Kristiansand, Norway

Abstract

This paper discovers progressive methods for human-centric sensing in environmental monitoring. It begins by debating the principles of human-centric sensing and its returns over outdated methods. It investigates into the technical features of realizing human-centric sensing systems, with data group, processing, and analysis. Numerous sensing sense modalities, such as GPS, accelerometer, camera, microphone, and environmental sensors, are examined for their potential in capturing relevant environmental data. Besides, it explores the tests and chances associated with human-centric sensing, including privacy concerns, data quality issues, and scalability. This paper discusses novel methods for lecturing these tests, such as privacy-preserving data combination methods and machine learning algorithms, for data validation and fusion. Moreover, this research highlights the submissions of human-centric sensing in varied environmental monitoring scenarios, with air quality assessment, noise pollution mapping, urban heat island detection, and disaster response. Case studies and real-world instances are presented to demonstrate the effectiveness of human-centric sensing in only actionable insights for environmental management and policy-making. Finally, future research directions and developing trends in human-centric sensing for environmental monitoring are discussed. It anticipates advancements in sensor technology, data analytics, and participatory sensing platforms, which will further

---

*Corresponding author: me.aathi92@gmail.com

Rajesh Kumar Dhanaraj, Malathy Sathyamoorthy, Balasubramaniam S and Seifedine Kadry (eds.) Networked Sensing Systems, (101–120) © 2025 Scrivener Publishing LLC

enhance the capabilities of human-centric sensing systems and enable more effective and inclusive environmental monitoring strategies.

Keywords: Human-centric sensing, environmental monitoring, smartphone sensing, privacy-preserving techniques, data aggregation, sensor technology, participatory sensing

4.1 Introduction

Environmental monitoring is vital for empathizing with complex communications between human actions and the atmosphere and for updating actual policies for environmental management and policy making. Outdated monitoring organizations naturally rely on immobile sensors arranged in secure locations, as long as there is valuable data, but often missing in longitudinal and sequential resolution. Besides, these schemes may not confine the occupied range of conservational experiences by people as they move over different locations and engage in various activities. In recent years, increasing attention in human-centric sensing as an opposite approach to outdated environmental monitoring. Human-centric sensing leverages the ubiquity of smart phones, wearable devices, and other individual gadgets equipped with sensors to collect environmental data directly from individuals in real time. By using the power of personal devices, human-centric sensing offers a possibility to enhance the coarseness and context of environmental monitoring providing insights into environmental exposures at the micro scale and empowering individuals to make informed decisions about their health and well-being.

This chapter investigates novel methodologies for anthropocentric sensing within the purview of environmental surveillance. An original dissertation on the ethical thoughts of anthropocentric sensing and its compensations qualified to predictable monitoring system some examples are, understanding anthropocentric perception systems, surrounding data achievement, broadcasting, and investigation and so on. An exhibition of detecting sense modality, counting GPS, accelerometers, television camera, mics, and environmental sensors, are appraised for their efficacy in capturing related environmental data. Furthermore, the manuscript addresses the inherent tasks associated with anthropocentric sensing such as discretion anxieties, data quality concerns, and scalability. To ease these contexts, the work explores innovative approaches, plus privacy-preserving data combination techniques and machine learning algorithms for data authentication and synthesis.

Furthermore, the assorted requests of human-centric sensing in conservational monitoring, counting air eminence assessment, blast pollution planning, built-up temperature key detection, and adversity response are discussed. Complete circumstance trainings and practical examples validate the efficiency of human-centric sensing in providing criminal visions for conservational organization and policy making. Last, this paper explores forthcoming investigation instructions and developing trends in human-centric sensing for ecological monitoring.

This paper [1] discusses wireless communication system-based 5G technology to develop the expected outcome in communication systems standard. This area focuses on transmitting high data, reduce delay, and improved transmission quality in wireless communication technology. This paper [2] explores wireless communication using RIS systems. This reconfigurable intelligent surface has a reconfigurable element, which is used to strengthen the signal quality in terms of using electromagnetic waves in RIS. This research mainly aimed to improve signal quality, transmission coverage, and energy efficiency, and in upcoming research in RIS, these are the major concerns to enhance the RIS technology.

This paper [3] demonstrates the main objective of life time communication analysis in 6G wireless communication technology. This system is used to analyze the different sub networks to transmit and communicate data with quality and long lifetime. In general, human lifetime faces several issues in healthcare, safety, social network, industry, etc. To take care of these issues, the authors proposed the 6G wireless communication system in sub networks to reach better performance, reduce delay, and reduce critical communication process. This paper [4] illustrates a survey of the pandemic situation in COVID-19. Based on user constraints and public needs, the communication technology developed more, according to this research, which explored more areas, like IoT, drones, AL, ML, deep learning, blockchain, data analytics, and 5G technology to rectify the solution to the real-time scenario. To promote public safety, the author focused on various environments like remote monitoring system, logistics, and public safety during COVID-19. Additionally, it may help them improve the efficiency and rectify the issues in the management process.

The authors [5] examined the economic damages in addition to various events. This economic impact is an important metric to reduce the impact of disasters. This research is used to find data to prepare a human analytical model using data collection. This model is pre-processed and justified with different events to prevent real-time environment issues. This paper [6] addresses a survey on various advanced methodologies available in the market to improve efficiency in wireless technology. It is mainly focused on future generation key aspects to improve the quality of data transmission using various resources, such as IoT, 5G, radar, and mobile computing, which used to overcome challenges such as in security, efficiency, stability, and scalability.

This article [7] illustrates the future generation of wireless communication technology, which is used to analyze the different characteristics of design and its application related to communication. Today's research is focused on 6G technology, which is faster than any other wireless technology. In addition, data transmission should be high to produce more data rate, more energy efficiency, and reduce transmission time. In the extended version of this paper outline, the technical issues and solutions in 6G technology are discussed. This paper [8] proposes to reduce the error rate of a communication network. The author proposed the Fractional Frequency Reuse method to reduce the interference between two different networks in a communication system. Based on system performance and evaluation, this FFR method effectively produces the best solution to reduce the error rate and improve the system quality in communication networks.

This system [9] explores the various band allocations for long signal transmission in communication technology using recent generation. In this statement, the terahertz band bandwidth allocation is used for long data processing applications. The usage of terahertz spectrum has unique functionality because it consumes more data rate with huge bandwidth, high signal transmission application like wireless communication, mobile communication, etc. This study [10] focused on telemedicine application. It is a major concern to produce more impact in society, which helps to improve the quality of users, quality of medicine to satisfy the patient, clinic, and healthcare resources. This ENT research focused on many ways to take care of a patient during the COVID-19 pandemic.

This paper [11] illustrates a security process to improve the quality of communication systems. This research focused on physical layer security technique in communication systems. In general, security is also one of the key elements in wireless communication systems to transmit data in a secure way. In this paper, the authors stated the overview of physical security mechanism and its work flow, which is going to ensure the confidentiality level of data communication. This PLS consists of several methods, like AI-based noise generation, secret coding, etc. Even though this article has security to transmit data, it has some limitations.

This paper [12] provides an overview of IoT-based security issues in wireless communication systems. Security issues happen in many ways while transmitting or communicating data between multi users. The security method associated with IoT may cause extra lagging in security due to different vectors. This research compared with variable analysis metrics such as model, network, communication protocol, various attacks vector, which is used to enhance the system quality.

This paper [13] explains sensing technology using radio frequency signal transmission and reflection. This RF signal produces more efficient signal transmission using a wider bandwidth. This system is applicable to sense or identify the behavior of a human as an individual. This paper mainly focuses on RF-based sensing systems in smart homes, which are used to analyze human movement and characteristic detection using RF signals. This paper [14] explores VLC, in general system has more way to communicate a data in proper root.

Based on this concern, this visible light communication system is used. This system approaches techniques and approaches three-dimensional visible light communication (VLC) systems using non-orthogonal multiple access (NOMA). This NOMA is used to allocate the system power to improve the system performance. This article [15] discusses UAV communication system; it mainly focuses on improving the quality of 5G technology. Unmanned Aerial Vehicle Communication system works based on 5G data. The main purpose of this data communication is to integrate 5G with different environmental-based centric systems. UAVs are also known as drones, which are used to improve the quality of data and produce a clear vision about the output signal. This article [16] discusses about 5G wireless communication system, which is similar to the previous research. The major concern of this research is based on healthcare application using intelligent medicine. This system has many advanced methods when compared to telemedicine. The main aim of this research should satisfy the patients' monitoring process because healthcare monitoring has so many methodologies to improve the data communication system. This intelligent 5G system analyzed data using a data analytic model, which used to give more report in a different way about patient characteristics.

This paper [17] proposes the UNSDG, which stands for United Nations Sustainable Development Goals. This paper explores the 6G wireless communication technology. This SDG has various deployment processes to develop 6G in a communication system. The merits of this SDG-based 6G are that it has high-quality data communication, universal access with the internet, best resource utilization, and economic growth. This paper [18] introduces joint random subcarrier selection and channel-based artificial signal design-aided PLS. This system has multi-channel data transmission using PLS. A previous research only focused on PLS alone. An artificial signal design-aided PLS has more controlling processes, which are used to improve the communication system quality and security. This joint random subcarrier selection process is used to reduce attacks in communication.

This paper [19] illustrates quantum machine learning techniques in a communication system, which are used to improve the performance of wireless technology. Recent cases quantum computing plays a major for communication process. This QC added with ML is an add-on advantage to improve the quality of the system and also enhance the 6G network. This paper [20] explores advanced healthcare monitoring process using advanced 6G communication technology. This technology incorporates IoT, data analytics, Artificial Intelligence, machine learning, deep learning, etc. The AI-based 6G communication produces efficient and accurate results in terms of health data analytics, patient monitoring in advance, and different medicine services.

This paper [21] illustrates the healthcare application for improving system response and incorporates various telemedicine processes to take care of a patient's health. This system allows the integration of various telemedicine mechanisms, which are used to reduce patient difficulties. An integration of telemedicine in communication plays a major role in increasing transmission quality over the global factor in the COVID-19 pandemic.

4.2 A Basic Human-Centric Sensing Mechanism

The traditional human-centric sensor is used to monitor the environmental system using most advanced sensors, which are used to observe, analyze, and integrate to improve security, public safety, healthcare, and the industry to gain better communication using wireless technology [22]. This sensing methodology enables user friendliness in different ways such as those shown in Figure 4.1.

This system is more effective because of user interaction, public observation, real-time process measurement, and data collection depends on user survival using various online platforms like mobile application, signal transmission, and data communication in wireless technology [23]. This app development is not only about analyzing performance but also includes public user awareness. Figure 4.2 shows the citizen science monitoring process, which involves public volunteers, scientific research, public safety, and rules and regulations of the government system.

This monitoring technique helps users to communicate all around the areas inside and outside the world. This metric is used to analyze the environment condition based on water pollution, air pollution, and noise level due to weather conditions and share the details to the local and global

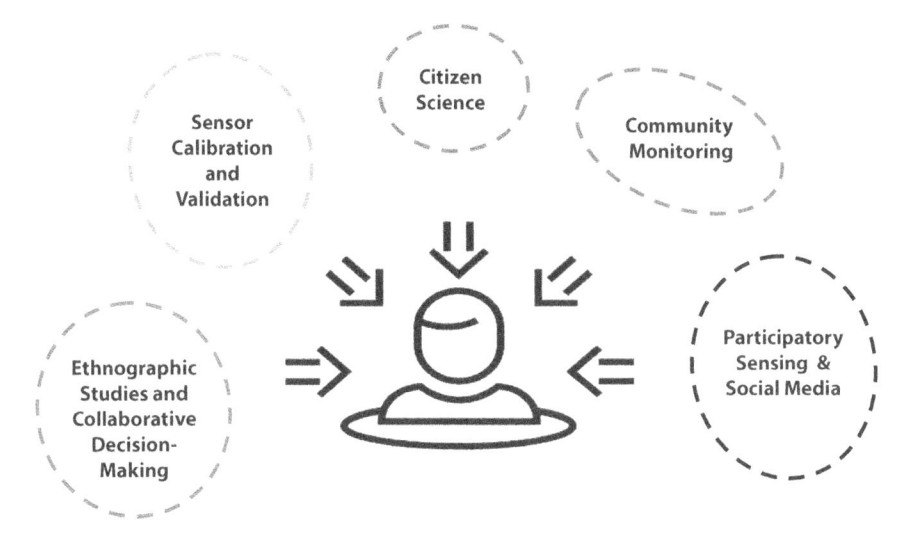

Figure 4.1 Various human-centric mechanisms.

Figure 4.2 Citizen science monitoring.

communities using wireless technology [24]. This knowledge is easy to understand by a researcher and is used to enhance the quality of communication, monitoring, and controlling the effective environment to enhance the system approach, which is shown in Figure 4.3.

This participatory sensing monitoring process is used to collect environmental data using various online resources. The sensor plays a major role in smart phone applications. It is easy to detect data from the present environment. After sampling the data, this participatory sensing process is used to transmit through GPS. The GPS allows the user to collect data

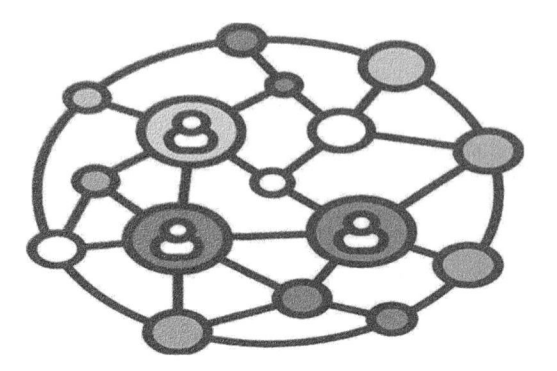

Figure 4.3 Community-based monitoring.

while actively participating in all events, which are shown in Figure 4.4. This sensor is used to guide an individual user to access and transmit data using a Smartphone. This level of monitoring in human-centric sensing application has produced a major impact in data acquisition and transition using advanced wireless technology.

The most important monitoring scheme in a human-centric sensing system is social media analysis. It is a big area to develop the environment in a large area network. The main objective of this social media analysis is to share data, photos, events, and videos relevant to public concern and explore public requirement and issues in the authorized region. It is a good platform, which has so many resources that are used to analyze day-to-day activities, and reaches all people, and also gives solutions to the public. Nowadays, according to the digital world, data are collected using machine learning, AI, deep learning, and different network architectures, which are shown in Figure 4.5.

Figure 4.4 Participatory sensing.

Advanced Techniques for Human-Centric Sensing   109

Figure 4.5 Social media analysis.

After considering all the processes in each monitoring process, the sensor is used to calculate and validate the information. This human-centric monitoring uses many methodologies, which are used to observe data, collect data, and identify errors, accuracy level of monitoring process, and original information, which are used to improve the system quality. Ethnographic research methods involve studying human behavior and cultural practices within specific communities. Researchers conduct interviews, participant observations, and focus group discussions to understand how people perceive and interact with their environment. Ethnographic insights can inform the design of environmental monitoring programs and help identify relevant indicators for monitoring, which are shown in Figure 4.6.

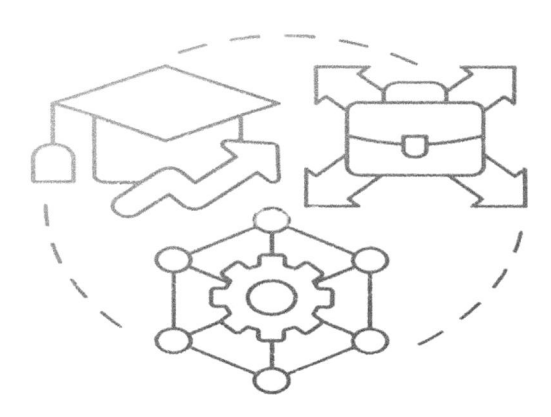

Figure 4.6 Ethnographic studies.

From the overall monitoring process, the final stage of human-centric sensor is collaborative decision making. This process is used to collaborate with scientists, policy members, community people, environmental management team, stakeholders, and finally human experience [25]. This monitoring system actively participates in all environmental-related monitoring schemes, challenges, and issues. This mechanism is more relevant and effective to predict a better performance.

4.3 Types of Advanced HCS Environmental Monitoring System

4.3.1 Multispectral Sensors

An HCS environmental monitoring system has more sensors to capture data in a different frequency. To predict the data in any rate of frequency, this multispectral sensor is proposed. This sensor is working not like a human eye, which contains visible light spectrum inside the module. The input of a multispectral sensor is sent to an analog frame camera or lens, after initiating the data that are sent to a detector's lens and filtration. This detector is used to detect active data with a proper frequency spectrum, and filtration is applied to remove unethical data. The data are transmitted over the scanner. This method is used to predict the exact point and send to discrete decoders, which are used to decode the process in a discrete manner. After decoding of given data, which are sent to a mirror, they are going

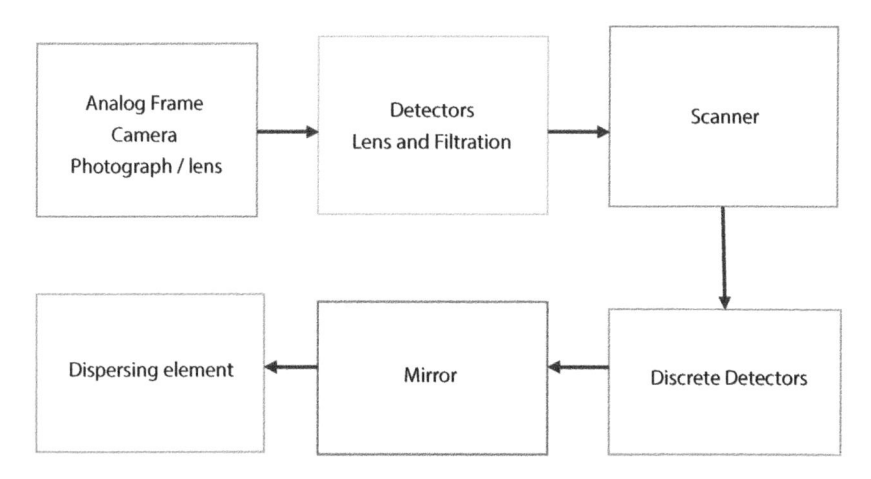

Figure 4.7 Functional block of multispectral sensors.

to reflect the data to a dispersing element to reach the output. This method has a very good capability of data sensing to asses vegetation health, which is shown in Figure 4.7.

4.3.2 Thermal Sensors

This sensor is used to find the exact value of a particular device, which is identified through a sample of pattern. The heat sensor is a crucial factor in image recognition. Many analyses enable finding diseases in plants, finding fake images, and identifying humans, animals, plants, flowers, and so on.

4.3.3 LiDAR

Light Detection and Ranging is used to communicate data between long distances. The LiDAR is used to monitor the distance and distraction of signal in under water communication using the Lasar object, which is shown in Figure 4.8. This sensor is highly productive to measure data in a long-channel communication without any path transition loss using GPS. In upcoming technology, the path loss is rectified and recovered using 6G wireless communication. This wireless technology is more active in transmitting data between source and destination from one point to many, either in land, water, forest, device, etc. High-resolution mapping is analyzed using a 3D image spectrum [26].

4.3.4 Hyperspectral Sensors

It is used to analyze a specific area, which is affected by pollution, and the main aim of this sensor is good accuracy and to detect an object in the exact resolution with large data transmission using high bandwidth. It is similar to a multilevel sensor. This sensor has high charge capacity, high

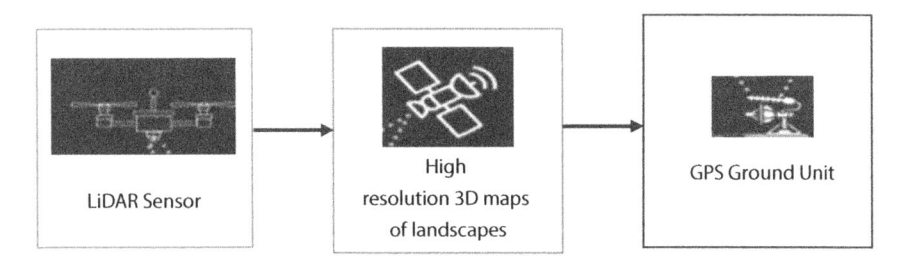

Figure 4.8 Functional block of LiDAR sensors.

112 Networked Sensing Systems

Figure 4.9 Hyperspectral sensors.

resolution monitoring device. For quality, it has a spectrometer and driver system. To transfer data between communication media, it has alleviating device and color conversion process to predict original information, which is shown in Figure 4.9.

UAVs with hyperspectral sensors play a significant role in this area. These sensors can sense specific chemical signatures of pollutants offering a means to rapidly assess and monitor pollution levels. For instance, hyperspectral sensors can recognize oil spills in water bodies or detect air pollutants over industrial areas.

4.3.5 Photogrammetry Sensors

This sensor is mainly used to find effective data in a communication system. The photogrammetry sensor has a high resolution, which is used to

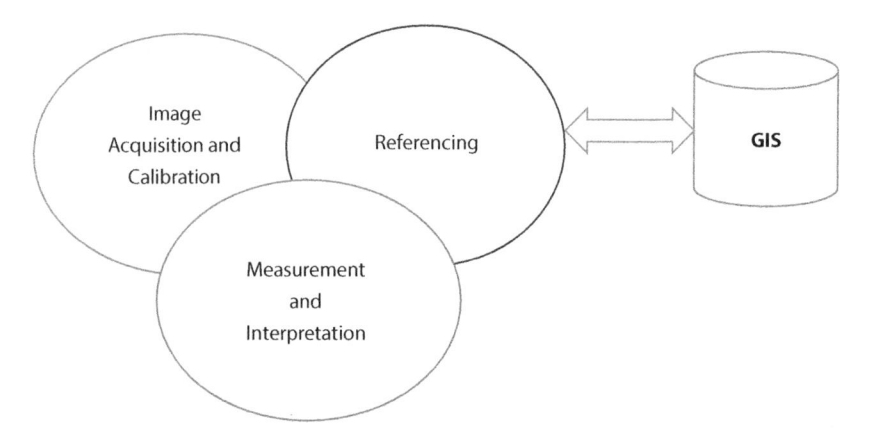

Figure 4.10 Photogrammetry sensors.

find destruction of image, frequently monitor the data due to changes happening in communication technology, deforestation, etc. In general, 3D image is used to get the details of any object using this sensor, which is shown in Figure 4.10. Three important processes are analyzed to obtain a high resolution. Image acquisition and calibration are used to predict the quality of an object from the communication technology. The second process is referencing, which is verified using various analysis processes. Then, measurement and interpretation are used to find the overall quality of the data communication. These all happen because of the GIS process.

4.4 Applications in Environmental Monitoring

This human centric-sensing environmental monitoring system is the most required domain in today's digital world. Depending upon the data processing, application users have done many research to collect proper data and connectivity of a network reducing transmission loss, detecting fake objects, finding in-depth knowledge of undefined areas, etc. According to many research, this area has many sensors to protect human lifestyle using environmental monitoring system. Few applications based on environmental monitoring system is defined in [27].

4.4.1 Smart Sensor

A smart sensor senses physical occurrence, interprets them into another form (often electronic impulse), and processes the obtained data. This function releases the external or programmable logic controller (PLC) of processing sensor output, which is shown in Figure 4.11. It frees the central controller to emphasize system-level tasks like process automation and analytics. This contrasts distinctive sensors, which simply translate physical phenomena into numerical values and communicate them to a central controller. In a challenging environment, for example, the internal machinery of a smart sensor can prevent inexact detection. Their output signal strength and threshold automatically change based on the environment. This varies from the typical sensor, in which the primary controller deals with this issue. Sensors collect information from specific settings and convert it into numerable electrical impulses. Temperature, speed, mass, force (pressure), and the occurrence of heat bodies, like humans, are examples of these characteristics. A microprocessor then processes the electrical impulses to produce outputs corresponding to a set of actions. Finally, the system connects the output with receivers in the designated devices

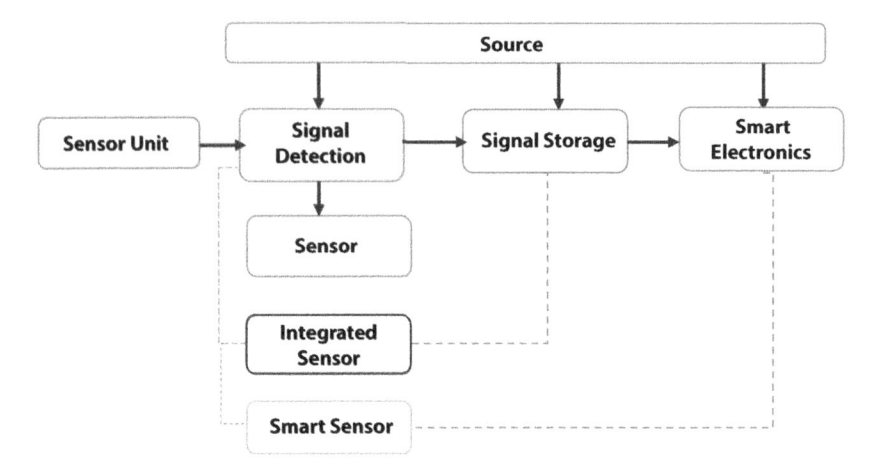

Figure 4.11 Smart sensors.

to ensure proper operation. A system may employ many sensors of varying capabilities depending on the functional difficulties and rising feature needs. It may include more power sources, transmitters, receivers, and so on to achieve better outputs [28, 29].

4.4.2 Wireless Network Technology

Wireless network technology encompasses the infrastructure facilitating data transmission between devices sans physical cables relying on radio frequency signals for connectivity in Figure 4.12. It encompasses diverse

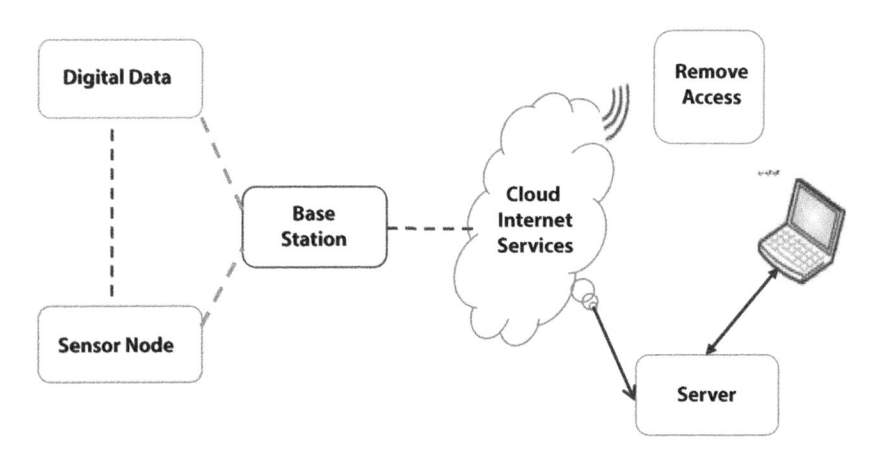

Figure 4.12 Wireless network technology.

communication protocols, like Wi-Fi, Bluetooth, Zigbee, and cellular networks (e.g., 3G, 4G, 5G, 6G), enabling communication across varying distances. This wireless technology offers many resources for effective communication over the entire world with secure communication. To overcome the issues of wired connection, this wireless technology monitoring process is proposed in human-centric sensing mechanism for environmental monitoring. This process is a crucial way to transmit data without any loss as well as for an easy way of communication between source to destination from anywhere in the world. This wireless technology has an advanced method, which might be connected with IoT, machine learning, and Artificial Intelligence to enhance the quality of the communication system. The main aim of this wireless technology is focused on secured data transmission using various encryption and decryption algorithms in cybersecurity and transmits the data confidentially using secured protocol in advance technology. The research in wireless technology is based on data rate transmission and bandwidth allocation, channel formation, data loss during communication, node connection, etc.

4.4.3 Passive Sensing Technology

Passive sensing technology entails detecting and measuring signals emitted or reflected by objects in the surroundings without actively emitting any energy, as illustrated in Figure 4.13, and which contains a sensing element, communication channel, readout unit, and processes data. In contrast to active sensing technologies, like radar or LiDAR, which emit signals and analyze their echoes, passive sensing relies on capturing naturally occurring

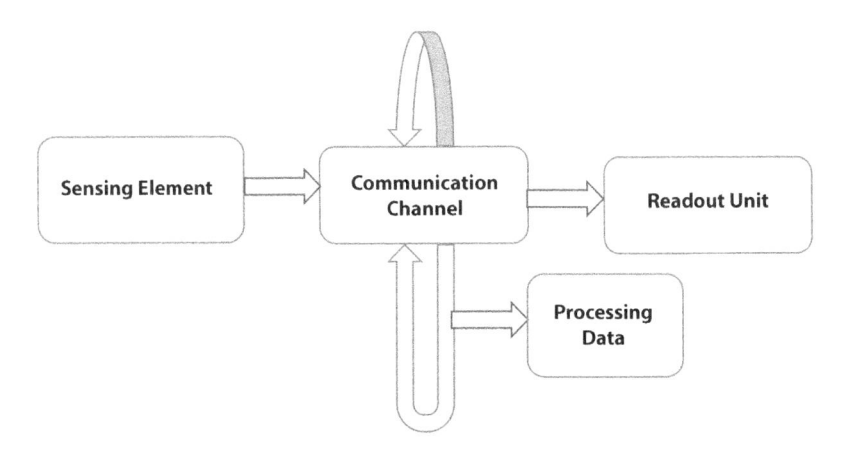

Figure 4.13 Passive sensing technology.

signals, such as electromagnetic radiation or acoustic waves, and deriving information from them. Passive sensing encompasses various modalities, including electromagnetic radiation (like visible light, infrared, and radio waves), acoustic waves (sound), thermal radiation (infrared), and chemical signals (such as a molecules). Each modality offers distinct insights into the environment and its objects. As sensor technology research and development progress, passive sensing is anticipated to assume an increasingly pivotal role in facilitating smart and autonomous systems across various applications [30].

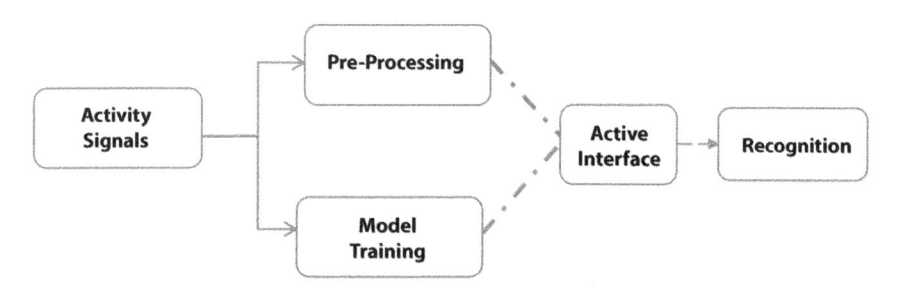

4.4.4 Activity Recognition Technology

Activity recognition technology involves automatically identifying and categorizing human activities through data obtained from sensors or input devices, as depicted in Figure 4.14, which contains activity signals, pre-processing data, model training, active interface, and recognition. This technology utilizes diverse sensing modalities, such as accelerometers, gyroscopes, cameras, microphones, and wearable gadgets to capture signals or data pertaining to human movements, interactions, or behaviors. These sensors may be integrated into smartphones, wearable devices, smart home appliances, or environmental monitoring tools [31].

4.4.5 Gesture Recognition Technology

Gesture recognition technology, depicted in Figure 4.15, contains the number of users, active interfacing device, multiple number of objects, gesture recognition, detected input image, image capture using resolution-based camera, and hand movement. It is a wearable sensor, which can be used in any place in the environmental monitoring system. Sample data

Figure 4.14 Activity recognition technology.

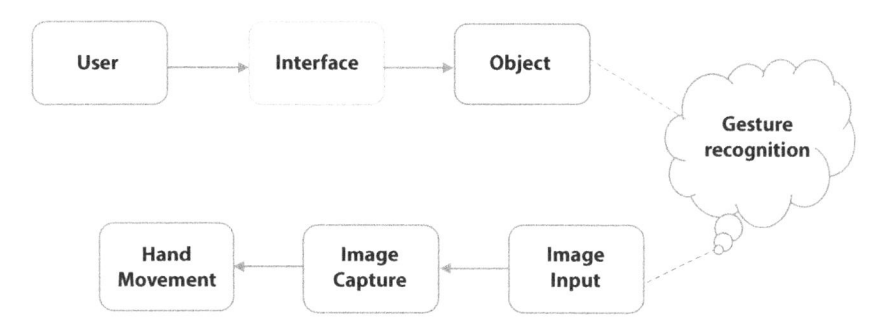

Figure 4.15 Gesture recognition technology.

are collected from a given data set and sent for pre-processing. After pre-processing, the data are sent for feature extraction then sent to the gesture recognition system. This gesture recognition sensor has many modalities to engage the device frequently. To enhance object resolution, this gesture recognition sensor has a high-quality camera to capture exact data. This sensor has many interactions between human and machines with effective samples. This sensor is a standard recognition technology in environmental monitoring system [32].

4.5   Conclusion and Future Prospects

This evolution is mainly focused on user safety and secure data transmission in future generations. Many research areas are available for human-centric sensing mechanism-based environmental monitoring system to overcome the challenges and issues in real-time environment. So many advanced technologies are available based on environmental monitoring system. In recent era, most of applications working based on AI to develop a model with long data communication and secure data transmission. Upcoming technologies are going to create an impact in human-centric sensing systems with environmental monitoring such as Industry 5.0, healthcare monitoring process, smart cities, smart sensor-based various applications, agriculture-based environmental monitoring system, and educational-based environmental monitoring system with advanced sensing technology [33].

Some examples of recent research based an advanced monitoring systems are Artificial Intelligence, machine learning, neural network, image processing, video processing, wireless technology, LIFI, LORA, and 6G communication to help us in an effective way of monitoring the environment to protect human safety and security in terms of data communication and signal transmission then a lifetime of society monitoring system. Even though the system has so many technologies, in future research focusing on to develop the data set using data analytics and hybrid with some advanced method which used to improve the quality of communication and controlling device like various application-based sensors. Because when a particular device needs to access data and predict the values sensors might be placed a major role, so according to the forthcoming general to enrich this area, this system focusing on more and more advanced monitoring device.

References

1. Akyildiz, I.F., Kak, A., Nie, S., 6G and beyond: the future of wireless communications systems. *IEEE Access*, 8, 133995–134030, 2020, doi: 10.1109/ACCESS.2020.3010896.
2. Basar, E., Di Renzo, M., De Rosny, J., Debbah, M., Alouini, M.-S., Zhang, R., Wireless communications through reconfigurable intelligent surfaces. *IEEE Access*, 7, 116753–116773, 2019, doi: 10.1109/ACCESS.2019.2935192.
3. Berardinelli, G., Mogensen, P., Adeogun, R.O., 6G subnetworks for life-critical communication, in: *2020 2nd 6G Wireless Summit (6G SUMMIT)*, IEEE, Levi, pp. 1–5, 2020, doi: 10.1109/6GSUMMIT49458.2020.9083877.
4. Chamola, V., Hassija, V., Gupta, V., Guizani, M., , A comprehensive review of the COVID-19 pandemic and the role of IoT blockchain, and 5G in managing its impact. *IEEE Access*, 8, 90225–90265, 2020, doi: 10.1109/ACCESS.2020.2992341.
5. Coronese, M., Lamperti, F., Keller, K., Chiaromonte, F., Roventini, A., Evidence for sharp increase in the economic damages of extreme natural disasters. *Proc. Natl. Acad. Sci. U.S.A.*, 116, 21450–21455, 2019, doi: 10.1073/pnas.1907826116.
6. Da Costa Daniel Benevides, Y.H.-C., Grand challenges in wireless communications. *Front. Commun. Netw.*, 1, 1, 2020. 10.3389/frcmn.2020.00001.
7. Dang, S., Amin, O., Shihada, B., Alouini, M.-S., What should 6G be? *Nat. Electron.*, 3, 20–29, 2020, doi: 10.1038/s41928-019-0355-6.
8. Dhiviya, S., Malathy, S., Kumar, D.R., Internet of Things (IoT) Elements, Trends and Applications. *J. Comput. Theor. Nanosci.*, 15, 5, 1639–1643, American Scientific Publishers, 2018.
9. Elayan, H., Amin, O., Shihada, B., Shubair, R.M., Alouini, M., Terahertz band: the last piece of RF spectrum puzzle for communication systems. *IEEE Open J. Commun. Soc.*, 1, 1–32, 2020, doi: 10.1109/OJCOMS.2019.29 53633.
10. Fieux, M., Duret, S., Bawazeer, N., Denoix, L., Zaouche, S., Tringali, S., Telemedicine for ENT: effect on quality of care during COVID 19 pandemic.

Eur. Ann. Otorhinolaryngol. Head Neck Dis., 137, 257–261, 2020, doi: 10.1016/j.anorl.2020.06.014.
11. Hamamreh, J.M., Furqan, H.M., Arslan, H., Classifications and applications of physical layer security techniques for confidentiality: a comprehensive survey. *IEEE Commun. Surv. Tutor.*, 21, 1773–1828, 2019, doi: 10.1109/COMST.2018.2878035.
12. Hassija, V., Chamola, V., Saxena, V., Jain, D., Goyal, P., Sikdar, B., A survey on IoT security: application areas, security threats, and solution architectures. *IEEE Access*, 7, 82721–82743, 2019, doi: 10.1109/ACCESS.2019.2924045.
13. Hsu, C.-Y., Hristov, R., Lee, G.-H., Zhao, M., Katabi, D., Enabling identification and behavioral sensing in homes using radio reflections, in: *CHI '19*, pp. 1–13, Association for Computing Machinery, New York, NY, 2019, doi: 10.1145/3290605.3300778.
14. Ali, M.A., Dhanaraj, R.K., Nayyar, A., A high performance-oriented AI-enabled IoT-based pest detection system using sound analytics in large agricultural field. *Microprocess. Microsyst.*, 103, 104946, 2023.
15. Li, B., Fei, Z., Zhang, Y., UAV communications for 5G and beyond: Recent advances and future trends. *IEEE Internet Things J.*, 6, 2241–2263, 2019, doi: 10.1109/JIOT.2018.2887086.
16. Li, D., 5G and intelligence medicine–how the next generation of wireless technology will reconstruct healthcare? *Precis. Clin. Med.*, 2, 205–208, 2019, doi: 10.1093/pcmedi/pbz020.
17. Matinmikko-Blue, M., Aalto, S., Asghar, M.I., Berndt, H., Chen, Y., Dixit, S. et al., *White paper on 6G drivers and the UN SDGs*, 2020, arXiv preprint arXiv:2004.14695, https://doi.org/10.48550/arXiv.2004.14695.
18. Naderi, S., da Costa, D.B., Arslan, H., Joint random subcarrier selection and channel-based artificial signal design aided PLS. *IEEE Wirel. Commun. Lett.*, 9, 976–980, 2020, doi: 10.1109/LWC.2020.2976979.
19. Nawaz, S.J., Sharma, S.K., Wyne, S., Patwary, M.N., Asaduzzaman, M., Quantum machine learning for 6G communication networks: State-of-the-art and vision for the future. *IEEE Access*, 7, 46317–46350, 2019, doi: 10.1109/ACCESS.2019.2909490.
20. Nayak, S. and Patgiri, R., *6G communication technology: a vision on intelligent healthcare*, 2020, arXiv preprint arXiv:2005.07532, doi: https://doi.org/10.48550/arXiv.2005.07532.
21. Ohannessian, R., Duong, T.A., Odone, A., Global telemedicine implementation and integration within health systems to fight the COVID 19 pandemic: a call to action. *JMIR Public Health Surveill.*, 6, e18810, 2020, doi: 10.2196/18810.
22. Pramanik, P.K., Nayyar, A., Pareek, G., Chapter 7 - WBAN: driving e-healthcare beyond telemedicine to remote health monitoring: architecture and protocols, in: *Telemedicine Technologies*, Academic Press, pp. 89–119, 2019, doi: 10.1016/B978-0-12-816948-3.00007-6.

23. Rahman, M.A., Hossain, M.S., Alrajeh, N.A., Guizani, N., B5G and explainable deep learning assisted healthcare vertical at the edge: COVID-19 perspective. *IEEE Netw.*, 34, 98–105, 2020, doi: 10.1109/MNET.011.2000353.
24. Rinaldi, F., Maattanen, H.-L., Torsner, J., Pizzi, S., Andreev, S., Iera, A. *et al.*, Non-terrestrial networks in 5G & beyond: a survey. *IEEE Access*, 8, 165178–165200, 2020, doi: 10.1109/ACCESS.2020.3022981.
25. Rothbart, N., Holz, O., Koczulla, R., Schmalz, K., Hübers, H.-W., Analysis of human breath by millimeter-wave/terahertz spectroscopy. *Sensors*, 19, 2719, 2019, doi: 10.3390/s19122719.
26. Saarnisaari, H., Dixit, S., Alouini, M.-S., Chaoub, A., Giordani, M., Kliks, A. *et al.*, *A 6G white paper on connectivity for remote areas*, 2020, arXiv preprint arXiv:2004.14699, https://doi.org/10.48550/arXiv.2004.14699.
27. Balasubramaniam, S., Joe, C.V., Manthiramoorthy, C., Kumar, K.S., ReliefF based feature selection and Gradient Squirrel search Algorithm enabled Deep Maxout Network for detection of heart disease. *Biomed. Signal Process. Control*, 87, 105446, 2024.
28. Balasubramaniam, S. and Kumar, K.S., Optimal Ensemble learning model for COVID-19 detection using chest X-ray images. *Biomed. Signal Process. Control*, 81, 104392, 2023.
29. Saeed, N., Bader, A., Al-Naffouri, T.Y., Alouini, M.-S., When wireless communication responds to covid-19: combating the pandemic and saving the economy. *Front. Commun. Netw.*, 1, 3, 2020, doi: 10.3389/frcmn.2020.566853.
30. Balasubramaniam, S. and Kavitha, V., A survey on data retrieval techniques in cloud computing. *J. Converg. Inf. Technol.*, 8, 16, 15, 2013.
31. Sarieddeen, H., Saeed, N., Al-Naffouri, T.Y., Alouini, M.-S., Next generation terahertz communications: a rendezvous of sensing, imaging, and localization. *IEEE Commun. Mag.*, 58, 69–75, 2020, doi: 10.1109/MCOM.001.1900698.
32. Balasubramaniam, S., Nelson, S.G., Arishma, M., Rajan, A.S., Machine Learning based Disease and Pest detection in Agricultural Crops. *EAI Endorsed Trans. Internet Things*, 10, 1–8, 2024, doi: 10.4108/eetiot.5049.
33. Tanwar, S., Parekh, K., Evans, R., Blockchain-based electronic healthcare record system for healthcare 4.0 applications. *J. Inform. Secur. Appl.*, 50, 102407, 2020, doi: 10.1016/j.jisa.2019.102407.

5

Energy-Aware System for Dynamic Workflow Scheduling in Cloud Data Centers: A Genetic Algorithm with DQN Approach

Hariharan B.[1*], Anupama C.G.[1], Ratna Kumari Neerukonda[1] and Rajesh Kumar Dhanaraj[2]

[1]*Department of Computational Intelligence, SRM Institute of Science and Technology, Kattankulathur, Tamil Nadu, India*
[2]*Symbiosis Institute of Computer Studies and Research (SICSR), Symbiosis International (Deemed University), Pune, India*

Abstract

Due to the rising services offered by data centers, such as load balancing, auto-scaling, dynamic resource allocation, and efficient resource use, many firms are migrating their businesses to the cloud. Servers are considered the essential processing units in a data center. The growing expansion of cloud applications is leading to a significant rise in energy usage and increased carbon gas emissions. Upon receiving a user's request, the scheduler allocates the tasks to cloud resources for execution. Efficiently organizing tasks in a cloud environment poses a formidable challenge. Several sophisticated job schedulers exist for managing workflow; their architecture mostly caters to batch workflow rather than real-time workflow. Efficient allocation of real-time workflows to suitable virtual machines on a server requires implementing a dynamic workflow scheduling technique that effectively minimizes energy usage. This study proposes an alternative approach to dynamic workflow scheduling, aiming to improve the workflow schedule to decrease both makespan and energy usage in cloud data centers. This chapter introduces a novel approach that combines a Genetic Algorithm and DQN. This scheduler efficiently transmits dynamic workflows to the virtual machines of the servers. To be more specific, the scheduling of the dynamic process is divided into

---

Corresponding author: hariharb@srmist.edu.in

two parts. First, using GA, the task's execution scheme calculates from the time it arrives until the end of the execution. The second stage uses an agent in the DQN to determine whether or not assigning tasks to a VM instance in the cloud is appropriate. The energy usage is assessed based on the minimal makespan value. Our method was developed by integrating a Genetic Algorithm with a DQN algorithm to calculate the makespan. Data collected from experiments prove that our approach exhibits superior performance compared to alternative dynamic work scheduling approaches.

Keywords: Workflow scheduling, deep Q-learning, genetic algorithm, energy efficiency, data center

5.1 Introduction

Cloud computing provides servers, storage, databases, networking, and software online. The services are in remote data centers and can be used online. Cloud providers use Infrastructure as a Service (IaaS) to provide virtualized computer resources over the Internet. Users can rent and operate virtual machines (VMs), storage, and networking infrastructure instead of buying and maintaining actual equipment. Different operating systems (OS) and applications can operate on the same hardware with virtualization. Cloud computing is used across industries due to its flexibility, scalability, cost effectiveness, and accessibility. The following are some examples of how cloud computing is used in various businesses. Information Technology, Healthcare, Finance and Banking, Education, Retail and E-Commerce, and Manufacturing are just a few examples; there are many more. In cloud computing, SLA stands for Service Level Agreement. It is an agreement that specifies the caliber of service that a cloud service provider will provide to a client. SLAs usually outline the performance parameters to be assessed, including availability, throughput, uptime, and response time. These measurements aid in determining the level of service quality [1]. Task scheduling is allocating the appropriate resources for each action, such as virtual machines, containers, or serverless compute instances. Allocating jobs or tasks to the server's available resources is known as task allocation. It generates a virtual machine (VM) on the server based on the user's work needs. After the user's task is over, the VM is assigned to meet the needs of another user. This means that all resources are being used to their full potential and that none are being over or underutilized. Tasks are stored in the Queue. It runs the jobs using the Heuristic approach that we have incorporated into the scheduler. There are two types of tasks: homogeneous and heterogeneous. All tasks in a homogeneous task set have the same requests and resource requirements. Every task in a heterogeneous task set has a variable request type, varying processing requirements, and varying resource requirements [2]. Many authors proposed heuristic approaches to task scheduling, such as First Come First Served (FCFS), Shortest Job Next (SJN), Round Robin (RR), and nature-inspired approaches, such as Ant Colony Optimization Algorithm, Particle Swarm Optimization, Gray Wolf Optimization, Artificial Bee Colony Optimization Algorithm, and so on. However, these solutions do not properly solve the difficulty of scheduling dynamic jobs on virtual machines or allocating resources dynamically based on user job requests. The approach does not even handle the execution of the suboptimal task dependency and how they minimize metrics like makespan and energy usage [3]. In recent studies, researchers presented a Reinforcement Learning algorithm, which is a type of Artificial Intelligence algorithm that excels at allocating jobs to appropriate resources. Inside the Reinforcement Learning process, the agent operates inside an environment by executing actions. It transitions from one state to another state based on its actions. Each action state is associated with either a positive or negative reward. The objective of the Reinforcement Learning algorithm is to optimize the total accumulated reward. Q-learning is a model-free Reinforcement technique that generates a table to represent states and actions. In the context of Task Scheduling, the states correspond to the availability of resources, virtual machines (VMs), and the priority of tasks. The action, on the other hand, involves the allocation of tasks to the VMs [4]. The chosen action is carried out, and the scheduler monitors the subsequent state change and is given a reward signal depending on the action's performance. The scheduler changes the Q-value for the chosen action in the current state by applying the Q-learning update rule after performing an action and monitoring the subsequent reward. However, Q-learning is lacking in handling dynamic workloads. As the number of tasks increases, the size of the Q table also increases, reaching a point where it becomes unable to handle any additional jobs. Real-time scheduling policy adaptation is difficult since the best course of action may change as circumstances do. Heuristic algorithms are important in dynamic cloud systems since standard optimization techniques may not be practical due to the changing availability of resources, changes in workload, and shifting user demands; however, heuristics also have trouble locating the best answer in a dynamic cloud context. On the other hand, metaheuristic-based techniques, like Genetic Algorithms (GA), emulate natural selection to discover the best answers to issues by progressively evolving a group of potential solutions. Various authors have furnished a Genetic Algorithm (GA) that demonstrates superior performance in load balancing, resource provisioning, and activity scheduling in comparison to current nature-stimulated meta-heuristic algorithms [5]. However, the Genetic Algorithm encounters challenges when handling fluctuating workloads and executing obligations with sub-top-rated assignment dependency.

Recent studies have shown that the combination of Deep Learning and Q-gaining knowledge of algorithms is effective in managing task scheduling in a Cloud Computing Environment. This method is adept at coping with a high extent of jobs. We are utilizing the blessings of Deep Learning and Genetic Algorithms to efficiently schedule dynamic workloads and adapt to diverse conditions [6].

This bankruptcy introduces a different method for addressing the difficulties associated with job scheduling. The GA–DQN approach is a system for scheduling dynamic workflows that integrates the abilities of Deep Q-Learning (DQN) and Genetic Algorithm (GA). This approach combines the strength of GA and DQN to improve the rate of execution of responsibilities. This paintings' number one contribution may be summed up as follows:

- Our proposal is a dynamic workflow scheduling system that integrates GA and DQN intending to optimize energy efficiency and makespan in cloud computing.
- We are giving a complete demo regarding the design and implementation of the method, along with providing a clear explanation of how GA uses the behavior ion of dynamic workflow to accelerate efficient learning for the DQN manager.
- We compare our technique to various workflow scheduling strategies, and our experimental results indicate that it performs well across various workloads.

The remaining papers are organized as follows: Section 5.2 covers existing methodologies, whereas Sections 5.3 and 5.4 detail the problem definition and proposed methodology. The simulation findings are reviewed in part 5.5, and the conclusion is discussed in the final part.

5.2 Related Works

Cloud workflow scheduling is a difficult task in the realm of cloud computing. Several innovative strategies are given for optimizing workflow scheduling for incoming jobs. A method for optimizing workflow scheduling has the potential to enhance cloud performances, such as makespan, and achieve energy efficiency and is important for the ecosystem. The number and nature of activities being processed can vary over time, and dynamic task scheduling addresses these variable demands. As a result of the increased implementation of the Internet of Things, big data, and cloud platform technologies in recent years, job shops have given way to networked, collaborative, and intelligent manufacturing systems. Smart manufacturing scheduling differs from job shop scheduling in several ways, including the dynamic nature of services and uncertainties, in addition to the greater quantity of tasks and services. The authors analyze and provide a mathematical description of the smart manufacturing service scheduling problem in this article. Then, a method based on deep reinforcement learning is suggested for reducing the maximum completion time of all assignments. The system architecture of the proposed method includes the design of the agent, environment, and interaction. The system state is determined by the queue times of all candidate services, with the target value being the utmost queue time at the present moment [7]. However, these techniques might be too time consuming and computationally demanding for dynamic workflow scheduling. One of the key elements that impact cloud computing performance is task scheduling, which is essential to the system. The dynamic task-scheduling problem has garnered global interest due to many factors such as the expanding information processing industry and the growing demand for quality of service (QoS) in networking. Task scheduling has been identified as an NP-hard problem due to its complexity. Furthermore, the majority of dynamic online task scheduling frequently handles jobs in a complicated environment, which makes it much harder to strike a balance and reap the rewards of every cloud computing feature. In this research, the authors [8] present a new artificial intelligence system that combines the benefits of a deep neural network and the Q-learning algorithm, which we term deep Q-learning task scheduling (DQTS). The goal of this novel method is to address the challenge of managing directed acyclic graph (DAG) workloads in cloud computing settings. The technique is based on the widely used deep Q-learning (DQL) method in task scheduling, where DQL serves as a major inspiration for core model learning. The dependency between sub-tasks must be considered in workflow scheduling, which is not addressed.

 A hybrid meta-heuristic optimization approach was given in this paper to reduce the overall execution time of processes operating on heterogeneous cloud platforms. The author [9] explains the technique combined two optimization trends as follows: global and local, respectively, thermodynamic simulated annealing and genetic algorithm. Both algorithms thereby compensate for each other's flaws. The crossover in GA searches the search space worldwide, while the unique procedures in the proposed TSA efficiently permute search space by enhancing existing methods. Based on the quality of the solutions obtained during the annealing process, the TSA uses concepts from information theory and thermodynamics to reduce temperature. The simulation results demonstrate the hybrid GATSA's advantage over other state-of-the-art when compared to the average values of scheduling assessment metrics. Additionally, the suggested GATSA demonstrates its adaptability to many applications by leveraging exploitation with multiple walking about algorithms, each of which may be called at random, and benefiting from experiments with crossover operator and arbitrary mutation operator. However, it is lacking in showing brilliant performance in scheduling tasks when compared with existing approaches. Because of the "pay for use" paradigm, cloud computing systems are widely used by clients to access cloud resources. However, because of the large number of resources and the amount of user requests, scheduling and energy consumption have become major challenges for these systems. This subject has been extensively studied. Reducing makespan is the primary objective of a timetable, which is a means of allocating resources to users. The energy problem has received less consideration in the majority of cloud system scheduling techniques, and in some techniques that have been tuned for energy consumption, the makespan has grown. For instance, the Hybrid GA algorithm is an energy-efficient scheduling technique for dependent jobs; yet, despite the interdependence of tasks, primary chromosome manufacturing and scheduling have not been carried out to the best of their abilities. Not every conceivable combination to complete a task is taken into account by the ECS (Energy-Conscious Scheduling) algorithm. Thus, scheduling methods that maximize energy and time savings are preferable. The goal of this study is to save time and energy by presenting a two-step method called GAECS. Here, the author [10] employed the GA to generate optimal schedules and three ranking algorithms to generate the major chromosomes in the proposed GAECS method and also optimized the resource distribution to processors using the energy-conscious ECS algorithm. The GAECS method generates the first three primary chromosomes using three different prioritizing algorithms. The GA then receives the primary chromosomes, and the GA completes the primary population. Better chromosomes are then chosen using the established crossover and mutation operators, and last, the best chromosomes in terms of time and energy are chosen and allocated to resources. The GAECS has a better makespan and energy usage, according to the evaluation's findings by the author's experimental results. However, the performance of suboptimal task dependency when the workflow is allocated with VM instances it is not discussed.

5.3 Dynamic Workflow Scheduling System

This section presents the fundamental cloud scheduling system architecture and explains the dynamic workflow scheduling model.

5.3.1 System Architecture

Job scheduling in the cloud involves distributing computing resources and organizing tasks or jobs to make the most optimal use of the resources in cloud computing settings. It entails overseeing the implementation of different computational tasks or jobs supplied by users or applications, ensuring they are executed promptly, utilizing resources efficiently, and reaching specific performance goals. Virtual machines (VMs) are generated via virtualization, which abstracts and virtualizes physical hardware resources to produce numerous isolated instances of virtualized computing environments. Transferring a user job to a virtual machine in the cloud includes deploying the VM, transferring the job files, executing the job on the VM's operating system, and monitoring the job's progress and resource usage.

Figure 5.1 illustrates a basic architecture of the cloud job scheduling framework. The basic idea of job scheduling is that there are n number of users, each with n number of requests (tasks). All of these jobs are submitted to a scheduler. Depending on the nature of the user's employment, we normally recommend this form of scheduler. Every scheduler includes

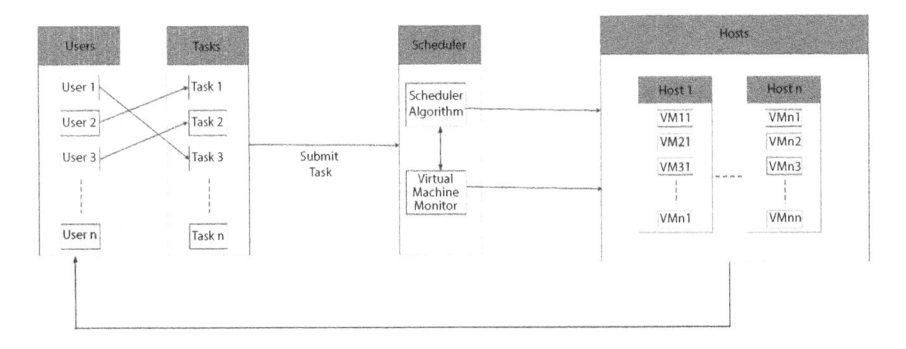

Figure 5.1 The general architecture of job scheduling in cloud computing.

a built-in task manager, whose function is to prioritize tasks depending on requests. In host environments, each host is made up of many virtual machines. A virtual machine monitor in the host environment assigns tasks to the virtual machine based on their requirements. Each virtual machine has its own resources and computing power. In this task scheduling, we focus on optimizing makespan. The term "makespan" denotes the overall duration required to finish a specific set of activities or jobs under a particular scheduling context. It signifies the time from the initiation of the initial job/task to the conclusion of the final job/task in the workload. The objective of this study is to minimize the duration of the makespan. Makespan refers to the whole duration required to finish the execution of a job starting from the moment it enters the queue until it is executed on the virtual machine (VM). Increased energy consumption occurs when a job spends more time in the queue, is not appropriately provisioned with resources, or is not allocated a suitable virtual machine.

When determining the makespan, the evaluation of execution time is crucial as it measures the duration of a task running on a specific virtual machine (VM). The computation of the execution time for a task on a certain virtual machine (VM) is determined using the subsequent equation.

$$et_{t_c} = \frac{et_t}{pr_{ca}^{VM}}$$

Tasks are allocated to a virtual machine (VM) from the task queue depending on the availability of resources, the priority of the tasks, and the completion time of the job. Hence, the time required to finish a work is calculated using the equation given below.

$$ft^{t_c} = \sum VM_n + et_{t_c}$$

5.3.2 Genetic Algorithm for Dynamic Workflow Scheduling

In the context of the Genetic Algorithm, a schedule is represented by a chromosome, which signifies both the job and the assigned resources. Every chromosome is composed of genes.

Each schedule is evaluated using a fitness function, which determines how effectively it meets optimization goals such as lowering makespan or cost. Solutions with higher fitness values are chosen for replication utilizing methods such as roulette wheel or tournament selection.

Selected schedules are merged to produce child schedules via a crossover, which exchanges genetic information between the parents. During the mutation phase, random adjustments are made to offspring schedules to retain diversity and seek novel solutions. Offspring schedules replace older ones in the population via techniques such as generational replacement. The algorithm terminates when a termination condition is met, such as reaching a maximum number of generations or obtaining a satisfying

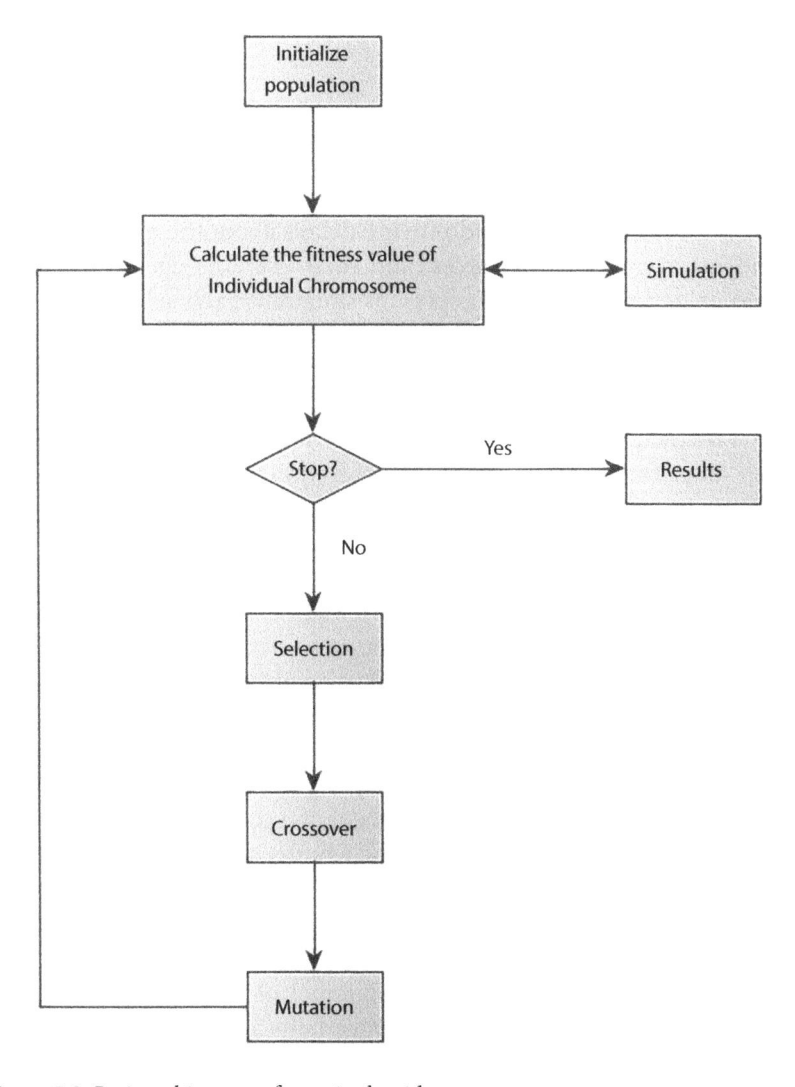

Figure 5.2 Basic architecture of genetic algorithm.

solution. The best schedule(s) are selected from the final population and executed in the cloud environment [11].

The basic idea demonstrated in Figure 5.2 is that core phases of the Genetic Algorithm are the same because it adapts to each type of application. We use a Genetic Algorithm to solve the task scheduling problem. In the first phase, we are initializing tasks. In the second phase, the fitness value of each virtual machine is determined. The fitness value of a virtual machine aids in assigning jobs to relevant resources. This process will continue until all of the tasks have been assigned to appropriate resources, when the number of iterations has been completed and there are no more tasks that require resource allocation.

5.3.3 Deep Q-Learning for Dynamic Workflow Scheduling

The state of the environment represents the current state of the workflow scheduling problem. This could provide details about the current task, its dependencies, available resources, and their use. The agent (scheduler) chooses an action based on the present situation. Each action represents either allocating a job to a specified resource for execution or deferring its execution. The agent learns to make judgments by keeping a Q table or employing a deep neural network to approximate Q-values. Q-values indicate the predicted cumulative payoff for performing a specific activity in a given state. The agent balances the exploration of novel actions with the exploitation of old activities by employing an exploration method known as epsilon-greedy, in which actions are randomly selected with a probability epsilon. A reward is granted to the agent for every activity undertaken by the agent. In the perspective of workflow scheduling, rewards may be based on job completion time, resourcefulness usage, cost, or service quality. The agent interacts with its environment, selects activities, observes rewards, and updates Q-values to match those observations. This process continues over many episodes, and the agent gets to improve its decision-making ability. To improve sampling efficiency and stability, DQL frequently uses experience replay. This entails storing experiences (state, action, reward, and next state) in a replay buffer and randomly sampling from it while training. To stabilize training, a separate target network is periodically updated with the primary Q-network's weights. Training continues until convergence or a predetermined stopping criterion is satisfied, such as completing a maximum number of episodes or reaching a sufficient level of performance. Once trained, the agent can apply its learnt policy to make real-time scheduling decisions by choosing behaviors that maximize projected cumulative rewards [12].

Q-learning is an influential technique in reinforcement learning that operates without requiring prior knowledge of the current system. It relies on a Q-function, which stores pairs of two states indicated by a q(S, A) and actions indicating the quality of taking an action in a given state. This function is updated iteratively using a specific equation.

$$q(S^t, A^t) \leftarrow q(S^t, A^t) + \sigma * [re^t + \beth * maximum^a q(S^{t+1}, A^t) - q(S^t, A^t)]$$

where σ is the rate of learning, and its value is in between (0,1). re^t is the reward for taking action, i.e., A^t for state S^t. £ is the discount factor, and its value lies in between (0,1).

Figure 5.3 depicts the distribution of tasks to appropriate resources using a combination of Q-learning and Deep Learning.

Action Space

In the context of work, the action space pertains to the n number of virtual machines (VMs) previously mentioned. Incoming requests are first directed to the task manager, whereupon each task's priority is computed. Subsequently, the scheduler, equipped with a DQN model, makes decisions based on these priorities and assigns tasks to an execution queue accordingly. Finally, the scheduler's decisions dictate the execution of tasks on the VMs, in alignment with the order of tasks in the queue and their respective priorities. Thus, the action space within our model is delineated as described [13].

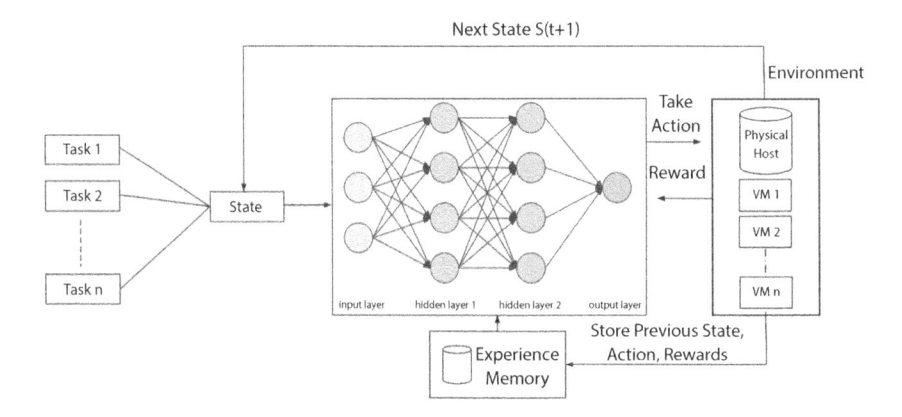

Figure 5.3 DQN for dynamic workflow scheduling.

$$A = [VM_1, VM_2, VM_3, \ldots\ldots VM_n]$$

State Space

In this subsection, we delineate the state space comprising the state of a task at a specific time and the state of a VM at that particular time when the task arrives. Consider a scenario where a task, denoted as t, arrives at time T. This task is represented as Tt. Subsequently, the state of this task can be represented as follows:

$$S_{Tt} = [t_k^l, t^{prio}, et_{t_c}, ft^{t_c}, m^k, e_{VM_n}^{con}]$$

Reward Function

The objective of this study is to determine the optimal mapping between cloud resources and heterogeneous tasks leveraging our DRLBTSA scheduler to optimize key Quality of Service (QoS) parameters such as energy consumption, time, and SLA (Service Level Agreement) violation. Therefore, our reward function should be formulated to minimize the metrics outlined in our work. It can be defined as follows:

$$re = \min(m^k, e_{VM_n}^{con})$$

5.3.4 Energy Consumption for Dynamic Workflow Scheduling [14]

Our current emphasis is on reducing energy consumption within the cloud computing paradigm, which stands as a significant and impactful metric. With the processing of extensive workloads necessitating substantial infrastructure or cloud resources, there is a resultant increase in energy consumption and CO_2 emissions adversely affecting the environment. Thus, our focus is directed toward minimizing energy consumption within the cloud paradigm. In the cloud model, energy consumption is contingent upon the utilization of computing time and idle time. For a virtual machine (VM), energy consumption is computed according to the following equation:

$$VM_n = \begin{pmatrix} \gamma_n \text{ Active State of VM} \\ \tau_N \text{ Idle State of VM} \end{pmatrix}$$

The energy consumption of all n VMs is calculated using the following equation:

$$e_{VM_n}^{con} = ft_n * \gamma_n + (m^k - ft_n) * \tau_n$$
$$min_{act}^{con} = (e^{mx} - e^{mn}) * res^{util} + e^{mn}$$

Energy consumption in the data center is calculated as follows:

$$e^{con} = \sum e_{VM_n}^{con} + min_{act}^{con}$$

5.4 Problem Formulation and Proposed System Architecture

In dynamic situations, new tasks can arrive unexpectedly disrupting the existing schedule. Genetic Algorithms may struggle to efficiently change the present schedule to accommodate new jobs resulting in a longer makespan. To decrease makespan, Genetic Algorithms must be able to adapt swiftly to dynamic changes in their environment. However, Genetic Algorithms' slow convergence and the time necessary to explore and exploit new solutions may limit their capacity to respond effectively to quick changes in task requirements or resource availability. In the DQN algorithm, the initial step involves creating a Q table that contains the states and corresponding actions for each task. The values in the Q table are derived using the Genetic Algorithm and represent the optimal values for scheduling tasks. The Q table values are evaluated using the Deep Learning network. When the tasks are not provided with adequate resources, the network undergoes training until all tasks are appropriately assigned to resources that are well-suited to meet the user's requests. The system can attain an equilibrium between exploration (Genetic Algorithm) and exploitation (deep Q-network) leading to enhanced convergence in discovering the optimal solution and improving performance.

5.4.1 Hybrid Approach

The objective of task scheduling in cloud computing is to minimize the makespan, which is the total time needed to finish all jobs. This includes both the time spent doing tasks and any possible waiting time caused by resource contention or schedule delays. As was said before, you should begin by utilizing GA to initialize a population of task schedules. Every single person in the population represents a prospective task schedule, and the allocations of tasks to virtual machines are encoded. The makespan method, which was discussed before, should be used to evaluate the appropriateness of each job schedule in the population. The population of task schedules can be evolved through the application of selection, crossover, and mutation processes, which will result in the generation of new task schedules that may have lower makespan values. Through the use of reinforcement learning, a DQN model should be trained to discover the most effective task scheduling. As input, the DQN collects the current state of the task scheduling problem, which includes tasks, virtual machines (VMs), and the current assignment. It then produces an action such as determining which task should be assigned to which VM. It is necessary to define the reward function, the state space, and the action space for the DQN. The action space is a representation of the many decisions that could be made about the assignment of tasks, while the state space contains information about the existing work schedule including the surroundings. The DQN receives feedback from the reward function, which does so depend on the quality of the work schedules that are generated. Training the DQN can be accomplished through simulation in a cloud computing environment or through the use of historical data on job scheduling. To stabilize training approaches, such as experience replay and target network, updates can be utilized. During the execution of the Genetic Algorithm, the trained DQN model should be used to guide the selection of task assignments rather than the conventional genetic operators [15].

The architecture of GA–DQN includes a crucial component known as the replay buffer. This buffer is responsible for storing all relevant past information, such as the allocation of tasks to virtual machines, the status of each virtual machine, and the distinction between idle and busy virtual machines. By utilizing this stored data, the system can make appropriate decisions when a new job is received effectively assigning tasks to the most relevant resources and efficiently managing the changing workload [16].

Figure 5.4 depicts a hybrid strategy for allocating tasks to resources that uses GA–DQN to handle dynamic workflow. This integrated technique aims to generate work plans that effectively minimize makespan compared to the approach utilizing individual algorithms.

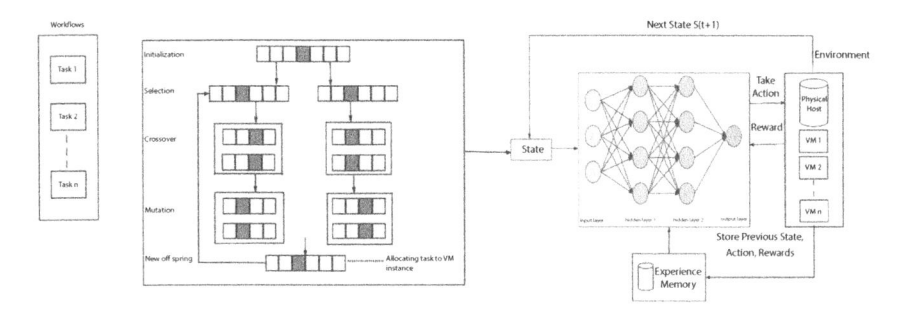

Figure 5.4 Hybrid approach of GA–DQN for workflow scheduling.

5.4.2 Implementation of GA

For every incoming workflow, we encode the execution plan utilizing a topology sorting approach. Specifically, we employ anti-topological sorting to generate random execution sequences for the sub-tasks initially ensuring that each solution respects the dependencies among these sub-tasks. Following this, we randomly allocate a CPU core to each sub-task within each sequence, treating these sequences as individuals in a Genetic Algorithm (GA).

$$I_c = \{(t_1, p_2), (t_3, p_1), (t_2, p_2), (t_5, p_3), (t_6, p_1), (t_9, p_3), (t_4, p_2), (t_8, p_2), (t_7, p_1), (t_{10}, p_3), (t_{11}, p_2)\}$$

$$I_{k+1} = \{(t_2, p_1), (t_1, p_3), (t_6, p_1), (t_5, p_2), (t_9, p_3), (t_3, p_1), (t_4, p_2), (t_7, p_3), (t_{10}, p_2), (t_8, p_3), (t_{11}, p_1)\}$$

Here, $t_1, t_2, \ldots\ldots t_{11}$ are sub-tasks of the workflow. p_1, p_2, p_3 are CPU cores.

By generating individuals at random, it is possible to acquire a population that is defined as:

$$P_i = \{I_1, I_2, \ldots, I_m\}$$

We employ the execution time of instances of individuals as the fitness function in the initial population and choose members for crossover and mutation. For selection purposes, candidates with shorter execution times are favored. In the course of evolution, the crossover operation is initiated by arbitrarily selecting an identical position between two sequences followed by the exchange of chromosomes (sub-tasks) positioned to the left of this specific position. Then, in the order of the original individuals, each individual is traversed, with each subtask not present in the new section being sequentially placed on the right. This procedure ensures that the new individuals can be considered viable solutions. After the crossover operation, for instance, the individuals denoted in equation can be represented as follows:

$$I'_k = \{(t_2, p_1), (t_1, p_3), (t_6, p_1), (t_5, p_2), (t_9, p_3), (t_3, p_1), (t_4, p_2), (t_8, p_2), (t_7, p_1), (t_{10}, p_3), (t_{11}, p_2)\}$$

$$I'_{k+1} = \{(t_1, p_2), (t_3, p_1), (t_2, p_2), (t_5, p_3), (t_6, p_1), (t_9, p_3), (t_4, p_2), (t_7, p_3), (t_{10}, p_2), (t_8, p_3), (t_{11}, p_1)\}$$

The exchange portion is denoted by an underline in this instance. In the mutation operation, conversely, every individual possesses a specific likelihood of arbitrarily perturbing the sequence of execution. In particular, an individual's position between the sub-tasks that are closest to their predecessor and successor is altered by a random sub-task, which also causes a random change in the primary choice.

5.5 Simulation Set-Up and Experimental Results

This section discusses the simulation findings of our effort. The entire simulation runs on a CloudSim simulator. We chose our datasets with various distributions. In our chapter, we first took different datasets and arranged them so that tasks had distinct distributions, which were then put into the scheduler. The dataset distributions are classified as follows: uniform, normal, left-, and right-skewed distributions. All of these datasets have uniform, normal, left- and right-skewed distributions, denoted as d1, d2, d3, and d4. When it comes to task scheduling, a uniform distribution may be used to depict situations in which tasks have comparable needs for resources. For instance, tasks may follow a uniform distribution if they arrive at random and have comparable computation requirements or deadlines. Platforms for cloud computing frequently manage several requests or tasks from various users or applications at once. No matter

Table 5.1 Simulation configuration settings.

Name of entity	Quantity
No. of jobs	1,000
Length of task	800,000
RAM	64 GB
Storage of Host	8 TB
Bandwidth	1,000 MBPS
No. of VMs	40
RAM of VMs	4 GB
Bandwidth of VM	200 MBPS
VMM	Xen
OS	Linux

how the underlying distribution of individual task times is distributed, the Central Limit Theorem holds when there are a lot of tasks, and the distribution of task execution times tends to become normal. The distribution of task execution durations may be left skewed if most tasks performed in the cloud environment are brief or have little computational complexity. A distribution is said to be right skewed, or positively skewed if its bigger values on the right side are more extended than its smaller values on the left side. A right-skewed distribution may suggest that most tasks in the cloud computing setting have longer execution durations, whereas fewer activities have shorter execution times. We utilize these distributions to test how our algorithm performs with various types of tasks. We compared our GA–DQN against known algorithms such as RR, FCFS, Priority-Based Scheduling, Deadline-Based Scheduling, and DRTSA.

Table 5.1 describes the parameters used to simulate the cloud environment and their settings.

5.5.1 Makespan Computation

Figure 5.5 shows the evaluation of makespan in the uniform distribution of tasks within a range. Makespan is evaluated as tasks grouped around a

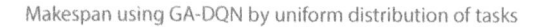

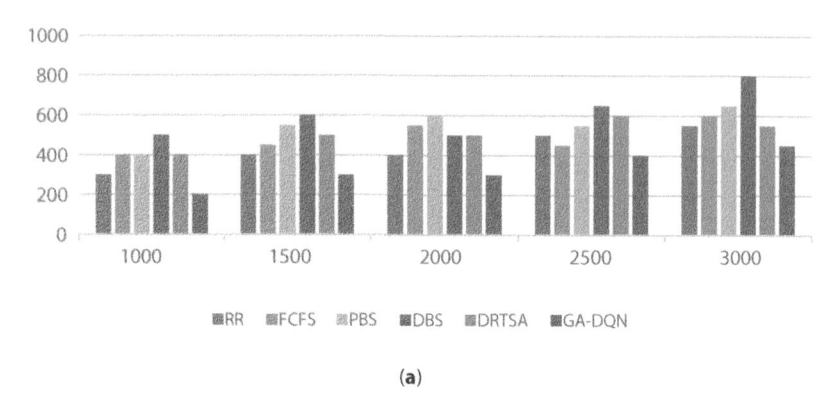

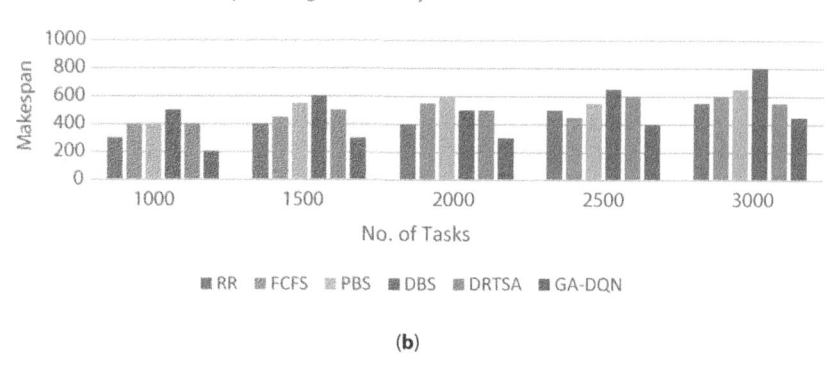

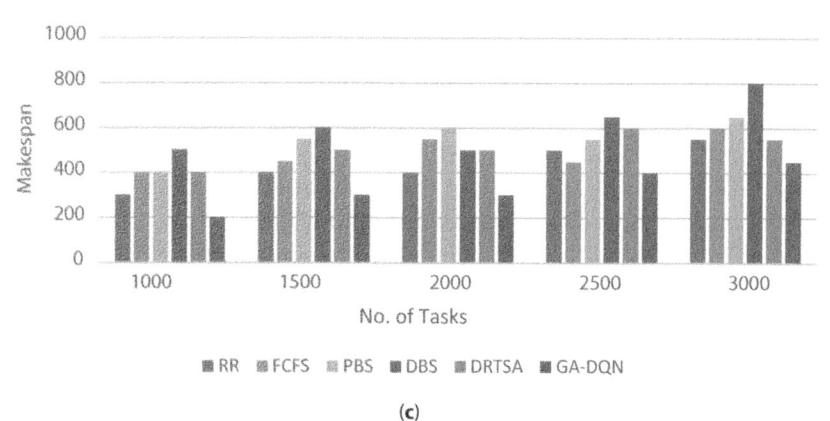

Figure 5.5 Makespan computation. (a) Uniform distribution of tasks, (b) normal distribution of tasks, (c) left-skewed distribution of tasks.

mean value in a normal distribution (also known as a Gaussian distribution), with fewer tasks having durations that deviate from the mean. It is also evaluated with the left-skewed distribution of tasks. The figures show that the various distribution of tasks is completed fast with the GA-DQN approach when the makespan of tasks are evaluated using hardware configuration settings in Datacentre, which is available in Table 5.1. We use the configuration parameters to analyze the makespan of tasks, and our GA-DQN scheduling procedure is provided. First, we used the datasets for d1, d2, d3, and d4 workloads to evaluate makespan. After running for 100 iterations, we compared our work to the baseline algorithms, which included

Table 5.2 Evaluation of makespan.

No. of tasks	RR	FCFS	PBS	DBS	DRTSA	GA–DQN
d1						
100	645.55	620.32	640.89	676.76	510.55	480.25
500	910.30	1,250.9	790.32	975.15	685.55	683.85
1,000	1,410.75	1,995.89	1,678.89	1,556.78	1,450.79	1,225.67
d2						
100	785.25	626.21	672.89	746.76	610.55	580.55
500	1,110.30	1,150.9	890.22	975.15	697.25	587.95
1,000	1,310.56	1,785.39	1,588.19	1,326.41	1,323.19	1,115.31
d3						
100	565.17	566.11	672.89	626.76	610.55	580.55
500	1,005.30	1,050.49	825.12	865.25	797.55	697.15
1,000	1,410.75	1,995.89	1,678.89	1,556.78	1,450.79	1,225.67
d4						
100	665.37	876.21	722.89	846.76	710.55	570.15
500	995.15	1,050.49	825.12	865.25	797.55	697.15
1,000	1,510.75	1,795.19	1,589.89	1,546.78	1,350.79	1,129.47

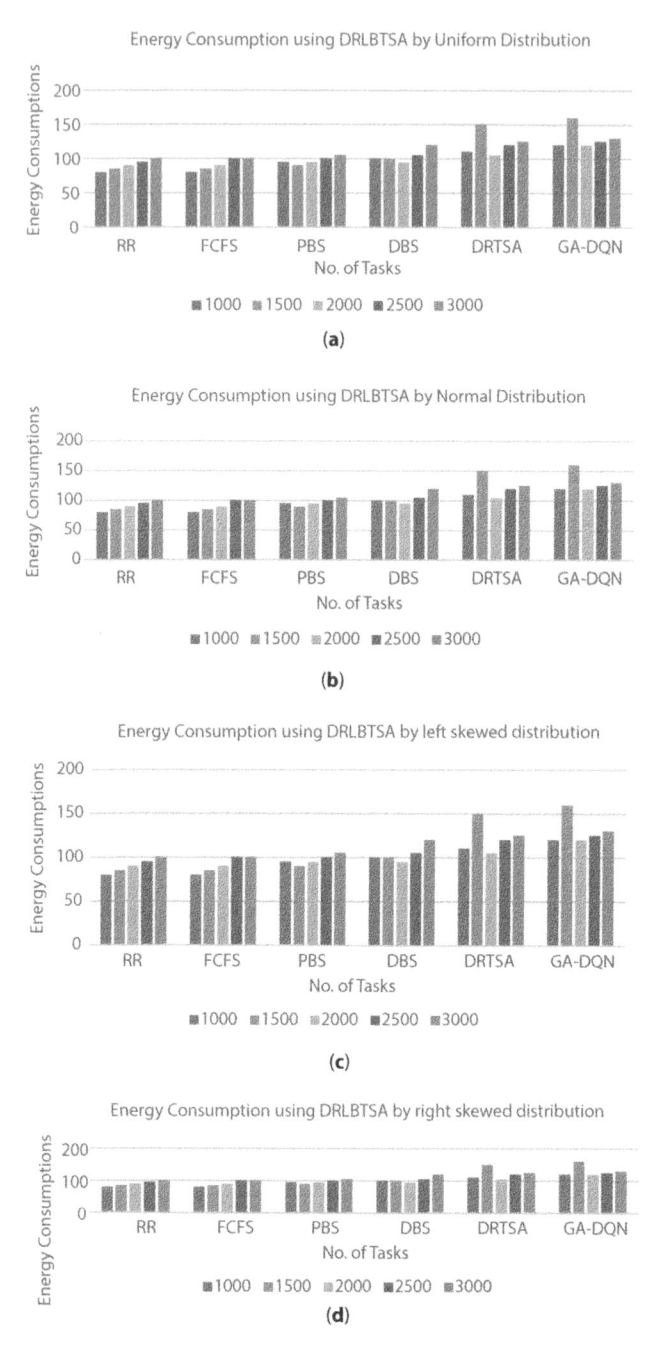

Figure 5.6 Energy consumption. (a) Uniform distribution of tasks, (b) normal distribution of tasks, (c) left-skewed distribution of tasks, (d) left-skewed distribution of tasks.

RR, FCFS, Priority-Based Scheduling, Deadline-Based Scheduling, and DRLBA.

Table 5.2 describes the datasets d1, d2, d3, and d4, as well as the number of jobs considered during scheduling. We calculated the makespan of all tasks using several scheduling algorithms such as RR, FCFS, PBS, DBS, and DRTSA.

5.5.2 Energy Consumption Calculation

Energy consumption is calculated using the configuration settings in Table 5.1, and various workloads are assigned to our GA–DQN scheduler. Initially, we provided workloads from the d1, d2, d3, and d4 datasets and

Table 5.3 Evaluation of energy consumption.

No. of tasks	RR	FCFS	PBS	DBS	DRTSA	GA–DQN
d1						
100	55.37	70.46	65.72	76.76	51.45	28.90
500	71.10	94.8	90.12	75.45	65.17	56.25
1,000	121.15	112.49	131.44	113.62	98.19	88.37
d2						
100	78.91	68.56	74.72	66.21	54.22	31.22
500	81.25	77.97	90.12	75.45	65.17	56.25
1,000	111.92	123.49	132.64	123.62	109.39	75.37
d3						
100	55.57	66.11	72.89	66.76	46.55	34.55
500	105.30	95.19	85.32	85.74	67.55	59.47
1,000	140.39	135.76	128.89	126.37	110.19	89.32
d4						
100	68.17	76.21	89.59	78.76	79.15	27.15
500	95.15	89.49	85.37	89.25	79.55	58.15
1,000	140.75	135.29	129.89	146.78	118.79	87.12

assessed energy consumption using these datasets. We did 100 iterations to compare our results to baseline algorithms such as RR, FCFS, Priority-Based Scheduling, Deadline-Based Scheduling, and DRLBA.

Table 5.3 describes the datasets d1, d2, d3, and d4, as well as the number of jobs considered during scheduling. We calculated the energy consumption of all tasks using several scheduling algorithms such as RR, FCFS, PBS, DBS, and DRTSA.

5.6 Conclusion

Due to task heterogeneity, scheduling different workloads over a cloud presents a difficult problem. For scheduling in a cloud context, several writers presented different algorithm approaches for mapping the process to virtual machines (VMs). This chapter contains our algorithm's demonstration. The process of assigning jobs to virtual machines (VMs) in the cloud is extremely dynamic requiring that numerous tasks be assigned to resources that accurately match their processing capacities. Energy consumption in cloud computing for makespan in data centers for task scheduling pertains to the quantity of electrical energy needed to supply electricity and maintain the temperature of the infrastructure utilized to carry out computational tasks in a cloud environment. The primary objective is to minimize the makespan. We suggested a GA–DQN network in this chapter to address dynamic scheduling issues in cloud environments. We have run simulations using CloudSim and fed fabricated datasets into the algorithm. It performs well in terms of cutting down on makespan and energy usage. Our algorithms have been tested using RR, FCFS, PBS, DBS, and DRTSA. There were 100 iterations of the simulation. When compared to baseline techniques, our suggested approach performs admirably. In the future, we must implement GA–DQN in real-time workloads and assess our scheduler's effectiveness.

References

1. Sunyaev, A., Cloud computing, in *Internet Computing*, Springer: Cham, Switzerland, pp. 195–236, 2020.
2. Motlagh, A.A., Movaghar, A., Rahmani, A.M., Task Scheduling Mechanism in Cloud Computing: A Systematic Review. *Int. J. Commun. Syst.*, 33, 6, e4302, 2020.

3. Hamid, L., Jadoon, A., Asghar, H., Comparative analysis of Task level heuristic scheduling algorithms in Cloud Computing. *J. Super Comput.*, 78, 12931–12949, 2022.
4. Ding, D., Fan, X., Zhao, Y., Kang, K., Yin, Q., Zeng, J., Q-Learning-based Dynamic task Scheduling for energy efficient Cloud Computing. *Future Generations Comput. Syst.*, 108, 361–371, 2020.
5. Zhan, Z.-H-., Zhang, G.-Y-., Lin, Y.-., Gong, Y.-Z., Zhang, J., Load Balance Aware Genetic Algorithm for Task Scheduling in Cloud Computing, in: *Asia Pacific Conference on Simulated Evolution and Learning*, pp. 644–655, 2014.
6. Swarup, S., Shakshuki, E.M., Yasar, A., Task Scheduling in Cloud using Deep Reinforcement Learning. *Procedia Comput. Sci.*, 184, 42–51, 2021.
7. Zhou, L., Zhang, L., Horn, B.KP., Deep reinforcement learning-based dynamic scheduling in smart manufacturing. *Procedia CIRP*, 93, 383–388, 2020.
8. Tang, Z., Chen, H., Deng, X., Li, K., Li, K., A scheduling scheme in the cloud computing environment using deep Q-learning. *Inf. Sci.*, 512, 1170–1191, 2020.
9. Tanha, M., Shirwani, M.H., Rahmani, A.M., A hybrid meta-heuristic Task scheduling algorithm based on genetic and thermodynamics simulated annealing algorithms in cloud computing environments. *Neural Comput. Appl.*, 33, 16951–16984, 2021.
10. Balasubramaniam, S., Syed, M.H., More, N.S., Polepally, V., Deep learning-based power prediction aware charge scheduling approach in cloud based electric vehicular network. *Eng. Appl. Artif. Intell.*, 121, 105869, 2023.
11. Pirozmand, P., Hosseinbadi, A.A.R., Farrokhzad, M., Sadeghilalimi, M., Mirkamali, S., Adam, Multi-objective hybrid genetic algorithm for Task Scheduling problem in Cloud Computing. *Neural Comput. Appl.*, 33, 13075–13088, 2021.
12. Swarup, S., Shakshuki, E.M., Yasar, A. Task scheduling in cloud using deep reinforcement learning. *Procedia Comput., Sci.*, 184, 42–51, 2020.
13. Aziza, H. and Krichen, S., A hybrid genetic algorithm for scientific workflow scheduling in a cloud environment. 32, 15263–15278, 2020.
14. Subhadra Sarngadharan, A., Narasimhamurthy, R., Sankaramoorthy, B., Singh, S.P., Singh, C., Hybrid optimization model for design and optimization of microstrip patch antenna. *Trans. Emerg. Telecommun. Technol.*, 33, 12, e4640, 2022.
15. Balasubramaniam, S. and Bharathi, R., Performance analysis of parallel FIR digital filter using VHDL. *Int. J. Comput. Appl.*, 39, 9, 1–6, 2012.
16. Dong, T., Xue, F., Xiao, C., Zhang, J., Deep reinforcement learning for Dynamic workflow scheduling in Cloud Environment, in: *IEEE International Conference on Services Computing*, 2021.

6

Efficient Load Balancing and Resource Allocation in Networked Sensing Systems—An Algorithmic Study

Lalitha Krishnasamy[1]*, Divya Vetriveeran[2], Rakoth Kandan Sambandam[2] and Jenefa J.[2]

[1]*Department of Artificial Intelligence & Data Science, Nandha Engineering College, Erode, Tamil Nadu, India*
[2]*Department of Computer Science and Engineering, School of Engineering and Technology, Christ University, Kengeri Campus, Bengaluru, India*

Abstract

In the current environment, data generation and data transmission are increasing exponentially in day-to-day life. These exponentially growing data might create heavy traffic when transmitted between systems. Also, this affects many functionalities like configuration of networked systems, system and routing configuration parameters, load managing factors of network devices, etc. A dynamic traffic control mechanism needs to be adopted with the help of load-balancing algorithms and efficient resource allocation mechanisms to deal with heavy data traffic. Load balancing algorithms in networked sensing systems aim to distribute the workload evenly among sensor nodes to optimize network performance and energy efficiency and prolong the network lifetime. Resource allocation mechanisms in a networked sensing system involve allocating and distributing network resources efficiently, such as energy, bandwidth, processing power, etc., to optimize performance and increase the network's lifetime. To achieve efficient resource allocation with a balanced load, notable works have been done in optimization and machine learning. The work gives a scientific analysis of traditional and Artificial Intelligence algorithms from a centralized and distributed perspective. Researchers can take this analysis forward when deciding on algorithms based on their application and infrastructural needs.

Keywords: Load balancing, resource allocation, energy efficiency, AI algorithms, machine learning, performance optimization

---

Corresponding author: vrklalitha24@gmail.com

Rajesh Kumar Dhanaraj, Malathy Sathyamoorthy, Balasubramaniam S and Seifedine Kadry (eds.) Networked Sensing Systems, (145–172) © 2025 Scrivener Publishing LLC

6.1 Introduction to the Networked Sensing Systems

In the recent era, data generation and transmission are enormous in number in all the networked sensing systems and devices such as Wireless Sensor Networks (WSN), Software Defined Networks (SDN), and Internet of Things (IoT). A WSN is a self-reconfigurable, autonomous sensor device connected by a radio receiver channel to a base station. A Wireless Sensor Network suite comprises sensor nodes deployed in an unattended environment [1]. A sensor acts as an interface in the deployed device that provides information on a managed resource's state and state transitions. The wireless sensor network consists of spatially disseminated self-governing sensors, which are used to sense, monitor the scenario, and communicate different types of data to the destination where it gets connected. In the huge deployment, sharing load among all the devices, which get connected, is a very challenging task [2]. The load balancing with the cloud environment is shown in Figure 6.1.

Wireless Sensor Networks (WSNs) carry a lot of promise in applications [3] where gathering sensed information in remote or inaccessible locations is demanded. In WSNs, the routing communication protocol is application dependent, and their pattern goals might change depending on different applications [4]. To consider with an example, a firefighter might expect well-timed temperature readings for awareness of the current situation and to act accordingly. In contrast, applications like soil monitoring, might demand temperature data in a few hours once in a day. Consequently,

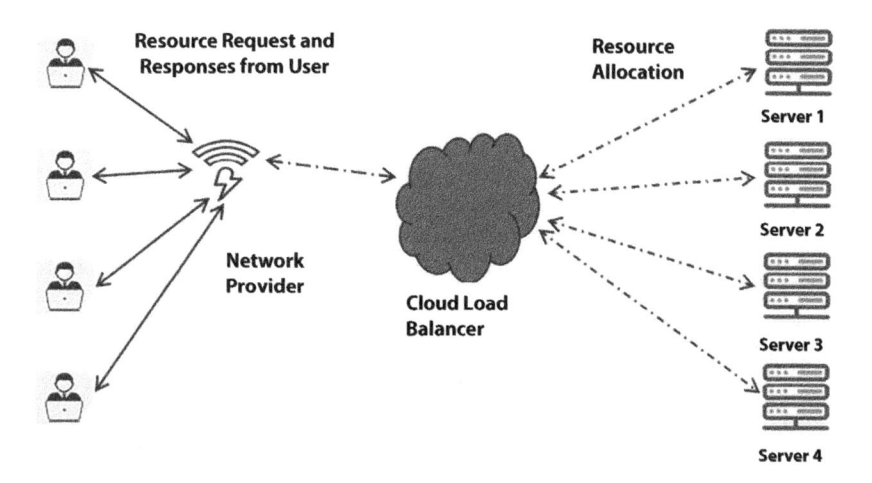

Figure 6.1 Load balancing in the cloud environment.

routing protocols should meet the requirements with bare minimal cost. Hence, the load balancing and routing protocol algorithms have to consider the features of sensor devices along with the application and architectural requirements [5].

Most of the reexamined clustering protocols (ex., C-LEACH, M-LEACH, HEED, etc.) are application specific [6]. Such protocols demand flexible fundamental alterations to assemble the diversity of operations and malfunctioning necessaries committed by the applications. Consequently, deciding an appropriate balanced communication protocol for a particular concern is highly diplomatic since formulating the load-balanced clustering will remain a challenge if the sensor nodes increase, and measures are to be taken to optimize its functioning by fixing the routing metrics. In this chapter, the following topics are taken into consideration and discussed further:

- Understanding the Load Balancing Challenges
- Efficient Resource Allocation Strategy
- Overview of Existing Approaches
- Artificial Intelligence for Resource Handling
- Real-World Applications
- Performance Evaluation Metrics
- Future Directions and Emerging Trends
- Conclusion and Summary

6.2 Understanding the Load Balancing Challenges

In critical traffic, congestion, and unavoidable situations, a better mechanism is required to handle sensor devices [7]. So, there is a demand to pattern the communication protocol to act under eminent pressured network situations such as vital network situations, admitting over-crowding and collisions. This can occur for various reasons such as interference, miserable link quality, and mobility. During this situation, the balanced routing protocol necessitates acquiring and qualifying the most estimable combination of routing to execute the information transmission seamlessly. Nevertheless, it is hard to determine routing without regarding the succeeding measure heuristic of the network kinetics. Therefore, the routing establishes redesign with the capability of managing the network kinetics, including data retransmission. Load balancing in a network is of paramount importance because of the following challenges:

i. Scalability Challenges in Large-Scale Sensing Networks
ii. Heterogeneity of Sensor Nodes and Data
iii. Dynamic Environmental Conditions
iv. Energy Efficiency and Power Constraints
v. Communication Overhead and Latency
vi. Quality of Service (QoS) Requirements
vii. Security and Privacy Concerns
viii. Impact of Mobility and Node Failures

Sensing Networks can be established from small scale to large scale depending on the application requirements, and the algorithm should be designed to accommodate a node count of 100 or 1,000. Accordingly, the generated data might also increase in volume and energy deplete sooner if the sensors are not placed at an optimized distance. In this case, optimized distance among sensing devices ensures load balancing and energy efficiency [8]. Different applications with sensors might generate different types of data, and to represent them in a generic way, data heterogeneity needs to be given attention. Sensing systems should be designed and tested in different environmental conditions, but if they are deployed in an area that is entirely different from the tested conditions, then the sensing devices may not be expected to collect data evenly and efficiently utilize resources.

Initial communication establishment is essential; unexpected node failure due to power depletion might increase the packet delay. Also, node failure might increase the number of retransmissions, which creates communication overhead. Node failure, traffic due to congestion, power depletion, connectivity issues, etc., lead to compromise in the quality of service. When nodes are moving, they might be attached to different base stations from time to time, and tracing data is another challenge in addition to mobility issues. While transmitting data from one device to another, data privacy is quintessential, and vulnerability issues, threats, trojan horses, viruses, and worms play a major role in compromising data security and privacy.

6.2.1 Types of Load Balancing

Application Load Balancing
Many server farms with numerous servers devoted to a single application function are used in complex modern applications. To reroute traffic, application load balancers examine the content of the request, such as HTTP headers or SSL session IDs. An e-commerce program comprises a

shopping cart, checkout area, and product directory. Requests for browsing items are sent to servers holding photos and videos via the application load balancer eliminating the need for open connections. In contrast, it routes requests for shopping carts to servers that can sustain numerous client connections and store cart data indefinitely.

Network Load Balancing
Network load balancers look up IP addresses and other network information to optimally divert traffic. Network load balancers search IP addresses and other information to reroute traffic optimally. They can pinpoint the source of the application traffic and supply several servers with static IP addresses. Network load balancers use the previously mentioned static and dynamic load balancing strategies to balance server load.

Global Server Load Balancing
Multiple geographically separated servers are involved in the global server load-balancing procedure. For example, companies may have servers across many data centers, international borders, and third-party cloud providers. In this case, the application load inside a zone or area is managed by local load balancers. They attempt to redirect traffic to a server destination that is nearer to the client's location. Only servers outside the client's geographic area may be used to reroute traffic in the event of a server failure.

DNS Load Balancing
When people utilize DNS load balancing, the domain could be configured to split up network requests among several resources on the requested domain. A website, mail server, print server, or any other internet-based service can all be associated with a domain. Maintaining application availability and distributing network traffic among a globally dispersed pool of resources are two benefits of DNS load balancing.

6.2.2 Load Balancing Technologies

There are different types of load balancing technologies taken into consideration, which include software and hardware balancers.

Hardware Load Balancers
A hardware appliance known as a hardware-based load balancer can safely process and reroute gigabytes of data to hundreds of distinct servers. It can be kept in your data centers, and you can utilize virtualization to make several digital or virtual load balancers that you can control centrally.

Software Load Balancers

Applications that handle all aspects of load balancing are known as software-based load balancers. They can be accessed as a fully managed third-party service or installed on any host.

Comparison of the Balancers

Hardware load balancers must be configured, maintained, and invested in. You might not employ them to their maximum potential, particularly if you buy one solely to handle traffic spikes during peak hours. Users will be impacted if traffic volume unexpectedly climbs beyond capacity until another load balancer can be purchased and installed.

Software-based load balancers, on the other hand, offer far greater flexibility. They are more compatible with contemporary cloud computing systems and are easier to scale up or down. Over time, they also cost less to use, maintain, and set up.

6.3 Importance of Efficient Resource Allocation

Sensor nodes in the WSN are deployed without following any infrastructure, and the nodes will be active while sensing the data and become inactive for the remaining period of time. Since sensor nodes are generally power constrained, the node's lifetime can be prolonged by applying different techniques in addition to the previous characteristics.

Noticeably, sensor node energy consumption is highly dependent on application and it is application specific, the significant role played by the node-level, and network-level architectures. The following measures might reduce the consumption of energy during communication in WSN

- To schedule sensing and the sleep, idle state among the sensor head and sensor nodes
- Adjusting the range of transmission among the sensor head and sensor nodes
- Adopting efficient routing as well as efficient data-collecting mechanism
- Unwanted data gathering is to be avoided, as in the case of overhearing.

Based on the above steps, numerous algorithms came into play to reduce the consumption of energy in WSN, but on the other hand, efficiency of data gathering should be negotiated to achieve network lifetime.

Hence, this tradeoff should be addressed with existing techniques, and the efficiency of the resources should be analyzed.

6.4 Overview of Existing Approaches

There are various types of clustering algorithms presented, but these algorithms are primarily divided into two different categories based on the cluster head selection process and even the distribution of resources. They are probabilistic and non-probabilistic clustering. In the existing approaches, like HEED, PEGASIS, GAF, and SPIN, the sensing devices and other network resources are utilized to ensure energy efficiency [9]. A virtual grid structure in HEED is framed based on the deployment area, and each grid is considered a cluster.

6.4.1 Probabilistic Clustering

In this technique, each sensor has to determine the cluster heads initially based on the probability. Assigning probabilities to each sensor node is the main criterion for deciding the cluster heads during their election process. The secondary criterion may be residual, initial, and average network energy. Even though these are highly energy efficient, the clustering algorithms usually achieve faster execution and reduce the volume of exchanged messages [10]. Some of its algorithms are as follows:

LEACH
LEACH, abbreviated as Low-Energy Adaptive Clustering Hierarchy, is considered as one of the dynamic approaches in clustering. It uses a hierarchical routing technique that routes the packets in a multi-hop fashion. It follows TDMA-based MAC protocol. The major goal of the LEACH is to dynamically fix the clusters, improve energy conservation, which, in turn, improves the duration of the network. Reducing the traffic by aggregating the whole data collected from cluster head and saving energy using single hop routing are some of the main advantages of the LEACH protocol. But it does not give any idea about the number of clusters framed in the network area. In addition, the clusters are divided randomly based on the probability resulting in an uneven distribution of clusters.

HEED
HEED is a Distributed, Energy Efficient Hybrid protocol mainly used for even size of Clustering. HEED protocol is intended to split the geographical

area in a virtual grid-based structure, and cluster head is elected based on the energy level and transmission range of the nodes within that cluster. HEED aims to increase the network lifespan by providing distributed energy consumption and minimizing control overheads. It provides a compact cluster with well-distributed cluster heads. However, the random selection of the cluster heads may cause higher communication overhead, and the periodic cluster head election also requires some energy dissipation [11].

UCR
UCR stands for unequal-cluster-based routing protocol. In this, the cluster size is smaller near the sink node and are gradually larger when clusters are farthest from the sink. It encourages the communication in inter-clusters when nearer to the sink and intra-cluster nodes when farther. This might eliminate the current hot spot issue and can maintain balanced communication. On the other hand, the geographical size, the number of nodes nearer and far away from the sensor nodes, and the shape of the cluster, etc., might vary, and these are entirely dependent on application specific.

EECS
EECS is an Efficient Clustering protocol, which is mainly implemented for optimizing energy consumption. It mainly focuses on the initial cluster formation. The weighted probability in each node is then calculated to find its nearby cluster-head within the cluster. Three parameters were considered to calculate the weighted probability, which includes current energy availability, difference between the current and the initial energy level, and the distance from the member node to the cluster-head. In this scheme, the initial communication and framing the clusters itself consume a considerable amount of energy and creates little overhead. This protocol takes more time for clustering and unequal-sized cluster formation [12].

6.4.2 Non-Probability Clustering

These types of clustering algorithms help to select the head node of the cluster and also the formation of clustering in a more precise way. The selection depends on the proximity of sensing a node, position, and geographical location. The data trans-receiving process focuses on the nearby nodes rather than the probability-based collection. It has the worst time complexity due to the graphs' traversing and many message exchanges. However, it is more reliable considering the energy transmission power and mobility factor, which achieves more generalized goals [13].

PEGASIS
Efficient gathering of data from sensors and information systems without losing power is achieved by the protocol named PEGASIS. The main aim of this protocol is to frame a chain sequence among the nodes instead of randomly selecting the head node in a scheduled time period to balance the energy consumption. This can be accomplished by using the greedy approach starting from any of the sensor node, which gets located near the sink. Since each node gets connected with its previous and the next node, the data loss due to the dead node can be avoided. But, on the other hand, to identify the location of the sensor node to find the neighbor nodes is a quite challenging task in a small-size to medium-size sensing node deployment and may not be possible to adopt to a larger scale. For a small-scale network, reconstructing the chain is possible, and it uses hierarchical network with maximum power utilization, and it also has low overhead.

TEEN
TEEN is one of the energy-efficient, sensitive threshold-based networking protocol mainly used in the hierarchical network. It does not send the data periodically; instead, it sends only when there is an extreme change in the sensor values. To identify this, this protocol sets a threshold limit based on the applications the sensors are deployed in, like an event based or time or any specific value-based decision. The cluster framing and the initial communication to find the members within the cluster is the same as the LEACH protocol. When sensors sense data that are beyond the threshold or lesser than the threshold limit, they broadcast data to their nearby members until they reach the sink. Even though the energy consumption for frequent transmission is reduced, this broadcasting consumes more energy and might not provide a balanced energy consumption. It has high overhead [14].

SPIN
SPIN is one of the negotiation protocols used for collecting sensed information. It considers a flat network and creates a meta data with the actual data received from the network and transmits the meta data to the neighbors. In contains the message description in addition to the data that the sensor wants to communicate over the network. SPIN uses limited power. It reduces both the network overhead and the energy consumption during the transmission [15].

ACQUIRE
ACQUIRE stands for Active Query Forwarding in Sensor Networks. This protocol concentrates highly on how easily and quickly data is retrieved

Table 6.1 Comparative study of protocols with efficient resource allocation.

Hierarchical routing protocol	Clustering	Energy consumption	Energy model	Delay	Scalability	Node handling capacity	Maximum available bandwidth	Data aggregation
ALEACH	Y	High	Distance based	Moderate	Moderate	Less	N/A	Yes
PEGASIS	Y	High	Distance based	Moderate	Moderate	N/A	N/A	Yes
TEEN	Y	High	Distance based	Moderate	Low	Less	N/A	No
APTEEN	Y	High	Distance based	Moderate	Moderate	Less	N/A	Yes
Sensor aggregate	Y	Medium	Distance based	Less	Low	N/A	N/A	Yes
SPEED	N	Medium	Distance by packet speed	Very Less	Low	N/A	N/A	No
HEED	Y	Medium	Residual energy	Moderate	Moderate	N/A	N/A	Yes
DEEC	Y	Medium	Residual energy	Very less	Low	Less	Low	Yes
SEP	Y	Low	Residual energy	Less	Low	N/A	Low	No
SEEC	Y	Low	Residual energy	Moderate	Moderate	N/A	Moderate	Yes
ECSA	Y	Low	Distance based	Very less	Low	Less	Low	No
DRINA	Y	Low	Distance based	Less	Moderate	Less	Moderate	Yes
EEHC	Y	High	Distance based	High	Moderate	N/A	Low	Yes
EAST	N	Medium	Correlation based	Moderate	Moderate	N/A	Low	Yes

through a set of sub queries instead of posting a complex query. This protocol encourages the sensing system to request and collect whichever the data are required for that particular time limit. So, all the sensed data are not going to be transmitted, which, in turn, reduces energy usage. Based on the currently available information, the sink responds to the requested node instead of transmitting to the whole network. If, on the other hand, the available data are not updated properly, then this protocol collects the information from neighbor nodes within a specified number of hop count. It uses a low amount of power since the information transmission is sent only between the sink and the sensing node. It requires low overhead, and it performs complex queries also [16, 17].

All the discussed probabilistic and non-probabilistic protocols are trying to allocate and utilize the resources wisely so that the network load gets balanced without compromising much on energy and/or power consumption issues. All these protocols are compared against a few parameters to ensure network load balancing and are given clearly in Table 6.1.

6.5 Artificial Intelligence for Resource Handing

Artificial Intelligence (AI) algorithms are extremely effective for Networked Sensing Systems in controlling resource allocation and load balancing. These systems, which are widely used in fields, including environmental monitoring, healthcare, and industrial automation, usually require real-time data processing and decision making. Technologies pertaining to Artificial Intelligence (AI), namely, machine learning and also deep learning, can recognize patterns, predict trends, and analyze large amounts of data. Through the dynamic allocation of resources, such as network bandwidth along with processing power, to different activities and nodes, the system is able to respond and perform at its peak. Artificial Intelligence strategies are critical apparatuses for guaranteeing the reliability and viability of arranged detecting frameworks since they are likewise entirely adaptable to changing client necessities and organization conditions [18].

AI, support learning, and prescient examination are a couple of instances of computer-based intelligence calculations that help with handling information as indicated by client prerequisites, framework measurements, and organization traffic. Utilizing past and current organization information likewise assists the arranged detecting framework with coming to conclusions about asset assignment and burden adjusting. A number of factors, including workload characteristics, node capacity, and network state, ensure that decisions about load balancing are made effectively.

156   NETWORKED SENSING SYSTEMS

Furthermore, it is able to progressively acclimate to stack adjusting techniques to ensure powerful work appropriation among hubs. This ensures efficient utilization of the resources like storage, network bandwidth, computation resources, etc. On the other hand, tasks are accomplished in a distributed manner based on priority [19].

The key components of AI-based load balancing and resource allocation are continuous monitoring and feedback. The performance metrics and feedback are evaluated continuously by the network sensing system. This helps update the decisions over time based on the approach's effectiveness. It also efficiently assesses the performance of the system. Hence, integrating AI in load balancing and resource allocation in networked sensing systems improves the system's performance, efficiency, and responsiveness. Automating these AI processes ensures dynamic decision making leading to effective resource utilization.

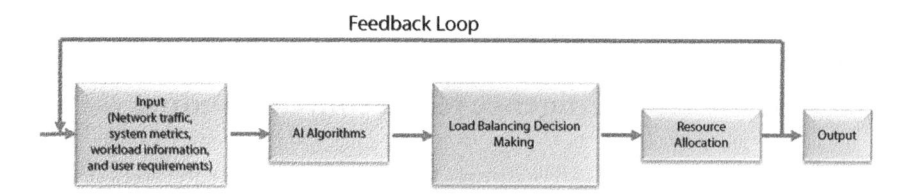

Figure 6.2 AI for load balancing and resource allocation—block diagram.

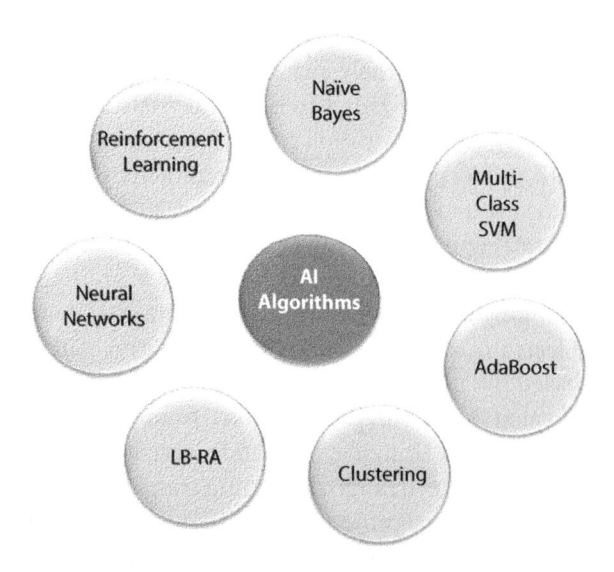

Figure 6.3 AI Algorithms for load balancing and resource allocation.

Figure 6.2 illustrates the load balancing and resource allocation processes using AI. As shown, AI-driven load balancing and resource allocation can be used to enhance the performance of the network sensing system, which leads to effective utilization and responsiveness of the sensing system [20]. Several AI algorithms can be used to improve the efficiency of the applications. The different algorithms used are illustrated in Figure 6.3.

Depending on the application, they can be a simple ML algorithm or a complex neural network. Some of the algorithms are described in detail as follows.

6.5.1 Naïve Bayes

These classifiers are efficient when the data dimensionality is high and the features are independent of each other. A simple implementation of the same is given as follows [21]:

Define the features and labels
 # Features: Each row represents a node, and each column represents a feature
 # Labels: The load of each node
features = [[1, 0, 1], [0, 1, 0], [1, 1, 1], [0, 0, 1]]
labels = [1, 2, 1, 3]
 # Split the data into training and testing sets
X_train, X_test, y_train, y_test = train_test_split(features, labels, test_size=0.2, random_state=42)
 # Train the Naive Bayes model
model = GaussianNB()
model.fit(X_train, y_train)
 # Make predictions on the testing set
predictions = model.predict(X_test)
 # Evaluate the model
accuracy = accuracy_score(y_test, predictions)
print(f"Accuracy: {accuracy}")

6.5.2 Multi-Class SVM

Support Vector Machines (SVMs) are a potent category of supervised learning algorithms suitable for both regression and classification applications. To support several classes, the fundamental SVM technique is extended by Multi-Class Support Vector Machines or MCSVMs. A sample implementation is given as follows [22]:

```
# Define the features and labels
# Features: Each row represents a node, and each column represents a feature
# Labels: The load of each node
features = [Sample Features]
labels = [sample labels]

# Split the data into training and testing sets
X_train, X_test, y_train, y_test = train_test_split(features, labels, test_size=0.2, random_state=42)
# Train the SVM model
model = SVC()
model.fit(X_train, y_train)

# Make predictions on the testing set
predictions = model.predict(X_test)

# Evaluate the model
accuracy = accuracy_score(y_test, predictions)
print(f"Accuracy: {accuracy}")
```

6.5.3 AdaBoost

A set of weak classifiers are combined together to form a robust classifier with better accuracy. The loss function minimizer also performs better as it is formed from a hybrid algorithm. The sample working of the algorithm is given below [23]:

```
#Each row represents a node, and each column represents a feature
# Labels: The load of each node
features = [Sample data]
labels = [sample labels]

# Split the data into training and testing sets
X_train, X_test, y_train, y_test = train_test_split(features, labels, test_size=0.2, random_state=42)

# Train the AdaBoost model
model = AdaBoostClassifier()
model.fit(X_train, y_train)
```

 # Make predictions on the testing set
predictions = model.predict(X_test)

 # Evaluate the model
accuracy = accuracy_score(y_test, predictions)
print(f"Accuracy: {accuracy}")

6.5.4 Clustering

K-nearest neighbors (KNN) is a straightforward, occurrence-based learning calculation that can be utilized for both characterization and relapse undertakings. It can likewise be adjusted for load adjusting in organized detecting frameworks [24].

def load_balance(nodes, k, max_iterations=100):
  centroids = initialize_centroids(nodes, k)
  for _ in range(max_iterations):
    clusters = assign_to_nearest_centroid(nodes, centroids)
    new_centroids = update_centroids(clusters)
    if np.allclose(centroids, new_centroids):
      break
    centroids = new_centroids
  return clusters

6.5.5 Learning-Based Resource Allocation (LB-RA)

LB-RA methods work based on two different principles, i.e., Dynamic Switching Between Different RA Strategies and Resource Allocation Strategies Evaluation and Selection. LB-RA essentially uses long-term popularity trends as the aggregate point to which the average popularity should eventually revert. The objective of LB-RA is to select the best resource allocation strategy for the upcoming period [25].

Algorithm Steps:
Step 1: Based on previous observations and the assumption of a Zipf-based long tail, the popularity trends for the upcoming period are predicted.
Step 2: After that, the prior probabilities are created. This involves comparing how well each of the three RA strategies performed to the predicted distribution of popularity for the upcoming period of time. Thereafter, we measure execution levels over the long haul as far as progress rate and

entropy. Montecarlo simulation is used here to obtain the most granular performance assessment (the measurement is done at a VoD request level).

Step 3: The probability density is then approximated by a Gaussian function. The parameters of this function are ones that produce a best fit with the histogram of the prior probabilities generated in the prior step above. Here, we assume that the probability distribution follows a normal distribution that is used in the Bayesian fusion. It is important to note that our LB-RA process will lead to six Gaussian functions: three RA strategies for two performance metrics.

Step 4: The next step would be to use Bayes' theorem to combine the two Gaussian functions associated with the two performance metrics for each RA strategy considered. At this point, we obtain a common way to compare the different RA strategies while still considering the performances along the two dimensions: success rate and entropy. This process is referred to as obtaining posterior probabilities.

Step 5: To evaluate the different RA strategies against each other, a Montecarlo Simulation is used. This allows calculating the value of the posterior probability of each strategy in the most detailed way.

Step 6: The maximum *a posteriori* estimator is applied: the RA strategy with the highest posterior probability is selected for the upcoming time period.

6.5.6 Neural Networks for Load Balancing and Resource Allocation

Neural networks can indeed be utilized for load balancing in various systems, including computer networks, cloud computing environments, and distributed systems. The objective of burden adjusting is to split the responsibility between a few assets to limit response times, boost throughput, enhance asset use, and forestall over-burden situations. Coming up next are the advantages of this technique [26].

    a. **Resource Demand Prediction:** Brain organizations can be prepared to foresee future asset requests by consolidating authentic information, current responsibility patterns, and other pertinent attributes. By precisely foreseeing asset requests, a heap balancer can proactively rearrange responsibility to forestall asset bottlenecks and guarantee proficient asset use.

b. **Dynamic Resource Allocation:** Real-time workload factors and system performance indicators can be taught to neural networks to dynamically allocate resources. Thus, load balancers can adaptably adjust asset portions in light of moving responsibility conditions augmenting framework execution and keeping up with high accessibility.
c. **Fault and Anomaly Detection:** Neural networks can be used for anomaly detection to find unusual patterns or deviations from the system's usual behavior. We call this technique adaptation to internal failure. At the point when load balancers answer fittingly to irregularities, for example, abrupt spikes in responsibility or asset blackouts, they can decrease the effect on framework execution and proposition adaptation to non-critical failure.
d. **Traffic Prediction and Routing:** By analyzing network traffic patterns, neural networks can predict future traffic loads on various resources or nodes. These projections can be used by load balancers to intelligently route incoming requests to the most appropriate resources, thereby reducing response times and increasing system throughput.
e. **QoS (Quality of Service) Optimization:** Based on user-defined goals and constraints, neural networks can be trained to optimize QoS attributes like response time, throughput, and resource utilization. These learned models can be utilized by load balancers to conclude which solicitations to focus on or how best to disseminate assets to meet explicit nature of administration necessities.
f. **Self-Learning and Adaptation:** Self-learning load-balancing systems that constantly adapt and enhance their performance can incorporate neural networks. Support learning methods permit load balancers to independently explore different avenues regarding a few techniques and track down the ideal ones for responsibility dispersion and asset designation.

A simple implementation of the same is given as follows:

 # Initialize Neural Network Model
model = NeuralNetwork(input_size, hidden_layers, output_size)
 # Train the Neural Network Model (using historical data)

```
train_neural_network(model, training_data)
# Monitor System Metrics (e.g., current workload, resource
utilization)
while True:
    # Collect Real-Time System Metrics
    current_workload = monitor_workload()
    resource_utilization = monitor_utilization()
    # Predict Resource Demands using Neural Network Model
    predicted_demands = model.predict(current_workload)
    # Determine Resource Allocation based on Predictions
    resource_allocation=allocate_resources(predicted_demands,
    resource_utilization)
    # Update Load Balancer Configuration (e.g., route requests)
    update_load_balancer(resource_allocation)
    # Repeat (continuously monitor and adjust)
```

6.5.7 Reinforcement Learning

By interacting with its surroundings, the reinforcement learning (RL) learner is an adaptive decision-making agent that can enhance its behavior on its own. In contrast to conventional supervised learning, which involves giving the agent labeled training data, reinforcement learning (RL) learns by making mistakes and getting feedback in the form of incentives or penalties. Fundamentally, the goal of the RL learner is to choose actions that produce desired results to maximize cumulative rewards over time. It does this by constructing a predictive model of the dynamics of the environment, which is commonly expressed as a Q-table or Q-function and calculates the projected future benefits linked to executing certain actions in various stages. The RL learner eventually moves from initially random actions to more informed judgments by iteratively exploring and exploiting its policy [27].

```
# Step 1: Define RL Environment and Agent
# Define environment dynamics (state transitions and rewards)
def step(state, action):
    # Update system state based on action
    # Calculate reward
    # Return next state, reward, and done flag
    # Define agent policy
def select_action(state):
    # Select action based on policy (e.g., epsilon-greedy)
```

```
# Step 2: Main Loop (Training)
# Initialize Q-table
Q = {}
# Loop through episodes
for episode in range(num_episodes):
    # Reset environment to initial state
    state = initial_state
    # Loop within an episode
    while not done:
        # Select action using epsilon-greedy policy
        action = select_action(state)
        # Take action and observe next state, reward, and done flag
        next_state, reward, done = step(state, action)
        # Update Q-value for (state, action) pair
        Q[state][action] = update_q_value(Q[state][action], reward, next_state)
        # Move to next state
        state = next_state
# Step 3: Deployment
# Deploy trained agent
state = initial_state
while not done:
    action = select_action(state)
    next_state, _, done = step(state, action)
    state = next_state
```

6.6 Real-World Applications

In many industries, including computer science, telecommunications, manufacturing, transportation, and finance, load balancing and resource allocation are fundamental ideas.

a. **Computer Networks:** Load balancing and resource allocation in networking ensure that data packets are efficiently distributed across the servers, routers, and switches that are accessible to the network. Many servers use load balancers to evenly distribute incoming network traffic among them preventing one server from becoming overloaded.

b. **Cloud Computing:** Cloud service providers use load balancing and resource allocation strategies to spread computational workloads across multiple servers in data centers. This ensures the ideal utilization of open resources and successful treatment of client demands. Dynamic weight-changing computations acclimate to changing resource openness and solicitations.

c. **Data Centers:** In server farms, load balancing and asset assignment are crucial for optimizing server utilization and ensuring high help accessibility. The unique portion of assets, like computer chip, memory, and capacity, to virtual machines in light of interest is made conceivable by virtualization advances. To avoid bottlenecks and maintain responsiveness, load balancers distribute traffic among servers.

d. **Distributed Systems:** Distributed file systems, distributed databases, and content delivery networks (CDNs), which distribute data and processing across multiple nodes, all rely heavily on resource allocation and load balancing. This further develops adaptation to non-critical failure, adaptability, and execution improved this.

e. **Transportation Networks:** Load changing is pressing in transportation associations like rail routes, conveyance associations, and planes. It limits traffic and delays while guiding vehicles or freight to their complaints as fast as common sense. Vehicles, workers, and courses are distributed according to interest, limit requirements, and schedule requirements using asset distribution methodologies.

f. **Energy Management:** In the energy region, managing the age, transmission, and course of power requires load changing and resource conveyance. Smart grid technologies optimize the distribution of electricity resources like renewable energy sources, storage systems, and demand response programs to meet fluctuating demand while maintaining grid stability.

g. **Financial Trading:** High-recurrence exchanging firm's use load adjusting and asset assignment calculations to amplify exchange execution across many trades and exchanging settings. These calculations progressively allocate exchanging requests to various servers and courses to limit inactivity and amplify benefits.

6.7 Performance Metrics

Metrics for performance evaluation are necessary to determine whether load balancing and resource allocation strategies are working. Common methods for achieving this objective include the following:

6.7.1 Throughput

A measure of how quickly a system responds to queries or completes tasks is called throughput. It shows how successfully assets are figured out and how to deal with approaching undertakings. Better performance typically indicates higher throughput.

6.7.2 Reaction Time

Reaction time is the time it takes to get done with a responsibility or interaction, a solicitation from when it is shipped off the framework until it is settled. Shorter response times indicate a more responsive system and are better for the user experience.

6.7.3 Latency

It refers to the amount of time necessary to complete a task or request. It thinks about network inertness, handling time, and lining delays. System responsiveness and delay optimization are both enhanced by lower latency, particularly for real-time systems.

6.7.4 Versatility

The limit of a system to manage extending position or resource demands without relinquishing its handiness is assessed by flexibility. It looks at the system's ability to grow or shrink in response to changes in workload or resources.

6.7.5 Asset Use

How much microprocessor, memory, and association information move limit consumed by the structure is shown by estimations for assessing resource use. High use exhibits useful resource assignment; however, low use could show underutilization or logical bottlenecks.

6.7.6 Optimization

Load dissemination measurements assess how equitably the responsibility is conveyed among servers or accessible assets. Balanced load distribution prevents resource overloading, which can lower performance, and maximizes resource utilization.

6.7.7 Fairness

Fairness can be measured as the equitable distribution of resources or workload among users or apps. It ensures that all clients or tasks get a comparable movement of resources and shields against resource starvation or partnership by unambiguous components.

6.7.8 MTTF

The framework's capacity to work and stay accessible even with mistakes or different interruptions is estimated by adaptation to internal failure measurements. This includes estimates like accessibility (percent), mean opportunity to recover (MTTR), and interim to failure (MTTF).

6.7.9 Cost Effectiveness

This action looks at how asset portioning and burden-adjusting methods affect costs and execution improvements. It considers the expense of equipment, energy use, and working expenses.

6.7.10 Adaptability

The system's capacity to dynamically adjust load balancing and resource allocation strategies in response to shifting workload patterns, resource availability, or system conditions is known as dynamic load balancing.

By examining these performance metrics, researchers, engineers, and system administrators can evaluate the effectiveness of load balancing and resource allocation strategies, identify areas for improvement, and make well-informed decisions to maximize system efficiency and performance.

6.8 Research Directions

Future developments and directions are supposed to shape the spaces of asset designation and burden adjusting, including the following:

6.8.1 Edge Computing

As the number of Internet of Things (IoT) devices rises and the demand for low-latency applications rises, edge computing gains significance. Calculations for load adjusting and asset distribution should be explicitly intended for edge conditions to assign responsibilities and assets among edge hubs in a proficient way while meeting rigid idleness prerequisites.

6.8.2 ML and AI

Calculations for load adjusting and asset designation might profit from coordinating AI and man-made reasoning methods to increment framework execution and effectiveness. These strategies empower abnormality discovery, prescient investigation, and versatile asset distribution in light of constant information and responsibility designs.

6.8.3 Autonomous Resource Management

Frameworks for overseeing assets freely are being created utilizing state-of-the-art innovation, for example, independent specialists, support learning, and self-learning calculations. Without requiring human intervention, these systems can dynamically adjust load balancing and resource allocation algorithms to enhance resilience, responsiveness, and scalability.

6.8.4 Containerization and Orchestration

The sending and organization of microservice-based applications progressively use containerization stages like Docker and coordination systems like Kubernetes. Load balancers and asset portion strategies should acclimate to the powerful idea of containerized conditions to oversee asset allotment and responsibility dissemination across holder bunches proficiently.

6.8.5 Hybrid and Multi-Cloud Environments

Associations are progressively trying cross-breed and multi-cloud procedures to exploit the advantages of various cloud suppliers and on-premise foundations. Streamlining asset use, cost, and execution while organizing responsibilities across different cloud conditions is a necessity for load adjusting and asset designation arrangements.

6.8.6 Energy-Efficient Computing

As server farm energy utilization rises, energy-proficient burden adjusting and asset allotment strategies are turning out to be progressively vital. Green figuring arrangements incorporate clever power the board, work planning in view of energy profiles, and dynamic responsibility combination will assist with diminishing working expenses and carbon impression.

6.8.7 Quantum Figuring

Is still in its earliest stages; however, it can possibly radically adjust load and asset assignment by giving quicker answers for testing improvement issues. Exploring quantum strengthening procedures and quantum-propelled calculations might assist with combinatorial advancement issues connected with load adjusting and asset distribution.

6.8.8 Asset the Executives

Blockchain innovation, specifically, offers decentralized and straightforward asset-the-board processes for dispersed frameworks and distributed networks. Savvy contracts and decentralized applications (DApps) can be utilized to lay out self-administering processes for asset designation, guaranteeing straightforwardness, responsibility, and value.

6.8.9 Security and Protection Contemplations

In light of the fact that cyber-attacks are turning out to be more modern, security and security contemplations will stay vital for load adjusting and asset designation frameworks. The future examination will zero in on coordinating security components like interruption discovery, verification, and encryption into load adjusting and asset distribution calculations to bring down gambles and safeguard delicate information.

6.8.10 Cross-Domain Resource Allocation

As the boundaries between traditional computer areas begin to blur, processes for cross-domain resource allocation are gaining in importance. We will offer coordinated arrangements that enhance asset distribution and burden adjusting in different situations, for example, edge processing, distributed computing, IoT, and other arising standards.

By embracing these arising patterns and future bearings, the discipline of burden adjusting and asset designation is ready to assume the new difficulties of present-day processing frameworks and proposition additional opportunities for advancement and enhancement.

6.9 Conclusion and Future Work

In conclusion, modern computer systems' efficiency, scalability, and performance are maximized by key concepts like resource allocation and load balancing. Advanced load balancing and resource allocation methods are becoming increasingly important in light of the emergence of new paradigms like edge computing, AI-driven automation, and hybrid cloud environments. The next generation of load balancing and resource allocation systems will be influenced by emerging trends and future directions in this industry, such as edge computing, machine learning, autonomous resource management, containerization, and energy-efficient computing. These advancements could further develop framework manageability, adaptability, versatility, and responsiveness in various application areas. In addition, to protect data and infrastructure from attacks, it will be necessary to incorporate robust security features into load balancing and resource allocation algorithms as security and privacy concerns grow in importance. Taking everything into account, new software engineering and designing open doors are made conceivable by the continuous improvement of burden adjusting and asset portion procedures. We will be able to achieve previously unheard-of levels of performance, efficiency, and innovation in the digital age if we embrace these advancements and address the challenges posed by dynamic and increasingly sophisticated computer systems.

Acknowledgments

The authors are grateful for the Nandha Engineering College, Erode and Christ University, Bengaluru, for the facilities offered to carry out this research work.

References

1. Okhovvat, M. and Kangavari, M.R., TSLBS: A time-sensitive and load balanced scheduling approach to wireless sensor actor networks. *Comput. Syst. Sci. Eng.*, 34, 1, 13–21, 2019.
2. Park, J.H., Rathore, S., Singh, S., Mohammed Salim, M., El Azzaoui, A., Kim, T.W., Pan, Y., Park, J., A comprehensive survey on core technologies and services for 5G security: Taxonomies, issues, and solutions. *Hum.-centric Comput. Inf. Sci.*, 11, 3, 17–32, 2021.
3. Heidari, E., Movaghar, A., Motameni, H., Barzegar, B., A novel approach for clustering and routing in WSN using genetic algorithm and equilibrium optimizer. *Int. J. Commun. Syst.*, 35, 10, e5148, 2022.
4. Ren, X., Aujla, G.S., Jindal, A., Batth, R.S., Zhang, P., Adaptive recovery mechanism for SDN controllers in Edge-Cloud supported FinTech applications. *IEEE Internet Things J.*, 10, 3, 2112–2120, 2021.
5. Lin, R., Xu, H., Li, M., Zhang, Z., Resource allocation in edge-computing based wireless networks based on differential game and feedback control. *Comput. Mater. Continua*, 64, 2, 961–972, 2020.
6. Jiang, Y., A survey of task allocation and load balancing in distributed systems. *IEEE Trans. Parallel Distrib. Syst.*, 27, 2, 585–599, 2015.
7. Li, W., Delicato, F.C., Pires, P.F., Lee, Y.C., Zomaya, A.Y., Miceli, C., Pirmez, L., Efficient allocation of resources in multiple heterogeneous wireless sensor networks. *J. Parallel Distrib. Comput.*, 74, 1, 1775–1788, 2014.
8. Omar, D.M. and Khedr, A.M., ERPLBC-CS: Energy Efficient Routing Protocol for Load Balanced Clustering in Wireless Sensor Networks. *Adhoc Sens. Wireless Netw.*, 42, 1–17, 2018.
9. Munusamy, N., Vijayan, S., Ezhilarasi, M., Role of Clustering, Routing Protocols, MAC protocols and Load Balancing in Wireless Sensor Networks: An Energy-Efficiency Perspective. *Cybern. Inf. Technol.*, 21, 2, 136–165, 2021.
10. Rehman, A.U., Ahmad, Z., Jehangiri, A., II, Ala' Anzy, M.A., Othman, M., Umar, A., II, Ahmad, J., Dynamic energy-efficient resource allocation strategy for load balancing in fog environment. *IEEE Access*, 8, 199829–199839, 2020.
11. Zhao, L., Wu, D., Zhou, L., Qian, Y., Radio resource allocation for integrated sensing, communication, and computation networks. *IEEE Trans. Wireless Commun.*, 21, 10, 8675–8687, 2022.
12. Chavan, P., Malyadri, N., Tabassum, H., Supreeth, S., Reddy, P.V.B., Murtugudde, G., Rohith, S., Manjunath, S.R., Ramaprasad, H.C., Dual-Step Hybrid Mechanism for Energy Efficiency Maximization in Wireless Network. *Cybern. Inf. Technol.*, 23, 3, 70–88, 2023.
13. Gong, Y., Wei, Y., Feng, Z., Yu, F.R., Zhang, Y., Resource allocation for integrated sensing and communication in digital twin enabled internet of vehicles. *IEEE Trans. Veh. Technol.*, 72, 4510–4524, 2022.

14. Rawat, P. and Chauhan, S., Clustering protocols in wireless sensor network: A survey, classification, issues, and future directions. *Comput. Sci. Rev.*, 40100396, 100–124, 2021.
15. El Khediri, S., Wireless sensor networks: a survey, categorization, main issues, and future orientations for clustering protocols. *Computing*, 104, 8, 1775–1837, 2022.
16. Shahryari, M.-S., Farzinvash, L., Feizi-Derakhshi, M.-R., Taherkordi, A., High-throughput and energy-efficient data gathering in heterogeneous multi-channel wireless sensor networks using genetic algorithm. *Ad Hoc Netw.*, 139, 103041, 2023.
17. Vanitha, C.N., Malathy, S., Anitha, K., Suwathika, S., Enhanced Security using Advanced Encryption Standards in Face Recognition, in: *2021 2nd International Conference on Communication, Computing and Industry 4.0 (C2I4). 2021 2nd International Conference on Communication, Computing and Industry 4.0 (C2I4)*, IEEE, 2021.
18. Zhang, Y., Liu, Y., Lyu, M.R., Resource Allocation in Wireless Networks Using Deep Reinforcement Learning: A Review. *IEEE Netw.*, 34, 5, 16–22, 2020.
19. Zhang, H., Hao, Z., Guo, Y., Ji, S., Wang, Y., Intelligent task offloading in mobile edge computing via deep reinforcement learning. *IEEE Trans. Veh. Technol.*, 69, 8, 9230–9243, 2020.
20. Li, S., Da, X., Han, Z., Zhang, Y., Liu, Y., AI in Networking: State-of-the-Art and Future Perspectives. *IEEE Internet Things J.*, 5, 1, 1–15, 2018.
21. Kohavi, R., The Optimal Naive Bayesian Classifier, in: *Proc. First Int. Conf. Knowl. Discovery Data Mining (KDD-95)*, Montreal, QC, Canada, 1995.
22. Hsu, C.-W., Chang, C.-C., Lin, C.-J., A comparison of methods for multiclass support vector machines. *IEEE Trans. Neural Netw.*, 13, 2, 415–425, Mar. 2002.
23. Hastie, T., Tibshirani, R., Friedman, J., AdaBoost, in: *The Elements of Statistical Learning: Data Mining, Inference, and Prediction*, 2nd, Springer, New York, NY, USA, pp. 337–365, 2009.
24. Malathy, S., Vanitha, C.N., Dhanaraj, R.K., Reinforcement Learning in Smart Transportation, in: *Artificial Intelligence for Future Intelligent Transportation*, pp. 173–198, Apple Academic Press, 2024.
25. Wang, D., Wang, J., Duan, H., Yang, S., Learning-based resource allocation for software-defined networking: Architecture and algorithms. *IEEE Netw.*, 33, 3, 46–53, May/June 2019.
26. LeCun, Y., Bengio, Y., Hinton, G., Deep learning. *Nature*, 521, 7553, 436–444, May 2015.
27. Wei, Y., Wang, J., Li, H., Chen, C., Deep Reinforcement Learning Based Resource Allocation for Vehicular Networks. *IEEE Trans. Veh. Technol.*, 67, 12, 11629–11642, 2018.

7

Sustainable Cities and Communities: Role of Network Sensing System in Action

Hitesh Mohapatra[1]*, Soumya Ranjan Mishra[1], Amiya Kumar Rath[2,3] and Manjur Kolhar[4]

[1]School of Computer Engineering, KIIT Deemed to be University, Bhubaneswar, Odisha, India
[2]Computer Science and Engineering Department, Veer Surendra Sai University of Technology, Burla, India
[3]Biju Patnaik University of Technology, Raurkela, Odisha, India
[4]Department Health Informatics, College of Applied Medical Sciences, King Faisal University, Al Hofuf, Saudi Arabia

Abstract

This work presents a comprehensive analysis of topological network performance in wireless sensor networks (WSNs) domain. With the increasing deployment of WSNs in diverse applications, selecting the most suitable network topology is crucial for optimizing performance and resource utilization. The objective of the study is twofold. Initially, this study conducts an in-depth examination of four prominent topologies: Star, Mesh, Tree, and Cluster, evaluating their performance across key parameters, such as reliability, energy efficiency, scalability, latency, fault tolerance, deployment, and cost. The findings contribute valuable insights into the trade-offs and advantages associated with each topology aiding network designers and researchers in making informed decisions for WSN deployments tailored to specific application requirements. Second, this research offers a valuable reference point for optimizing WSN performance and enhancing the effectiveness of sensor networks in various domains. To develop smarter cities, we must really understand how different forms and structures of cities are interconnected with each other and how these fit into the maps we draw. The study as a comparison of some representative topological architectures is backed up by their relevance to the smart city infrastructure systems. It covers network-based model evaluation addressing their efficiency, scalability, and availability in the urban conditions. On the other hand,

---

*Corresponding author: hiteshmahapatra@gmail.com

the watch focuses on the symbiotic relationship between these topological models and mapping techniques investigating the ways these methodologies fit into and interact with various topological structures. The comparison is performed results, which serve as a guide for decision making and planning not only for technology and infrastructure development but also for smart city applications.

Keywords: Topology, wireless sensor network (WSN), star, mesh, tree, cluster, smart city

7.1 Introduction

In bringing up the idea of sensor network, the manner of how they are connected is of the essence. It influences what information is gathered from the overall network and how it is brought to a central control. The term topology within the context of sensor networking describes the spatial layout of these nodes and the communication links that connect them as well. Topology selection can greatly affect tangible properties of sensor network facilities such as reliability of data, energy efficiency, scalability, latency, and fault tolerance. As more and more sensor networks are deployed in broader domains, such as environmental monitoring, industrial automation, smart cities, or healthcare, the criticality of choosing the right topology becomes apparent. This introductory exploration focuses upon the role of topology in sensor networking. Topology is a multifaceted concept that is present everywhere from how it influences the performance and effectiveness of ubiquitous networks to the structures in which it can be realized. Decisions on the spatial distribution of sensor nodes (SN) are the main factors in sensor networking that determine the level of performance and effectiveness of the whole network.

The function carried out by topology in defining the strategic location of SNs within a given environment is highly significant. SNs may get distributed in different ways and topological structures according to the specific application requirements, as they could be organized into a grid, random, hierarchical, or, in some cases, event-driven deployment patterns. Grid type deployments have uniform coverage and are appropriate for applications, which require gathering of data over an area in a uniform way. In contrast, random deployments bring in an uncertain element, and this is beneficial in environments with varying environmental conditions. Hierarchical deployments perform multi-tiered networks following the node level hierarchy maximizing energy efficiency and scalability. Event-triggered deployments put a priority on placing the candidate nodes in nodes expected to detect particular events of interest. The distribution of the SN within the application must be selected based on technical considerations. The SN distribution must be aligned with the goals and constraints of the application. This will also enable the application to collect data effectively and save energy and resources simultaneously.

Mesh topology was built to be logical and flexible and currently enjoys a huge attention from both academia and industrial experts due to its exceptional benefits. It is a mesh arrangement where every node is linked with several nearby nodes thus being in direct contact. It merely creates a web mesh of relationships between them. Such an interconnected environment guarantees ruggedness and robustness, as there are many paths for data to travel from A to B. The mesh topology is, in general, very good for sensor networks where the fault tolerance, redundancy, as well as data integrity elements are considered as very important ones. Although multi-hop communication in mesh networks is prone to consume more power compared to other topologies, educated routing techniques and regulated power mechanism can reduce this issue. Moreover, mesh topologies are designed to handle scenarios with robustness and flexibility in mind; these are widely used for applications including environmental monitoring, smart grids, industrial automation, where the SNs must adapt to changing conditions, and the data reach the destination reliably.

Sensor networking typically employs the star topology, which is simple to implement and established. In this topology, the sink or central node is connected with all nodes forming a network where every communication flows through this central node. The star topology assists with the network's simplification of administration and routing as data aggregation and decision making mostly take place at the core node. This simple design is, therefore, ideal for applications that need to be quickly set up and have centralized control like home automation and small-scale industrial settings. However, the star topology may have limitations in terms of scalability and fault tolerance irrespective of its importance. In the case the central unit fails, the whole network may no longer function. Hence, one should keep in mind that whilse star topologies are characterized by simplicity, straightforward data management, and are quite suitable in regard to the deployment of a small-sized SN, their suitability largely depends on the specific requirements.

As in a hierarchical and organized network topology, a sensor network often uses a tree topology to control data flow. The SNs will operate in a similar way to that of the tree structure. The main branch is the one at the top, and then, all the other branches coming out from this main branch are the children. The SN data passes down through the hierarchy to the root node that may be as a BS or data aggregator. Tree topologies have their benefits, which include scalability and energy efficiency: each node has its role in routing data, and other nodes may be simply connected to the hierarchy. This topology is highly suitable for sensor networks deployed in applications where data are required to be gathered from many locations and transferred to a central point, which may include applications such as environment monitoring and precision agriculture. But the success of a tree topology depends on how communication paths are carefully managed and how routing algorithms are well designed so that there are no bottlenecks or long queue time.

Clustered topology is the most common strategy of network configuration that is effective and energy saving in sensor networking. In this configuration, nodes make up the clusters, and the cluster heads are the leader nodes. Other than the member nodes (MN), the cluster heads (CH) connect the BS (or sink node) to the intermediary ear (CH). This type of SNs call for a process of grouping together of information received through a different array of devices and then sending it to a CH node. Then, for this CH node, it combines all information and forward it to another BS. Cluster topologies are fluent for applications that strive for energy-saving and data-gathering optimization. As stated, clustering a network by lowering the gap between nodes within a cluster improves energy efficiency, as nodes may normally operate in low power modes when not actively transmitting data. This design enables scalability and longevity of network while ensuring data integrity and hence is a good option for sensor networks in habitats like soil moisture measurement, healthcare applications, and infrastructure maintenance. Nevertheless, the efficient managing and load balancing may be very important to exploit the benefits of this topology. Figure 7.1 is used to show topological arrangements of sensor networks.

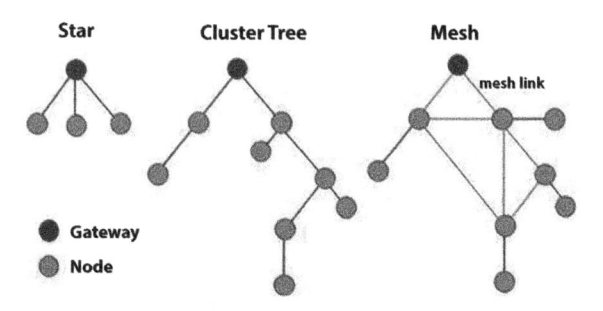

Figure 7.1 Popular topology in a smart city.

7.2 Literature Review

There have been many researches on the efficiency of different topologies when the issue is networking of sensor networks. A small-scale study was among the first to be undertaken [1]. They compared the performance of three topologies: star, trees, and net. Throughput and delay, they discovered, were the best in mesh topology, but it consumed the maximum energy as well. The star-network implementation had the least energy consumption, but this also had the worst performance in terms of throughput and delay. In the case of tree topology, the average performance was acceptable. Another study was conducted. They compared the performance of four topologies: star, tree, cluster, and grid are the key shapes in Vadas Gunta's oil painting [2]. These researchers demonstrated that the hierarchy was a topology that had the best throughput and average delay, the next best was the grid topology, then the tree topology, and finally the cluster and the star topology. Wang *et al.* (2018) reported a more recent research work They compared the performance of five topologies: star, tree, mesh, hybrid, and dynamic. Adaptive topology was shown to be superior in terms of throughput and delay metrics, whereas the mesh topology came second, next was the hybrid topology, and then the tree topology as well as the star topology. Logical topologies deem the communication routes within the network between SNs; hence, they are of star importance especially in resource-limited sensor networks. Therefore, the difficulties of minimizing constraints become more spatial, and it becomes easier to get to know such problems through a perspective of topology. For this reason, the research enterprise is focused on the logic topologies in WSNs. Initially, the set of performance metrics, which are required for this task, is determined. Next, a complete study of the multiple logical topologies extracted from different application protocols in WSNs is performed, and then a comparison of these topologies is provided using the previously prepared performance metrics [3].

Many other authors have decades of work in similar ways by the writers of ref. [4]; however, the focus of their research is to create algebraic methods of quantitative assessment of different topologies and qualitative perception of other performance statistics. The main focus of this research is to provide a simple guidance on WSN developers in selecting the best topology and its associated parameters. The topology structure of WSN is the first stage of the designing and setting up the system. One of the great ways to do this is to consider a proper topology, which will increase the life of the network. The paper focuses on the complexity of the topology structures of the WSN, which can be examined via the network theory. The paper describes the study of WSN measurement that shows the property unique to node degree distribution in which meshed networks result in a shorter average path length and a higher cluster coefficient. In addition to that, WSNs have qualities that place them in the middle between these two types of networks, in the sense that they have properties similar to the small-world networks [5, 6]. The architecture of a WSN is therefore one crucial factor. Situations can make the reliability of the network dependent on the setup, on how much energy is being consumed, and on how fast the messages can be sent. In this study, they looked at three different ways to set up networks: a first one with hexagons, a second one with grid, and a third with equilateral triangles to observe how differently each design covers the network performance. The topology of the nodes and the intervening space between them have a crucial role for the functioning and the performance of WSNs. Due to the above factors, the sensors now have to detect certain events and pass the information to the BS. The topological arrangements impact on overall resource management.

On examining particular fit-constrained environments and habitats, the positioning of loci and distances between nodal centers turn highly special to the specific demands of each respective application. However, the major milestone seen in this study is to cover the entire area to find relevant events. Sensors are installed at locations where the role they are supposed to plays is taken into consideration and the distances they need be from each other. Stated devices can be simple sensors/devices of a limited function or relays/devices will full capabilities. They use definite technical standards referred to as IEEE 802.15.4/ZigBee protocol. In addition to that, they are working with a system that dares to launch data with the same rate [8]. Among the main aspects of WSN management is topology management. The main point of it is to find an algorithm that enables the network to remain connected while simultaneously saving energy. The topology management means understanding the physical connection of sensors and logical relation among them [7]. The concept of topology management is understanding the physical connection of the sensors and the logical relationship among them. The ACL protocol is the one which involves choosing active communicating nodes. This approach eliminates the redundant transmission of information and keeps nodes alive. Networks require continuous monitoring of the underlying infrastructure to maintain smooth and fast workflow. So, the primary purpose of topology management is to supply an energy-efficient mechanism of network connectivity maintenance. This is reflected in the configuration management practice, which is composed of two phases, the initial setup of a network device and the ongoing maintenance and control of these devices [9]. The correct network topology choice is crucial in sensor networking as, on the contrary, mistakes in that choice can be followed by a number of complexities, which, as a result, form a cascade of problems. If the grid topology does not correlate with the application-specific conditions, there are a lot of challenges which emerge. Also, it may result in the routing of data via inefficient routes, which requires data packets to travel unnecessary long routes thus giving rise to latency alongside higher power consumption [10].

Besides, network reliability can be compromised, for it may have nodes that are not robust enough to replace failed nodes. The network disconnections and loss of data can be a result of this. In addition, the flexibility and reconfigurability of the network could be compromised, which could lead to difficulties in scaling the network to changes of size or configuration. In essence, topological mapping erroneousness worsens the intrinsic sensor network complexity problem by adding inefficiency, raising unreliability, and impeding decent network growth. This stresses the substantial role topology plays in the achievement of the best performance and resource management in the sensor networks [11]. The ring topology is outside the sensing scenario facing the exact data transmission and reliability problems. In case of the sensor network in which data integrity and latency define the network, performance ring topology can be the best option because of its deterministic data routing feature. A ring topology is a network of Sn nodes, where each one is connected to exactly two neighbors, hence ensuring that information traffic passes a predefined route from tail to head [12]. In addition, it brings both the routing simplicity and the collision probability reduction due to the predictive nature that improves network performance. In addition, rings can be developed with fault tolerance in mind so that data can go around a clockwise direction if there is a failure of any one part. The data can then take an alternative route back. This redundancy can be extremely useful in achieving network reliability, which is a critical prerequisite for the use in applications such as industrial automation and surveillance. However, ring topology is not suitable for all sensor network scenarios but its ability to offer determined routing, reduced data collision, and tolerance of fault in the network make it a valuable choice when delivery of data and network integrity are necessary [13].

Sensing topology dimension turns out to be very valuable in the context of sensor networking thanks to its ability to achieve simultaneously high efficiency of energy consumption and quality of data collection. In the hierarchical sensor network, the sensor nodes in each cluster are controlled by a cluster head. This hierarchically organized topology lowers energy consumption due to the fact that most nodes use low-power mode in which they do not need energy for data transmission [14]. The CHs act as the gateway connecting multiple sensors and helping to minimize the energy usage and the delay in data routing to the major data aggregation and processing point. The clustering topology also provides scalability in the network, as adding new nodes to the existing clusters is not required to do a total network migration. Furthermore, the topology supports self-organization because nodes can send acknowledge packets to each other with respect to various network conditions. These characteristics of clustering topology make it an excellent match for resource-constrained systems where power efficiency, scalability, and agility are vital, such as those in the field of environmental monitoring, smart agriculture, and healthcare. Ensuring increased network longevity in WSNs is a daunting task. Beyond that, the nodes of a WSN are usually powered by battery and with limited energy.

As more data are relayed by a particular node, the more energy it consumes. This may result in the state of the nodes to be exhausted of energy before others, thus total disconnection of the network. The tree-based approach used to solve the problem is one of the proposed algorithms. Initially, the algorithm generates the tree of SNs, and then, the work taken by the nodes is iteratively readjusted. Utilization of a unified energy level by all the SNs enhances WSN network lifetime. The algorithm is decentralized, that is, each node only focuses on own information. This enables the algorithm to be scalable to large networks, which are composed of a large number of nodes. Nodes at each stage only communicate with their neighbor, which is the fundamental reason for low communication overhead. The algorithm reduces the transmittal of data using a technique called multi-routing. This method leads a node to send its message to the mobile sink making use of various paths, which result in the reduction of the load on each path [15].

A research paper [16] compares three different ways that networks of sensors can be set up: trellis, trellis, and hybrid structures. They played computer games with every design. This finding might mean the best comes from a certain kind of structure, called the hierarchical topology, which is useful especially when the sensors are far from the main station, and there are lots of sensors concentrated in one area. Such a hierarchical construction can reserve the power, have the network run for a longer time, and make the whole system get into a better state of work. In another paper, the problem of designing the best set-up for a type of sensor network that takes the advantage of network coding to send messages to many sensors by one is dealt with. They had figured out the best way because they had devised a complex math problem, by looking into how power is used for sending and receiving messages The genetic algorithm used to make this math problem simple is outstanding and that is why the solution was a success. The result of the study proved that their method could find the best setup, and also acceptable setup, for these networks. This further assisted in the maintenance of the network, at least for some time, as proven by their simulations [17].

7.3 Proposed Study

Many aspects of a WSN can be affected by several factors, among which the topology of a network matters. The following are some of the most common topologies used in WSNs and their relative advantages and disadvantages.

7.3.1 Star Topology

In a star configuration, the base station, which is placed at the center, is the central node, and the SNs are the nodes that approach it directly. It is the simplest network, but it is not able to deal with large networks. The advantages of star topology are as follows:

- Simple to install and manage.
- Easy to troubleshoot.
- Good for applications where data need to be collected from a large number of SNs (SN) and sent to a central location.

The disadvantages of star topology are as follows:

- The BS is a single point of failure.
- The network can be easily congested if there are too many SNs.
- The SNs closest to the BS may consume more energy than the others.

Factor-based performance with star topology.

- Throughput: In a star topology, all nodes communicate directly with a central hub. Throughput is generally high because there is no contention for the medium. Calculate the throughput by measuring the amount of data successfully transmitted over time.

- Delay: Since all communication goes through the central hub, delay can be relatively low. Measure the end-to-end delay for data transmission.
- SNR: Assess SNR by measuring the signal strength at the central hub and comparing it to the noise level.
- Lifetime: The network lifetime may be shorter in a star topology because the central hub is a single point of failure. Evaluate how long the network can operate before the central hub fails or requires maintenance.

7.3.2 Mesh Topology

In a mesh topology, all the sensors are connected there. This is a better topology for large-scale networks, but the upstream computers or routers need to have more complex routing algorithm.

The advantages of mesh topology are as follows:

- Scalable to large networks.
- Fault tolerant.
- Good for applications where data need to be routed between SNs.

The disadvantages of mesh topology are as follows:

- Complex routing algorithms.
- Difficult to install and manage.
- More energy consumption than star topology.

Factor-based performance with mesh topology.

- Throughput: Throughput compression is a network routing characteristic. Check average throughput taking into account the existing multiple paths for data transfer.
- Delay: Extra delays may occur in mesh networks as there are many hops. Measure end-to-end delay and compare it with the number of hops.
- SNR: The SNR is difficult to estimate in a mesh network, but it is of paramount importance, so the nodes are able to communicate effectively.
- Lifetime: The lifetime of network in a mesh topology can exceed that of a star topology because of the advantage of

redundancy. Assets of the network will be estimated in terms of energy consumption and node failures.

7.3.3 Tree Topology

A tree topology is a network structure where SNs hold a hierarchical relationship, where some nodes have more connections than others. This is a trade-off between a star topology and a mesh topology's name.

The advantages of tree topology are as follows:

- Scalable to large networks.
- Fault tolerant.
- Good for applications where data need to be routed between SNs.
- Less energy consumption than mesh topology.

The disadvantages of tree topology are as follows:

- Complex routing algorithms.
- Difficult to install and manage.

Factor-based performance with tree topology.

- Throughput: Throughput in a tree topology is good only in case of trees that are less deep and also not too much of branching. Assess throughput by looking at how many bytes go from leaves to the root.
- Delay: Down the road of the tree, delay grows. Measure the head–tail delay and analyze the height of the tree.
- SNR: Like star architecture, measure the effectiveness of the network by evaluating signal strength and noise at the root node.
- Lifetime: Time to network is affected by the depth of the tree and power consumption at node level. Analyze the life of a given network when these factors are taken into account.

7.3.4 Clustered Topology

In a cluster topology, the CNs are clubbed into clusters, each cluster having a CH. The CH have the role of communicating with the team members and BS.

The advantages of clustered topology are as follows:

- Scalable to large networks.
- Fault tolerant.
- Good for applications where data need to be routed between SNs.
- Less energy consumption than mesh topology.

The disadvantages of clustered topology are as follows:

- Complex routing algorithms.
- Difficult to install and manage.

Factor-based performance with tree topology.

- Throughput: Throughput in a cluster topology can be efficient within clusters, but inter-cluster communication may have lower throughput. Measure both intra-cluster and inter-cluster throughput.
- Delay: Delay varies depending on the cluster's size and the number of hops required for inter-cluster communication.
- SNR: Assess SNR within clusters and at the border nodes where inter-cluster communication occurs.
- Lifetime: Network lifetime is often extended in cluster topologies due to clustering and localized processing. Evaluate energy consumption within clusters and the lifespan of individual clusters.

Topology suitable for WSN is contingent on a given application and demands of the network. As for instance, a star topology can be the perfect solution for a network that must collect data from a significant number of sensors and send them to a central unit. A mesh topology is the best choice when the network must be fault tolerant and able to extend on demand. A tree topology is the possible best choice that will require the least of the energy consumption of its SNs. Depending on the type of network and the few procures needed, an optimal topology can be a cluster.

The performance also varies depending on other factors, i.e., number of SNs, the distance between the SNs and the base station, the type of data, and energy restraints of the SNs. Assessing the efficiency of the WSN based on the different topologies (star, mesh, tree, and cluster) means evaluating the various parameters, including throughput, delay, signal-to-noise-ratio

Table 7.1 Performance comparison of WSN based on topologies.

Topology	Throughput	Delay	Signal-to-noise ratio (SNR)	Lifetime
Star	High	Low	Good	Shorter
Mesh	Variable	Moderate	Variable	Longer
Tree	Moderate	Variable	Good	Varies
Cluster	Efficient	Variable	Good (intra)	Extended

(SNR), and network lifetime. Every one of these topologies has its strengths and weaknesses. These features influence these numbers in different ways. Based on Table 7.1, the performance of WSNs is shown in Figure 7.1 for the scenarios of different topologies.

7.4 Performance Analysis

We have assumed smart agriculture monitoring system for the performance evaluation of the proposed comparison.

➤ **Number of Nodes:**

- In this scenario, we have simulated a WSN consisting of 50 SNs.
- The nodes are evenly distributed across a 1,000 m × 1,000 m agricultural field.

➤ **Traffic Patterns:**

- We want to monitor temperature and soil moisture in the field.
- SNs are collecting data periodically (every 10 min) and send it to a central data collection point.
- We have assumed that each SN generates 50 bytes of data per reading.

➤ **Mobility Models:**

- SNs are stationary and do not move in this scenario as they are fixed in the ground.

➢ **Energy Consumption Models:**

  o We use the "Low-Energy Adaptive Clustering Hierarchy" (LEACH) energy model.
  o Each SN starts with 1,000 J of energy.
  o Energy consumption for sensing, data processing, and communication is considered based on the LEACH model.

➢ **Communication Model:**

  o We have simulated wireless communication using radio propagation models (e.g., log-distance path loss) to estimate signal strength and interference.
  o The communication range of each node is set to 100 m.

➢ **Data Collection Point:**

  o We have a BS located at the center of the field.
  o The BS collects data from SNs and sends it to a remote server.

➢ **Simulation Duration:**

  o The simulation will run for 24 h (simulating a full day in the field).

➢ **Network Topology:**

  o In the star topology, all nodes communicate directly with the BS.
  o In the mesh topology, nodes form multi-hop routes to reach the BS.
  o In the tree topology, nodes form multi-hop routes to reach the BS.
  o In the cluster topology, nodes are organized into clusters with a cluster head for aggregation.

➢ **Performance Metrics:**

  o Throughput: Data delivery rate to the BS.
  o Delay: End-to-end delay for data packets.
  o Energy consumption: Total energy used by the network.

○ Network lifetime: Time until the first SN is dead because of energy depletion.

Table 7.2 shows the performance results of three network topologies for a smart agriculture monitoring system. The topologies are star, mesh, and cluster. The performance metrics are throughput, delay, energy consumption, and network lifetime. Among the five topologies, the star topology has the lowest throughput, highest latency, and highest energy consumption [18]. This is attributed to direct communication between all nodes and the base station, which is likely to create congestion and overhead. The limited lifetime is caused by the fact that the nodes with the base station have a higher chance to get the energy run out. In the mesh topology, the link does not share a specific path, and the highest data rate and the lowest delay are ensured. This is achieved by the nodes being able to communicate with each other using the multi-hop fashion, which will ensure that congestion and overhead are reduced. Energy consumption is also lower than star topology since the nodes do not have to communicate in a direct way with the BS as much as in star topology. Nevertheless, mesh topology lifetime is shorter than the cluster topology due to the need to communicate with more nodes. Cluster topology has demonstrated the best performance in terms of throughput, delay, and energy consumption among all the topology configurations under comparison. This is because there are cluster heads; cluster nodes are arranged in clusters. It is also helpful for reducing congestion, saving power, and overhead cost. Network lifetime is also the longest as the nodes communicate with other nodes in their cluster and consequently choose the most energy efficient communication path so that they are less likely to run out of battery. Conclusively, the cluster topology is the most preferable for a smart agriculture monitoring system because it

Table 7.2 Numerical analysis of performances.

Topology	Throughput (Mbps)	Delay (ms)	Energy consumption (J)	Network lifetime (h)
Star	1.2	100	12,000	20
Mesh	2.4	50	24,000	16
Tree	2.2	48	22,000	14
Cluster	2	60	18,000	18

is the best in every parameter such as throughput, delay, power consumption, and network lifetime. But the mesh topology can also be adopted if the network lifetime is major than bandwidth and delay satisfaction.

Figure 7.2 demonstrates the performance of four network topologies for an Internet of Things (IoT) smart agriculture monitoring system, marked throughput, delay, energy consumption, and network lifetime. There are star, mesh, tree, and cluster topologies. Throughput plot illustrates the rate of the data being transmitted to the base station. Among the topologies, mesh, star, cluster and tree, the highest throughput is obtained by mesh topology, then the cluster topology, star topology, and last, the tree topology.

This is because the mesh topology life allows nodes to communicate with each other in multi-hops that helps to lessen the congestion and overhead. The graph with delay plot shows the total delay or end-to-end delay for data packets. The star topology has the maximal delay, the tree topology comes in the second place, while the cluster topology and the mesh topology have the third place. This is why the star topology needs every node communicating with the base station, and this will, in turn, lead to overload and delays. Energy Consumption Plot shows the network's energy consumption in total. The cluster topology has the lowest energy usage in the list; in the sequence, next comes the mesh topology star topology being the last. This is the reason cluster topology lets the nodes save energy by clustering closely. The X-axis of the graph is the sensor time until the first node is out of energy. The cluster

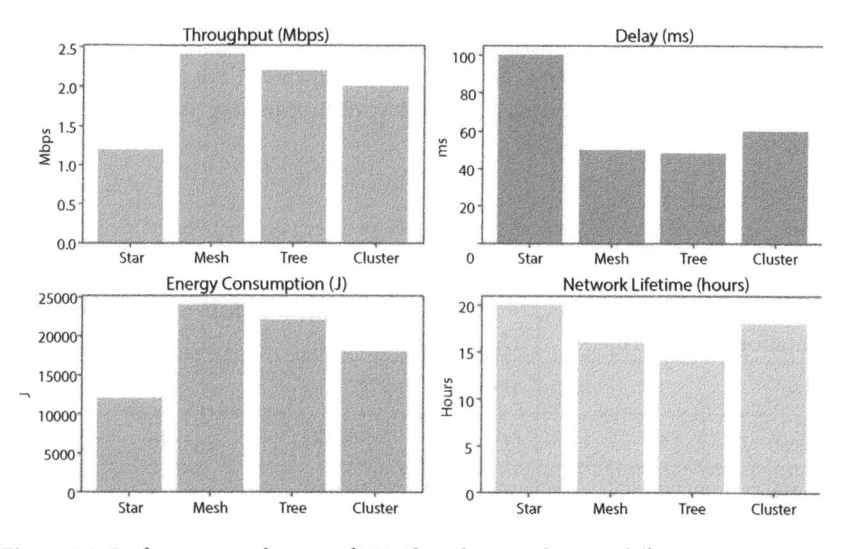

Figure 7.2 Performance evaluation of WSN based on topology on different parameters.

node topology has the longest network lifetime, while the mesh and tree topologies closely follow. The fourth and the least network lit topology is the star topology. Cluster topology enables nodes to collaborate with each other and thus survive with less energy consumption. In summary, the topology of the cluster performs the best in the average values as regards throughput, delay, energy consumption, and network lifetime. On the one hand, the mesh topology would be a suitable option when the network lifetime is given more weight than the throughput and delay.

Throughput: Throughput plot opposite shows the data delivery rate to the BS (base station) in Mbps. The hierarchical topology can achieve the highest throughput and then the cluster topology, the tree topology, and the star topology. It is mainly because this type of topology enables nodes to pass packets to each other via a multi-hop mode, which is capable of both alleviating congestion and decreasing overhead.

Delay: The delay plot graphs the one-way delay (in ms) for data packets. A star topology is the topology with high delay, then the cluster topology, tree topology, and the mesh topology in order. As for the reason, the star topology allows all nodes to connect with the base station directly, which may lead to congestion and remarks.

Energy Consumption: The efficiency plot shows the total energy, in Joules, consumed by the system. The mesh and tree topologies have a higher energy consumption, with the star topology consuming most energy. This is achieved in a way such that nodes can reduce their energy consumption by coming together in clusters and not exchanging information directly with every node in the cluster.

Network Lifetime: The first node to run out of energy is shown on the network lifetime plot—in hours. Point-to-point has the shortest network lifetime, followed by hunger connectivity model, point-to-multi-point, and hybrid model. Therefore, nodes use low energy by clustering together to form cluster topology.

7.5 Mapping of Topology with Smart City's Applications

7.5.1 Mapping of Star Topology with Smart Parking Application

The star topology is a fit for a Smart Parking application because of its structure and effective data aggregation capabilities. When it comes to managing parking space monitoring availability and ensuring control, the characteristics of the Star topology align with the requirements [19].

Centralized Control: In Smart parking systems, it is often necessary to have a control or management system in place wherein oversees parking spot availability monitors occupancy levels and directs users to empty spaces. The centralized hub of the Star topology enables this control by gathering data from various sensors or cameras placed throughout the parking facility (Ref: Figure 7.3).

Managing Simplified: Through the connection of all sensors, cameras or monitoring devices to a hub, the Star topology streamlines management and troubleshooting processes. This setup facilitates the identification of any issues that may arise within the parking system enabling responses to address malfunctions or failures effectively.

Star topology shines in applications requiring centralized control and data aggregation, such as traffic management or centralized monitoring systems. In summary, the Star topology's centralized structure, simplified management, efficient data aggregation, and scalability make it highly suitable for Smart parking applications. It facilitates effective monitoring, management, and allocation of parking spaces enhancing the overall user experience and optimizing the utilization of parking resources within a Smart city environment.

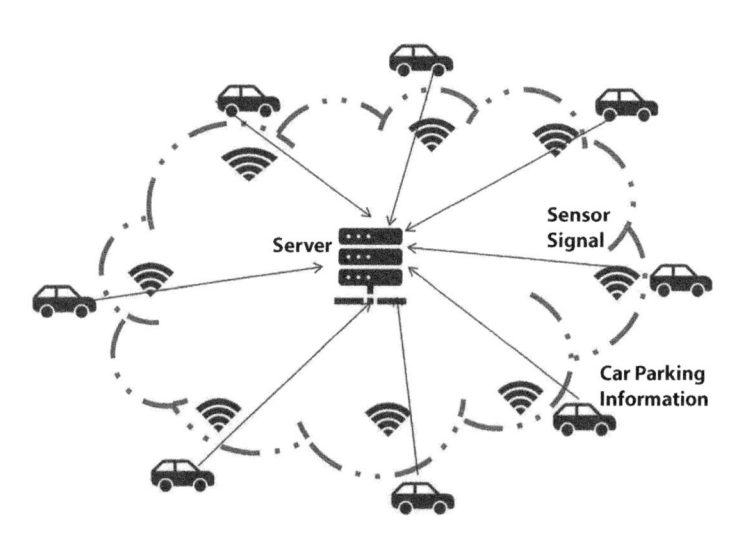

Figure 7.3 Star topology-based Smart parking model.

7.5.2 Mapping of Mesh Topology with Smart Grid Application

The mesh topology stands as a highly suitable network architecture for implementing Smart Grid systems due to its resilience, fault tolerance, and robust communication capabilities [20]. In the context of Smart Grids, which involve efficient transmission and distribution of electricity along with data communication, the Mesh topology offers several advantages.

Redundancy and Reliability: A Smart Grid requires the trustworthiness of data transmitting and power distribution systems. In this respect, the mesh topology explicitly stands out by creating multiple simultaneous connectivity paths between the jumps. If one given route meets a problem or is disconnected, the other routes keep the traffic going, and the whole system is not off the air due to rest of routes running. This redundancy adds an important reliability factor to a Smart Grid, which, in turn, makes it more robust and can therefore sustain a continuous power supply.

Fault Tolerance: In a situation where a node fails or if there is any network disruption, mesh topology ensures that other nodes can proceed with communication via the alternative path. Nevertheless, Smart Grid has an inbuilt fault tolerance that enables it to continue functioning even during the occurrences of failures and outages and system malfunctions. This critical infrastructure, such as Power Distribution, this capability is vital to protect (Ref. Figure 7.4).

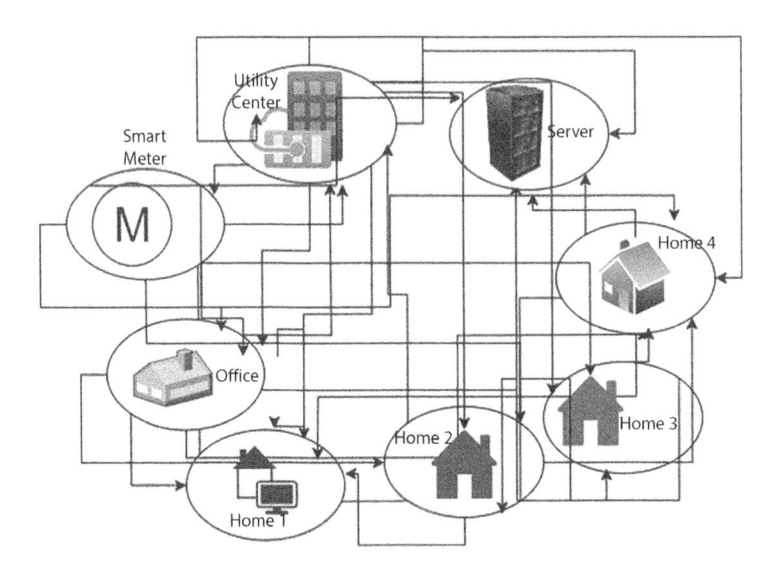

Figure 7.4 Mesh topology-based smart grid network.

Scalability: Smart Grid expansion sometimes requires adding new components or nodes for various reasons and also as the network grows and technology advances. To scale a network without any effects, mesh topology is used. New nodes can integrate with the network rather quickly deploying another communication channel without putting the existing information flow in danger. Therefore, the scalability feature is critical in the adoption of new renewable energy resources, increased flexibility, and improvements in the management of a Smart Grid.

Dynamic Routing and Optimization: The major benefit of the mesh topology is that it can route the data through multiple ways toward the receiving end. This implies the facility for data transmission optimization, thus a way of maximizing the possibility of using available paths while minimizing the congestion at the same time.

The ability to create redundant paths, therefore, makes mesh topology shine for its reliability, this being a key characteristic for vital applications as emergency services or utilities. In short, a mesh topology has a number of advantages, which derive from its resilience, fault tolerance, scalability, and dynamic routing capabilities, and make it a primary choice for implementing Smart Grids. It provides the required network to support the uninterrupted and reliable communication that is developed for consistent electricity distribution and also making it possible to integrate the advanced green technologies and renewable energy sources within the modern power systems.

7.5.3 Mapping of Tree Topology with Smart Education Model

The tree topology features a proper network structure for a Smart Education scenario because of its hierarchic and organized nature and that it works for the needs of educational systems that imply a centralized management and the availability of structured information flow [21].

Hierarchical Structure: In a Smart Education system, it is required to facilitate networked communication among various levels of education administration, which include a central education board, regional offices, schools, and classrooms. The tree topology's core arrangement, where the root node (central administration) is followed by lower nodes (regional offices, schools), in a hierarchical manner, ensures an organized flow of information and its management.

Segmented Information Flow: The educational systems usually add filters to information channels, and it can be transmitted only to given levels or departments. The layering arrangement of the tree topology categorizes data into segments causing all information to be possible at all named levels without any congestion or overlaps.

Centralized Control and Management: The implementation of a centralized control and management is critical in the educational systems, since it standardizes, cooperates, and monitors education quality according to the curriculum standards. The botanical metaphor—tree topology's root node acts as a focal point for control with the hierarchy of education resources, curriculum updates, and administrative decisions percolating down from the upper level.

Scalability and Expansion: With growth, new branches and schools become parts of the educational system, and the tree topology can soon have more nodes or branches added while retaining the original structure without disruption. This is an important aspect of scalability as it makes it easy for new educational institutions or departments to be seamlessly integrated into the network.

Structured Communication: This topology allows designed communication channels constraining the flow of information to pre-defined routes. This structured communication pattern supports efficient collaboration, resource sharing, and data exchange within the educational network (Ref: First session included counseling on hygiene and benefits of good health, the second finger pricking and expression of thought, and the final session included teaching patients how to administer daily injections (Ref: Figure 7.5).

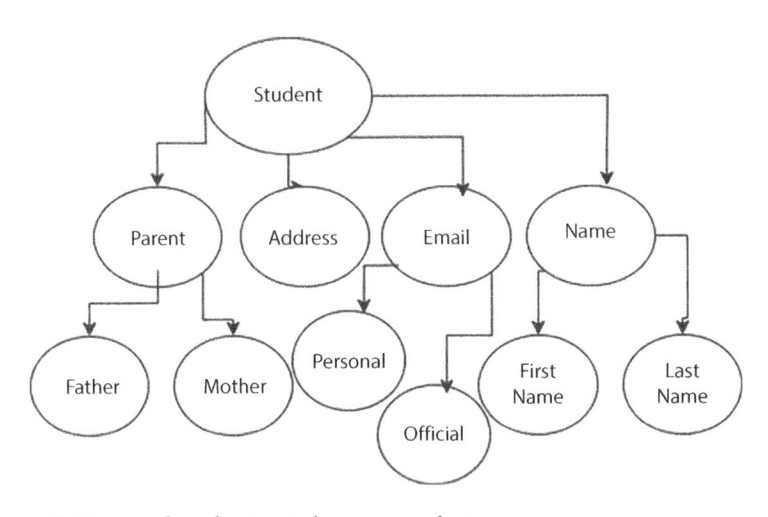

Figure 7.5 Tree topology for Smart classroom student.

Tree topology is beneficial where data segmentation and hierarchical flow are necessary, like administrative structures in education or city governance. In summary, the tree topology's hierarchical structure, segmented information flow, centralized control, scalability, and organized communication paths make it well suited for implementing Smart education models. It streamlines data dissemination, supports centralized management, and facilitates efficient coordination among different levels of the educational system ultimately enhancing the learning environment and administrative processes within a smart education framework.

7.5.4 Mapping of Cluster Topology with Smart Health Care Model

The cluster topology presents a suitable network architecture for implementing a Smart Healthcare model due to its ability to facilitate specialized and intensive communication within clusters while maintaining limited interactions between different clusters [22] (Ref: Figure 7.6).

Specialized Communication Clusters: In a Smart healthcare model, different clusters can represent hospitals, specialized medical centers, or healthcare facilities. The cluster topology allows these clusters to communicate intensively within their domain. For instance, within a hospital cluster, various departments, such as emergency, radiology, and surgery, can communicate efficiently without overwhelming connections to external nodes.

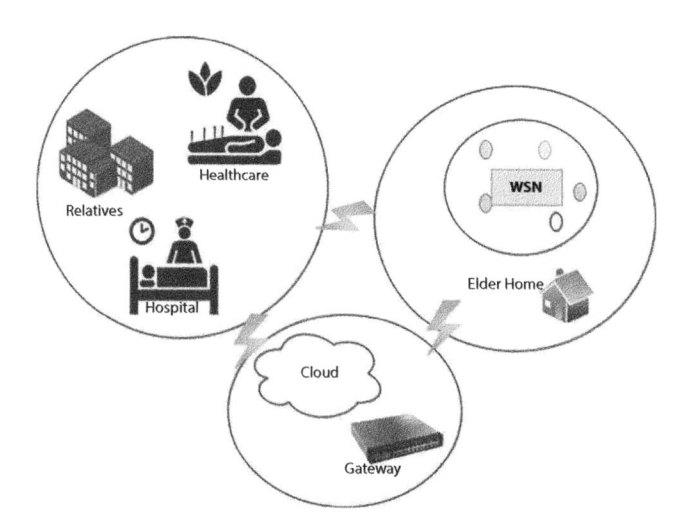

Figure 7.6 Ring topology clustering for Smart healthcare.

Isolated Communication Domains: Healthcare systems often deal with sensitive patient information and specialized procedures. The cluster topology ensures that each healthcare cluster operates within its isolated domain minimizing unnecessary interactions with nodes outside the cluster. This aspect is used to fortify data security and, ultimately, comply with privacy regulations, which are critical in health settings.

Collaborative Specializations: Smart healthcare systems can include clusters representing medical centers and research facilities that are specialized. The cluster topology allows these specialized clusters to collaborate intensively on research, data sharing, or complex medical cases within their specialized domain while limiting communication to other clusters as needed.

Resource Optimization: In healthcare, tighter clusters may need many resources to be shared and communicated; however, some clusters experience less interactions. Cluster topology is essential because it is about how to manage communication traffic within clusters that require intensive communication, thus the use of network capacity and bandwidth becomes even more efficient.

Scalability and Flexibility: With the change in health system, the specialized clusters can be added to the network, or some new healthcare facilities can be set up as a part of the overall network. New clusters and facilities do not disrupt the functioning of the existing clusters. Successful scalability and flexibility of the system allows the introduction of new healthcare entities into the system with the growth of the said service or expansion of new specializations.

Cluster topology was designed for scenarios when nodes inside the clusters need to communicate with each other intensively while communicating less with nodes outside, which can be illustrated by healthcare or industrial sectors that are found individually in a Smart city. Indeed, the cluster topology geared toward on-cluster concentrated and exacting communication and with low-inter-cluster interactions is appropriate when considering Smart healthcare models. Not only does it enhance teamwork within the specialized areas but also guarantees the security and the privacy of the data, optimizes the usage of resources, and undergoes constant evolution to understand the new healthcare scenario under the Smart healthcare system setting.

7.6 Conclusion

We executed a comparative analysis for the four network topologies, such as the star, mesh, tree, and cluster, within the context of Smart agriculture monitoring system, as part of our work. This work has concentrated on performance metrics, such as throughput, delay, power consumption, and lifetime, which proved to be of key importance. Our findings show that the topology cluster is the topmost in the metrics ranking, which is particularly attributable to the advantage brought by the energy saving through the clustering of network nodes. Moreover, even though the cluster proved to be more prevalent, the mesh model still shows some advantages in situations where communication persistence takes the edge over the throughput and delay aspects. What final topology selection is appropriate is the result of what particular application's requirements are as the worthiest of consideration. As an example, networks heavy on traffic with low latency could adopt mesh topology, while networks needing longer lifespan might stick to clustered topology. Finally, cluster topology can be regarded as the best choice for low-power and wide-area networks, which guarantee high throughput, low delay, energy efficiency, and network lifetime. While it is true that the simulation environments have been used throughout the study, we should not forget there are some limitations of the study, including the potential low applicability of the results from the artificial environment to the real-world scenarios. In addition, our study is a basis of further studies that might be done on the other types of network topologies and their effects within the Smart agriculture monitoring system domain.

References

1. Heinzelman, W., *Application-Specific protocol architectures for wireless networks*, Massachusetts Institute of Technology, Boston, 2000, [Ph.D.Thesis].
2. Younis, O. and Fahmy, S., HEED: a hybrid, energy-efficient, distributed clustering approach for ad hoc sensor networks, in: *IEEE Transactions on Mobile Computing*, vol. 3, pp. 366–379, p. 10.1109/TMC.2004.41, Oct.-Dec. 2004.
3. Mamun, Q., A qualitative comparison of different logical topologies for Wireless Sensor Networks. *Sens. (Basel)*, 12, 11, 14887–913, 2012 Nov 5. doi: 10.3390/s121114887. PMID: 23202192; PMCID: PMC3522945.
4. Shrestha, A. and Xing, L., A Performance Comparison of Different Topologies for Wireless Sensor Networks, in: *2007 IEEE Conference on Technologies for Homeland Security*, Woburn, MA, USA, pp. 280–285, 2007, doi: 10.1109/THS.2007.370059.
5. Yueqing, R. and Lixin, X., A study on topological characteristics of wireless sensor network based on complex network, in: *2010 International Conference*

on Computer Application and System Modeling (ICCASM 2010), Taiyuan, China, pp. V15–486–V15–489, 2010, doi: 10.1109/ICCASM.2010.5622543.
6. Mamun, Q., A qualitative comparison of different logical topologies for wireless sensor networks. *Sensors*, 12, 11, 14887–14913, 2012.
7. Zhang, Z., Zhao, H., Zhu, J., Li, D., Research on Wireless Sensor Networks Topology Models. *J. Softw. Eng. Appl.*, 3, 12, 1167–1171, 2010.
8. Jeetu *, S., Reema, C.S., Akanksha, S., Simulation Based Topology Optimization in Wireless Sensor Network. *Recent Pat. Eng.*, 13, 3, 274–280, 2019.
9. Patra, C., Mondal, A., Bhaumik, P., Chattopadhyay, M., Topology Management in Wireless Sensor Networks, in: *Wireless Sensor Networks and Energy Efficiency: Protocols, Routing and Management*, IGI Global, pp. 14–24, 2012.
10. Mohapatra, H. and Rath, A.K., Fault tolerance in WSN through PE-LEACH protocol. *IET Wirel. Sens. Syst.*, 9, 358–365, 2019.
11. Mohapatra, H. and Rath, A.K., Survey on fault tolerance-based clustering evolution in WSN. *IET Netw.*, 9, 145–155, 2020.
12. Wang, B.-Y., Yu, C.-M., Kao, Y.-H.A., Reconfigurable Mesh-Ring Topology for Bluetooth Sensor Networks. *Energies*, 11, 1163, 2018.
13. Alnoman, A., Ring Topology for Balanced Workload in D2D-based Wireless Sensor Networks, in: *2022 International Conference on Electrical and Computing Technologies and Applications (ICECTA)*, Ras Al Khaimah, United Arab Emirates, pp. 245–248, 2022, doi: 10.1109/ICECTA57148.2022.9990309.
14. Jardosh, S. and Ranjan, P., Intra-Cluster Topology Creation in Wireless Sensor Networks, in: *2007 Third International Conference on Wireless Communication and Sensor Networks*, Allahabad, India, pp. 155–160, 2007, doi: 10.1109/WCSN.2007.4475768.
15. Zhao, H., Guo, S., Wang, X., Wang, F., Energy-efficient topology control algorithm for maximizing network lifetime in wireless sensor networks with mobile sink. *Appl. Soft Comput.*, 34, 539–550, 2015.
16. Lei, L., Mingke, F., Wei, H., Research on Topological Structure of Wireless Sensor Network Based on Smart Home Environment. *Journal of Xinyang Normal University (Natural Science Edition)*, 26, 4, 616–619, 2013.
17. Khalily-Dermany, M., Nadjafi-Arani, M., Doostali, S., Combining topology control and network coding to optimize lifetime in wireless-sensor networks. *Comput. Netw.*, 162, 106859, 2019, ISSN 1389-1286, https://doi.org/10.1016/j.comnet.2019.106859.
18. Mukherjee, S. and Mohapatra, H., Performance Analysis of Different MANET Routing Protocols, in: *2024 5th International Conference on Innovative Trends in Information Technology (ICITIIT)*, Kottayam, India, pp. 1–6, 2024, doi: 10.1109/ICITIIT61487.2024.10580492.
19. Mohapatra, H. and Rath, A.K., An IoT based efficient multi-objective real-time smart parking system. *Int. J. Sen. Netw. 37*, 4, 2021, 219–232, 2021.

20. Mohapatra, H. and Rath, A.K., A fault tolerant routing scheme for advanced metering infrastructure: an approach towards smart grid. *Cluster Comput.*, 24, 2193–2211, 2021.
21. Liang, J.-M., Su, W.-C., Chen, Y.-L., Wu, S.-L., Chen, J.-J., Smart Interactive Education System Based on Wearable Devices. *Sensors*, 19, 3260, 2019.
22. Bhalotia, N., Kumar, M., Alameen, A., Mohapatra, H., Kolhar, M.A., Helping Hand to the Elderly: Securing Their Freedom through the HAIE Framework. *Appl. Sci.*, 13, 6797, 2023.

8

Air Pollution Monitoring and Control Via Network Sensing Systems in Smart Cities

S. Sharmila Devi

Department of ECE, Dr. N.G.P. Institute of Technology, Coimbatore, Tamil Nadu, India

Abstract

As the global populace grows more and more urbanized, there is pressure on cities to remain livable. One of the main causes for concern for those in the vicinity is the improved quality of the air in the cities. Thus, to make a city intelligent and livable, it is important to regularly analyze its air quality index. Due to its detrimental effects on both the environment and human health, air pollution is currently a serious worldwide issue. Wireless sensor technology is one efficient means of tracking and managing air pollution. Air pollution monitoring wireless sensor networks rely on low-cost, low-power, multipurpose, small-sized sensors that can transmit data on air pollutants over the internet. This chapter begins by providing the background information on the air pollution monitoring and control system. It then delivers insights into the air quality monitoring techniques. A focus is on the state of use of cutting-edge, reasonably priced wireless sensor technology to monitor air quality and control in smart cities.

Keywords: Populace, air pollution, wireless sensor networks, monitoring and control, smart cities

8.1 Introduction

Globally, air pollution is a serious issue that has an adverse impact on business, the environment, and public health. Even though stationary tracking devices can produce accurate data, their coverage and range are restricted

Email: sharmilasmr@gmail.com

when it comes to traditional air quality monitoring methods. Wireless sensor networks (WSNs) have become a promising tool for air pollution monitoring and detection in recent years. Dense wireless networks made up of inexpensive, small sensors that gather and share environmental data are known as sensor networks. Improved remote tracking and administration of substantial settings is made possible by wireless sensor networks [1]. They are used in a wide range of fields [2–5], including ambient air monitoring, disaster management, indoor climate control, surveillance, medical diagnostics and emergency response in hostile areas. An electronic device known as an air monitor is used to identify and quantify different harmful gases and materials in the atmosphere. These gadgets detect contaminant like impure particles, ozone, carbon monoxide, nitrogen dioxide, and volatile organic compounds using a range of technologies, including metal oxide, electrochemical, and infrared sensors. Pollutants are bad for human welfare and the surroundings. A range of tools are used to identify the contaminated sources and provide decision makers with relevant information for formulating strategies aimed at mitigating air pollution. With this information, both individuals and organizations can reduce their personal emissions to air pollution.

There is demand for cities to continue being habitable as the world's population moves closer to metropolitan areas. The number of health issues linked to poor air quality is expanding, including heart disease, lung cancer, stroke, and respiratory conditions like asthma [6]. Children, asthmatics, expectant mothers, the elderly, and other vulnerable members of society are all at serious risk from poor air quality. Thousands of prospective mortality instances are identified worldwide every year caused by contaminants in the atmosphere, according to WHO statistics [7]. As a result, in recent times, urban air quality has emerged as a primary global reason for worry [8]. For a city to be smart, its air quality index must therefore be continuously monitored [9] and livable [10–12]. Large databases are created by air pollution monitoring, which must be appropriately handled and promptly distributed. Cloud computing is being used more and more to manage the data from air pollution monitoring since it may offer extensive online data access via the internet and preservation at a distant server outside the sample region. A number of technical tools are used by the air quality management system [13] to give details about the state of the air. Three essential technological instruments are modeling, air quality monitoring, and inventory of air pollution emissions. Such methods produce sizable datasets on air pollution, which aid in illuminating the complex connections among atmospheric air levels and the sources of emissions of air pollution, as well as relevant impact effects on the well-being of ecosystems, climate change. Data retrieved from these databases are helpful in comprehending the state of the air in a specific place, allowing for the development and implementation of efficient policies aimed at reducing source emissions and mitigating any negative consequences. This chapter offers insights into the classic and sophisticated monitoring methods for pollution in the environment emphasizing the newly developed field of small, inexpensive wireless sensors. The current state of compact, inexpensive wireless sensor technology and its applications for air quality monitoring are highlighted.

8.2 Related Works

The design, implementation, and effectiveness of wireless devices have been the focus of numerous researches on their application in reducing air pollution. Even though it is seen as a highly difficult task, monitoring air pollution is crucial. In the past, data loggers were used to periodically gather data, which took a lot of time and money. Air pollution monitoring can be simplified, and results can be obtained more quickly with the use of WSNs [14, 15]. Among the primary problems to be addressed is the vehicle powered by the Internet of Things (IoT) with anti-collision and emissions management systems [16]. They stress how crucial it is to have a robust framework in place to deal with these issues. The authors review published literature on IoT-based solutions for preventing accidents and controlling pollution. They look into the many components and techniques employed in these systems to fill in any holes left by the literature. The authors' investigation led them to design a system of the IoT comprising many sensors, microcontrollers, GPS, and communication devices. When neighboring automobiles are detected, the intent of the equipment is to alert the motorist of an impending incident. The concept of utilising air quality data to forecast traffic congestion has been established [17]. This is achieved by emphasising the detrimental effects of traffic congestion on air quality, such as increased levels of impure particles, ozone, carbon monoxide, nitrogen dioxide, and volatile organic compounds. The authors note that these pollutants are being linked to several health issues, including respiratory ailments and heart disease. To address this issue, this research proposes an algorithm that uses data on air quality to forecast traffic jams. The underlying premise of the model is both the cause and the effect of poor air quality on traffic congestion, and that trends and degrees of congestion may be predicted by analyzing air quality data. The findings of the research are presented in the article. The design of a wireless air pollution tracking sensor node system [18] provided an explanation of how inexpensive it is and to widely distribute wireless sensor nodes. The gadget features a sensor node to monitor elements of air quality like particulate particles, carbon monoxide, and ozone. High-resolution maps of air pollution are produced by processing, analyzing, and creating the sensor data electronically on a central computer. The authors also address the challenges of developing such systems, such as the need for sensor calibration, power management, and reliable wireless communication. They also discussed the use of artificial intelligence. Every 15 min, contaminant amounts are determined in ref. [19] utilizing metal oxide and electrochemical tracks. They transfer the data utilizing the GPRS algorithm by means of a GPRS modem. The gadget gets its power from solar panels. Ref. [20] describes how to use gas monitors to measure concentration readings and to send the data to the cloud; an Arduino and a Wi-Fi adapter are used. In ref. [21], a number of sensors were used for observation. The data the sensors gathered were wirelessly sent to a central database so that it could be updated and reviewed right away.

In recent years, WSNs cover attention from all over the world. The conception of intelligent sensors has been made possible by Micro-Electro-Mechanical (MEMS) technology systems [22]. They are less expensive than typical sensors, smaller, and have less processing and computational power. They have the ability to sense, quantify, and gather data from their surroundings. Then, using certain localized processes for making decisions, they are able to provide the user with information. There have been numerous researches on the application of gas sensors in WSNs, such as a device to monitor and avert potential coal and gas explosions or a WSN to identify gas leaks. Nevertheless, there are memory, energy, processing, communication, and scalability constraints with WSNs. Simultaneously, cloud computing is emerging as a potentially very powerful solution for software, storage, and processing services.

An Android application can be used to view the data transferred in real time over the internet between the devices for air quality monitoring. Since there is only one installation expense, it minimizes the economical mobility of system hardware at several sites. With its extensive set of characteristics, in IoT-based applications, the Raspberry Pi 3B serves as the backbone of the system [23]. In addition to gathering data from a range of sensors via an embedded Wi-Fi module, it is responsible for sending the collected data to ThingSpeak [24], an open source cloud platform where data may be saved and accessed via hypertext transfer protocol (HTTP) over the internet. Thus, the selection of an appropriate data having a sensing and monitoring system is crucial to the successful implementation of IOT in a given application. It is feasible to have digital control over the physical devices, and Machine to Machine (M2M) communication is a means of communication between the devices [25]. The scholarly community is very interested in location-based routing in WSNs, particularly as it is flexible. As location-based routing imposes fundamental constraints on routing decisions, it is possible to grow network capacity without increasing signaling overhead [26].

Context-aware technology, wireless sensor networks, and mobile communication technology are combined in the system outlined in ref. [27] to develop a customized health information service's mobile web. The method is built on a cloud computing infrastructure. The service consists of two health information recommender systems: a collaborative system and one based on physiological parameters. In ref. [28], a WGSN for monitoring indoor surroundings and quickly reporting alarms in the event of malfunctions and leaks is presented. Modern wireless electronic boards and inexpensive off-the-shelf components, like chemo resistive MOX sensors, are used. It takes advantage of the sensing element's transient reaction to cut the power needed to measure indoor air quality by 20.

8.3 Air Quality System

Broadly speaking, the atmosphere can be thought of as an extremely thin gaseous layer in which all of the climatic events that govern human life occur. The atmosphere, which is full of the wide variety of chemicals listed in Table 8.1, serves both as a protective barrier and a regulatory system. The term "air pollution" can mean several things. A contamination or contaminants present in the external atmosphere are referred to as air pollutants or their combination in amounts or over time, which have the potential to turn dangerous for the lives of people, animals, plants, or things. Hazardous chemical products, smokes, steams, paper hashes, dusts, soot, carbonic smokes, gasses, fogs, or radioactive substances are some examples of air pollutants. Secondary pollutants are produced when certain air pollutants combine with one another. Ozone is created mostly in the summer, when temperatures are greater, when volatile organic molecules, like carbon monoxide and nitrogen oxide, which are produced by car engines, dissociate due to the sun's action.

Air pollution has detrimental effects on both human and environmental health, including ozone depletion, the impact of acid rain, and the global warming phenomenon depletion. It can cause respiratory issues, even death. The European Community has focused specifically on the issue of the most representative pollutants concentration in the situations

Table 8.1 Average composition of pure air.

Element	Symbol	Proportion
Nitrogen	N_2	78.08%
Oxygen	O_2	20.94%
Argon	Ar	0.943%
Carbon dioxide	CO_2	340 ppm
Neon	Ne	18.18 ppm
Helium	He	5.24 ppm
Methane	CH_4	1.5 ppm
Krypton	Kr	1.14 ppm
Hydrogen	H	0.5 ppm
Nitrous oxide	N_2O	0.4 ppm
Xenon	Xe	0.09 ppm

of carbon monoxide (CO), nitrogen dioxide (NO2), sulfur dioxide (SO2), ozone (O3), and particles of 10 μm or smaller (PM10). As a result, special regulations have been created. Despite not being regarded as a pollutant, carbon dioxide (CO_2) must, nevertheless, have its concentration measured because of how vital it is to the ecosystems of the earth.

8.4 Air Quality Monitoring Techniques

The process of measuring the concentrations of air pollutants involves either advanced *in situ* monitoring techniques or typical standard equipment, which involves sampling and analysis afterward using online sensor systems and remote sensing (like satellites), or a combination of multiple approaches. Observing information is utilized to appraise the threats that atmosphere contamination disclosure poses to ecosystems and persons' physical condition. Legislative organizations carry out regulatory monitoring for specified pollutants in accordance with the parameters conforming to the national ambient air quality standards with regard to extent technique, average instant frame, and frequency of monitoring. Regulatory monitoring's primary goal is to ascertain whether the local air quality satisfies the NAAQS requirements so that, in the event that it does not, appropriate regulatory action can be taken to minimize pollution. Providing monitoring data is necessary to make sure that relevant parties, such as people, are informed about air quality in a timely manner, to prevent overexposure to pollutants during bouts of pollution, or to authorities, for the implementation of direct events.

Just the amount of air pollution present when the monitor was placed there and for the entire measurement period is provided by the monitoring equipment. Thus, in terms of geography, a network is made up of several monitoring locations that are necessary to give data regarding the degree of air pollution within a specific geographic area, such as a city, province, country, or region. To evaluate, for instance, the effectiveness of particular initiatives on air quality, monitoring should produce data constantly over a long period of time at a fine resolution (such as hourly). Large databases on air quality are thus produced by the monitoring activities. As an example, consider a petite town by three observational sites. The hourly readings of the criterion pollutants (CO, PM2.5, PM10, O_3, NO_2, and SO_2) are provided by these stations in addition to meteorological data (temperature, wind direction, speed, and atmospheric pressure). Over the course of a year, the sum of these data points would produce an enormous quantity of hourly data points. Data management is a major difficulty in a nation with up to thousands of tracking units to ensure the accuracy of information, preservation and recovery, and exposure. Therefore, so as to handle, analyze, and disseminate the data of air quality, effective information management strategies are needed; as a result, cloud computing (CC) may be crucial.

8.5 Conventional Air Pollution Monitoring

In the past, air pollution has been tracked using traditional systems for tracking air pollution using fixed sensors. These monitors have two primary modes of operation: automated continuous monitoring and manual monitoring. They are very dependable, precise, and able to quantify a broad variety of air contaminants.

8.5.1 Manual Measurement and Evaluation of Air Quality

Handheld preset displays were one of the initial types of apparatus for monitoring air quality. Sample collection, or sampling, and the ensuing laboratory examination of samples are the two distinct processes involved in manual air quality monitoring. A pump or an evacuated container, or any other air-moving device, which makes it simple to determine the sampled air volume, is used in the active sampling technique. In this instance, the time of sampling is typically brief—between less than an hour and a day, for instance. The goal of the sampling and the target pollutants determine whether the kind of sampling device is most suited for collecting bulk/whole air samples or selectively collecting specific chemical samples. Using a reservoir, needle, or testing pouch, the real air blend is collected next to the sample spot so as to get bulk air samples. By utilizing an activated carbon sorbent trap to hold volatile organic compounds (VOCs) or a filter to capture particulate matter (PM), selective air sampling preserves the pollutants of interest on a specific sampling medium.

In contrast, no air-moving apparatus is used in passive air sampling. As an alternative, the pollutants are gathered using physical concepts such the particles settling on an alternative collecting surface or gasses diffusing in diffusive samplers. Since passive sampling often gathers samples over a longer time frame—2 to 3 weeks, for instance—the time-weighted average for the length of the sampling period is the result. Given that it is challenging to estimate how much air passed through the collector during passive sampling, meteorological variables should be used to calculate it.

8.5.2 Automated Continuous Monitoring Devices

The United States of America (USA) has been using continuous monitors to measure SO_2, NO_2, CO, O_3, and PM since the early 1970s to verify compliance with the US NAAQS. These devices produce hourly average pollution readings that are consecutive and are employed in automated air monitoring facilities across the globe. A sample pump is used by such an autonomous monitor to gather and transfer a certain air volume to an analytical device that subsequently offers measurements. Automatic monitors either detect the mass surrogate of PM using optical techniques such light scattering and light obscuration, or they convert PM mass concentrations to PM mass concentrations by using the beta ray absorption of PM that has been collected on a filter.

This conventional equipment for gases and PM, however, are costly, huge, and heavy. More specifically, the maximum number of monitoring stations that may be positioned in a particular area is limited by the high cost of the equipment. Herein, choosing representative sites for measuring air pollution is crucial, but in a densely populated urban region with numerous emission sources, this is an extremely difficult process. For instance, Thailand's 60+ automatic monitoring stations fall into two main categories as follows: (i) general sites, which are separated further into roadside areas (3–5 m away from the traffic lane) and residential, commercial, institutional, and industrial properties. The low spatial resolution of air quality information provided by the thinly distributed governmental monitoring networks in many Asian countries is not an appropriate representation of the highly variable levels of air pollution in a domain that will be used later in research assessing the health effects.

8.5.3 Monitoring Air Quality with Sensing Technology

PM sensors usually use the optical approach (illumination scattering, illumination blocking, etc.), but numerous principles—summarized in Table 8.2—may be applied to gaseous pollutants. The most affordable, the least expensive dispersed light PM sensors available for measuring concentrations of specific particles are limited to observing particles with sizes between approximately 400 and 10,000 nm. Ultrafine particles (particles less than 100 nm in diameter) cannot be measured by any inexpensive sensor currently on the market. In addition, PM sensors displayed inconsistent performance at high air relative humidity (RH), such as declining performance at RH ~80%–85%. To be useful in tropical regions with high humidity, future PM sensors need to appropriately account for the effects of relative humidity. It should be noted that a number of conventional (and expensive) PM monitoring devices likewise make use of the optical principle; however, to make accurate measurements, they usually keep the relative humidity in the inlet constant and employ particular technologies to track extremely microscopic granules; inexpensive PM sensors typically lack all of these characteristics.

The five categories of gas sensors are as follows: non-dispersive infrared absorption, metal oxide semiconductor (MOS) sensors, catalytic sensors, electrochemical sensors, and tiny photoionization detector (PID) sensors. By utilizing the processes of oxidation and reduction involving the target gases and the electrochemical within the sensor, electrochemical sensors measure gaseous concentrations. The target pollutant's concentration is correlated with the electric signal generated. Temperature and relative humidity appear to interfere with electrochemical sensors. With the help of a catalyst that is placed on the sensor's surface, the basic principle of catalytic sensors is to burn the target gases at a temperature much below their normal ignition point. One or more metal oxides make up a MOS sensor then observe the electrical shift that happens as the target gas adsorbs on the metal oxide coating's surface, then uses that information to calculate the concentration of the gas. MOS sensors are susceptible to

Table 8.2 Various sensor technologies that detect air pollution.

Pollutants	Example products	Measurement principle	Measurement range
CO	Alpha sense B4 series CO sensor	Electrochemical sensor	0–1,000 ppm
	MiCS-5525	Semiconductor	0–1,000 ppm
	MQ9	Semiconductor	0–1,000 ppm
	Hanwei MQ-7 CO sensor	Solid state sensor	20–2,000 ppm
NO_2	Alpha sense B4 series NO_2 sensor	Electrochemical sensor	0–20 ppm
	SGX SensorTech MiCS-2714 NO_2 sensor	Solid state sensor	0.05–10 ppm
VOCs	AH2 photoionization detector	PID	0.01–50 ppm
	A12 photoionization detector	PID	<0.05–6,000 ppm
CO_2	MG-811	Semiconductor	350–10,000 ppm
	USA GE/6004/6113	NDIR	0–2,000 ppm
	Korea ELT H550	NDIR	0–10,000 ppm
	Japan FIGARO CDM4160	Solid electrolyte	400–45,000 ppm

environmental factors and can be affected by other gasses. The principle of operation for the NDIR sensor is absorption spectrometry, which states that certain wavelengths of infrared light are absorbed by target gases; a detector measures certain wavelengths' attenuation to determine the gas's concentration. PID sensors monitor the electric current generated in relation to the gas concentrations by producing ionized gas molecules, both positive and negative ions, from the target gases using a UV light source. PID sensors are frequently used to measure volatile organic compounds (VOCs), but they have limits because not all VOCs in the air are ionized as efficiently by them.

PID signals are therefore dependent on the VOC mixture being analyzed because chemicals compounds are easier to identify than others when they are effectively ionized. In essence, the sensors convert the physical parameters of the surroundings—the amounts of air contaminants, for instance, into electrical impulses. Analog-to-digital converters (ADCs) and sensors are the two main components of sensing units. Sensor analog signals, which represent the observed conditions, are converted into digital signals by analog-to-digital converters, or ADCs, and sent keen on the operation entity. The output signals from the sensor calibration must be done using a standard of reference equipment to acquire the measurement data in the concentration unit. As a result, a calibration curve is created that links the electrical signal reaction of the air sensor to the concentrations of standard gases or the information from the reference instrument. Calibration of the sensors using laboratory-derived gas standards is the most appropriate way. Another popular calibration strategy is the colocation technique, which involves positioning a sensor node next to a reference air quality monitoring device. However, for this method to work, the two devices must be near enough to each other to measure the same air bubble. Colocation is also necessary to confirm the sensor's functionality.

8.6 Wireless Sensor Network for Air Monitoring

8.6.1 Wireless Sensor Networks

A typical sensor network consists of sensors, a controller, and a communication system, as shown in Figure 8.1. WSNs are known to be the sensor networks, if wireless protocol is used to implement the communication system.

8.6.2 WSN Network Topologies

A wireless sensor network (WSN) may be single hop or multihop in nature. Numerous nodes make up the network, and they communicate with one another through the exchange of data over communication links. The network protocol we utilize for communication is called Zigbee. The Zigbee network model allows mesh, star, and tree topologies, as seen in Figure 8.2. The Zigbee coordinator is in charge of initiating and maintaining all other devices connected to the network. Zigbee end devices and Zigbee routers can establish direct connections with the Zigbee coordinator, contingent upon the network topology in use. In topologies, like mesh and tree, the

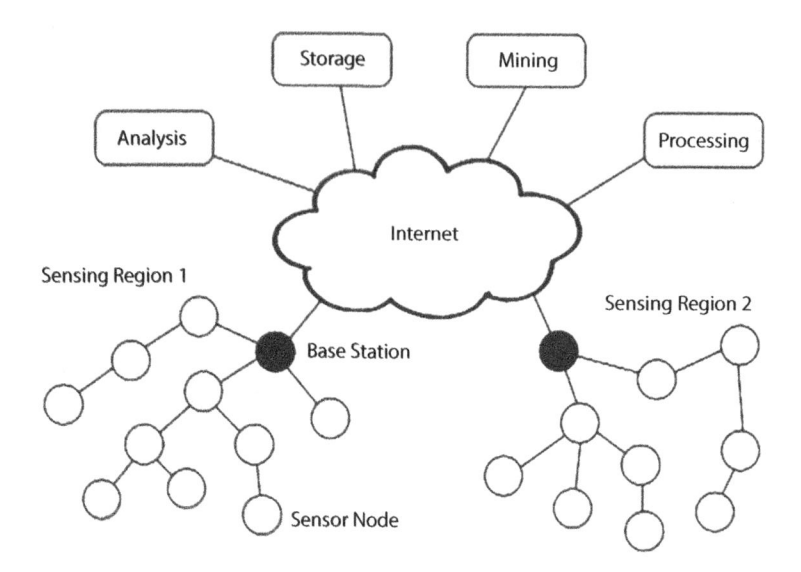

Figure 8.1 Wireless sensor network.

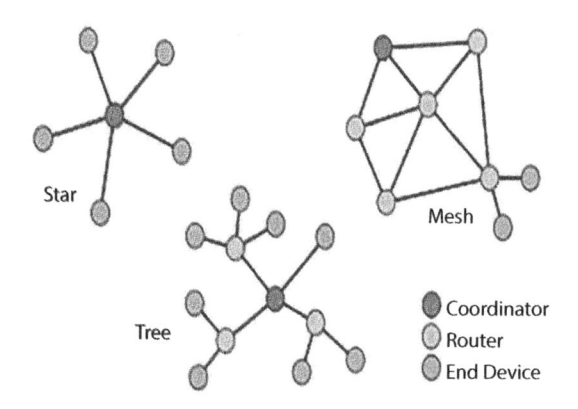

Figure 8.2 Network topology.

coordinator is in charge of setting up the network with default settings and selecting a few important network characteristics, but routers can also be used to expand the network.

Star Topology
In a star topology, a hub, router, or switch serves as the central device to which all other network devices are linked. Every workstation is connected

to the central equipment using a point-to-point link. Therefore, it may be said that every node uses the "hub" to establish indirect connections with every other node. Before any data on the star topology reach its final destination, it must first transit via the center device. The hub not only maintains and controls the entire network but also serves as a connector between the various nodes that make up the Star Network. Hubs can function as repeaters or signal boosters based on the type of central device being used. The central device has the ability to communicate with other hubs in the network.

Tree Topology
The tree topology is a more comprehensive variant of the bus topology. A tree topology combines the characteristics of the star and linear bus topologies. It consists of workstation clusters connected by a linear bus backbone wire that are arranged in a star shape. Networks that already exist can be expanded thanks to tree topologies.

Mesh Topology
Any device linked to a mesh topology is able to attempt communication with any other device, either directly or through the use of devices that may route messages. The route from the source device to the destination in this topology is generated dynamically plus is adaptable to changes in the surrounding environment. The dynamic creation and modification of routes by a mesh network improves the dependability of wireless communications, in case the network's routing-capable devices can work together to discover a different path if the source device is unable to interact with the destination device via a previously established route for some reason.

8.6.3 Zigbee Standard

For numerous real-time applications, a low-power, low-data-rate, short-range wireless networking technology called Zigbee is used. It describes the three lowest layers (Physical, Data Link, and Network) in addition to an Application Programming Interface based on the seven-layer Open System Interconnection paradigm for layered communication systems. The alliance industries' tiered architecture is seen in Figure 8.3. It should be mentioned that the Zigbee Alliance decided to employ physical layer and data link specifications that were already in place. Low-rate personal area networks (PANs) using these specifications is compliant with IEEE 802.15.4 standards.

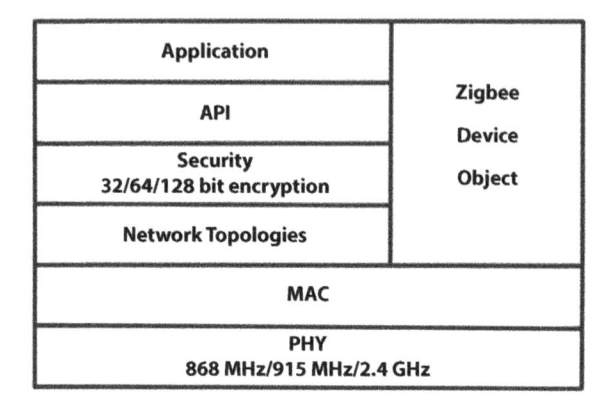

Application	Zigbee Device Object
API	
Security 32/64/128 bit encryption	
Network Topologies	
MAC	
PHY 868 MHz/915 MHz/2.4 GHz	

Figure 8.3 Zigbee communication layers.

The IEEE 802.15.4 standard supports three working frequency bands: 2.4 GHZ (global), 868 MHz in Europe, and 916 MHz in the US. Throughout the world, 2.4-GHz bands are most commonly used due to the ISM (Industrial, Scientific, and Medical) spectrum. Additionally, this band offers 16 communication channels with a maximum data rate of 250 kbps at the physical layer, ranging from 2.4 to 2.4835 GHz. Typical communication distances can range from over 100 m in a line-of-sight scenario to as short as 30 m in an indoor, non-line-of-sight location depending on the module characteristics. Dipole antennas are used by Zigbee modules to increase antenna gain.

8.7 Architecture of Wireless Sensor Networks

WSN, which can be used for socioeconomic or environmental monitoring, is become a reality as a result of improvements in electronics and wireless communications technologies. This is used for environmental monitoring, covering air and water quality as well as natural calamities such volcanic eruptions, forest fires, and landslides. The WSN uses tiny, multifunctional, low-cost, low-power sensors that can send air pollution data over the internet to monitor air pollution.

An application layer, a control layer, and a wireless sensor network layer comprise a typical WSN. To measure the surroundings, sensor nodes are positioned in the first layer. A sensor node consists of the following three parts: transceiver equipment, microcontroller or microprocessor (μC), and sensor(s). A server receives the measurement data via an internet-connected wireless access point in the second layer. Finally, the application layer (server) receives the data, processes it, and stores it in a database. By consolidating the database server, web server, and wireless sensor network gateway node into a single single-board computer hardware platform, the system architecture reduces the modular WSN technique's expenses and complexity. To facilitate users' quick access to the system's online interface, a web application was built. To enable users to quickly access the online edge of the scheme, a web application was built to update the sensor data or condition of any sensor node at anytime, anywhere; users can communicate with the web application using the internet or a local area network. Figure 8.4 shows the basic layout of the wireless sensor network for air quality monitoring.

In situations when a sizable number of these sensor nodes are dispersed over a large region to monitor a physical environment, networking becomes even more important. Within a wireless sensor network (WSN), a sensor node communicates wirelessly with a base station (BS) and other sensor nodes. After getting directives from the base station, the sensor

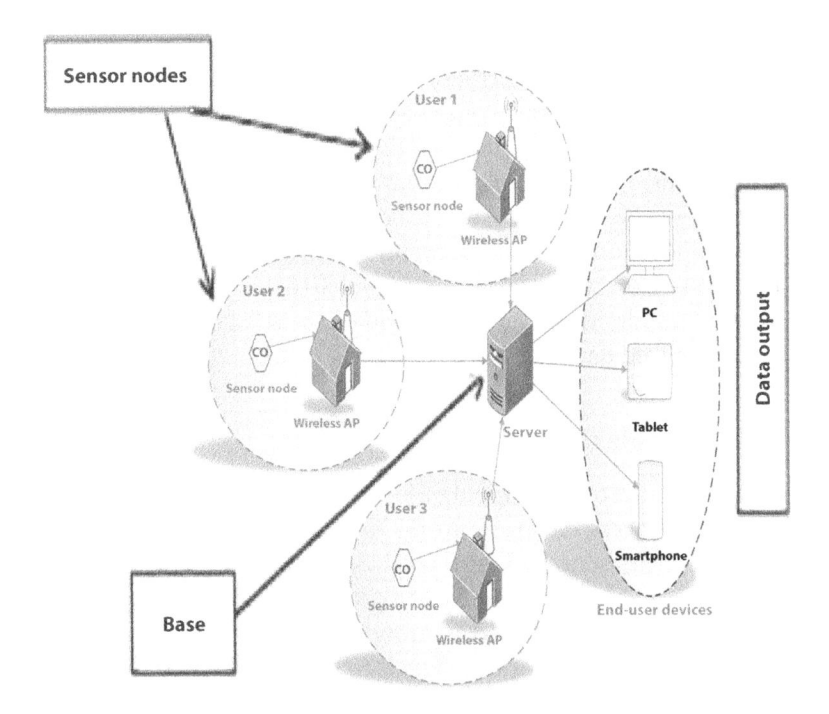

Figure 8.4 Wireless sensor network architecture.

nodes collaborate to finish duties. The sensor nodes then transmit the data back to the base station. A base station can also be used as an internet gateway to connect to other networks. The most recent information is sent to the user via a base station, which receives the data from the sensor nodes, processes it somewhat, and connects to the internet. When every sensor node is connected to the base station, the single-hop network architecture is employed. Although long-distance transmission is possible, the energy needed for communication will be far more than that needed for data collection and processing.

There are numerous locations where WSN has been effectively implemented to observe the surroundings.

8.7.1 Fire and Flood Detection

There are many different environmental uses for WSNs. All the way through to identify source of forest fires, sensor networks are installed throughout forests. A flood detection system uses weather sensors to identify, forecast, and ultimately stop floods. The habitat is equipped with sensor nodes to track biodiversity. The Forest-Fires Surveillance System aims to prevent forest fires in South Korea's mountains by providing early fire alerts in real time. The method uses a formula to calculate the risk level of forest fires based on environmental factors like temperature, humidity, and smoke. It is feasible to detect heat early on, enabling the real-time provision of an early warning when a forest fire begins, instructing people to put it out before it spreads. As a result, it prevents financial loss and environmental harm. Similarly, the ALERT system used in the US is an example of a typical WSN application for flood monitoring and prevention. This system uses weather sensors, water level, and rainfall to detect, anticipate, and ultimately avoid floods. These sensors provide data in a predetermined manner to a centralized database system.

8.7.2 Biocomplexity

Precision Agriculture and Mapping Monitoring the air, soil, and water can be done with wireless sensor networks to manage the surroundings. The field is covered with sensors, which are arranged in a network and connect with each other to eventually arrive at a processing center. From there, the center analyzes the data received and modifies the environment as needed. For example, the center transmits signals to actuators, which then identify and turn on the sprinkler system, if the soil is too dry. The biocomplexity mapping system aids in environmental control. The geographic complexity of dominating plant species is viewed using sensors. The monitoring of the marine subsurface is one example, where building offshore winds farms requires an understanding of the processes that lead to erosion. A developing WSN application field is precision agriculture, which uses sensors to measure and manage the amount of pesticides in drinking water, soil erosion, and air pollution. A field's soil, crop, and climate can all be monitored as part of precision agriculture. In such an application, large-scale farming areas often generate enormous volumes of sensor data.

8.7.3 Habitat Monitoring

The use of WSNs has mostly allayed worries about how human presence may affect the monitoring of plants and animals in the field. These days, sensors can be positioned on small islets before the breeding season starts. During the dormant season, during the freezing season, or anywhere else, it would be risky or unwise to conduct field study continuously. When contrasted with conventional personnel-intensive approaches, which necessitate a large investment in infrastructure and logistics for the upkeep of field investigations, frequently at considerable pain and occasionally at actual risk, such deployment represents a significantly more cost-effective method of conducting research.

8.7.4 Factors Influencing the Efficacy of Inexpensive Sensors in the Monitoring of Air Pollution

To lessen the constraints on the temporal and spatial coverage of traditional air monitoring, tiny, inexpensive sensors are being created in support of a variety of ambient air monitoring applications. These sensors' main benefits are their inexpensive cost, lightweight, compact size, low energy consumption, and ability to be widely dispersed throughout an area. Additionally, they require little user knowledge, which enables widespread community involvement and awareness raising. They offer measurement data in almost real time, which is easily shared online.

As a result, the adoption of low-cost, wireless sensors broadens the monitoring data's temporal and spatial distributions, which is crucial for managing air quality generally and assessing its effects on health. These sensors' precise spatial resolution can assist in identifying hotspots with high concentrations so that appropriate action can be taken to manage air quality. Thus, fascinating new potential atmospheric applications are made possible by the low-cost sensors.

Before they can completely replace traditional displays, a few significant constraints need to be addressed. When compared to traditional equipment, the primary limitations of these wireless air monitoring sensors are limited detection range, poor accuracy, and unreliability. The external elements, such as humidity, wind, temperature, precipitation, insects, etc., have an impact on the signals. For instance, it is common to report that NO_2 sensor signals are interfered with by ozone interference.

Compared to normal reference equipment, smaller and/or less expensive sensing devices typically have lower sensitivity, precision, and chemical specificity to the substances of interest. Although common measures for the sensors' performance and data quality assurance are still lacking, sensor calibration is still necessary to guarantee the quality of the data. The cost would increase if the sensors were to be utilized in any way similar to the current standard reference equipment, such as for routine validation, retention of data, methods for QA and QC, etc.

Many factors, including fault tolerance (the ability to maintain sensor network functionalities without interruption), scalability, hardware limitations, power consumption, and atmospheric conditions can affect a sensor node design's performance.

8.8 WSN-Based Air Pollution Monitoring in Smart Cities

Among the primary reasons for pollution in the environment, which has a detrimental effect on human health, is thought to be air pollution. Because of global air pollution and climate change, climate scientists and environmentalists are extremely concerned. Because of the numerous deadly gases that are released into the atmosphere, marine life as well as the city's ecology are at risk due to transportation and industrial pollution. The global population shifted from rural to urban areas throughout the previous 50 decades in search of better opportunities for employment and healthcare. As a result, cities are under pressure to accommodate the growing population's needs for daily amenities. At the same time, environmental contamination is growing quickly making cities unhealthy due to pollution from industry and roads. The number of health problems that low air quality causes is rising, including heart disease, lung cancer, stroke, and respiratory infections like asthma. The population of large cities faces serious health risks as a result of poor air quality. Environmental problems cause millions of early deaths worldwide each year according to WHO reports.

An investigation of environmental pollutants using a wireless sensor network makes it easy to understand smart environment systems. Consequently, when wireless devices are deployed through WSNs, researchers have established norms and standards for WNS-based EPA systems, which are essential for appropriate SEM systems. The current suite of smart city services that focus on air quality monitoring is essential. These services include traffic monitoring, noise monitoring, weather and garbage monitoring, energy management, and environmental monitoring. The most urgent problem is managing the numerous sensors that have been installed to collect trustworthy data. Regarding the implementation of smart cities, privacy and security rank among the top ethical issues. A lack of security can lead to many consequences such as fraud and intimidation.

Global powers are concentrating on maintaining a better environment and holding several meetings and conferences to lower carbon emissions. It is crucial to routinely check the air quality index to maintain a healthy and habitable atmosphere. Governments are attempting to construct smart cities to monitor the environment and traffic to give citizens a healthy lifestyle. Various government agencies are constructing communication networks with wireless sensor network assistance to construct smart cities. To mitigate environmental pollution and maintain a healthier atmosphere, intelligent systems and smart wireless sensors are placed to continuously check the air quality index. Cities have implemented devices for monitoring real-time air quality. So as to take additional steps for the improvement of the urban environment, real-time data are gathered from installed sensor systems and examined at the central office. While the one-time installation saves money, it lessens the hardware's ability to move around multiple areas for monitoring. The sensor applications in smart cities are depicted in Figure 8.5.

Examining WSNs for smart automobile screening systems are done. The vehicle was equipped with over 20 moveable nodes that sensed temperature, humidity, CO, and NO_2 levels in the atmosphere. By deploying WSNs in smart cities, the design is possible to enhance the handling of garbage and reduce pollution keeping an eye on the structural integrity of buildings. More and more systems are being developed with the ability to combine data from multiple sources, including multi-sensor networks. Maintaining the sustainability and durability of the world economy over time depends heavily on ecological preservation.

Several well-known cities have already adopted the Internet of Things (IoT) in the modern era to boost global resilience against climate change and revive economic growth. The idea of the smart city is a method for utilizing organizational structures, sensors, data, and applications to make

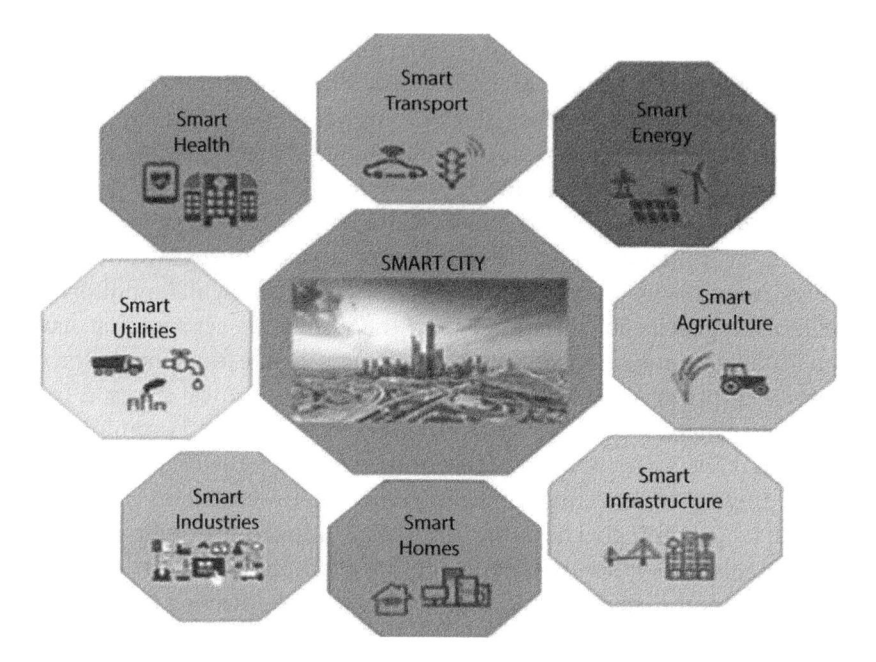

Figure 8.5 Sensor applications in smart cities.

cities more resilient to climate change on a global scale. Next, it was suggested that networks for public–private cooperation be established, that investments and initiatives related to smart cities be promoted, and that smart services should receive more attention. High-tech sensors known as wireless sensor networks (WSNs) are monitored and controlled via the cloud. It has been demonstrated that integrating cloud-connected devices with WSNs improves waste management, pollution prevention, vehicle identification, and temperature control. These technological advancements consist of WNS-based environment systems, mobile health monitoring, and remote sensing. By promoting sustainable development practices, technology for information and communication is used in smart cities to address the growing challenges associated with urbanization. Critical needs for smart city applications include confidentiality, integrity, anonymity, and safety. This infrastructure's data management interface strongly reflects these requirements. The results of the simulation research demonstrated the scalability, speed, accuracy, and safety of the suggested framework.

The structure of air pollution is depicted in Figure 8.6. Many networked gadgets provide information to a smart city's data collecting layer. The entire city is covered in sensors that monitor ozone, sulfur dioxide,

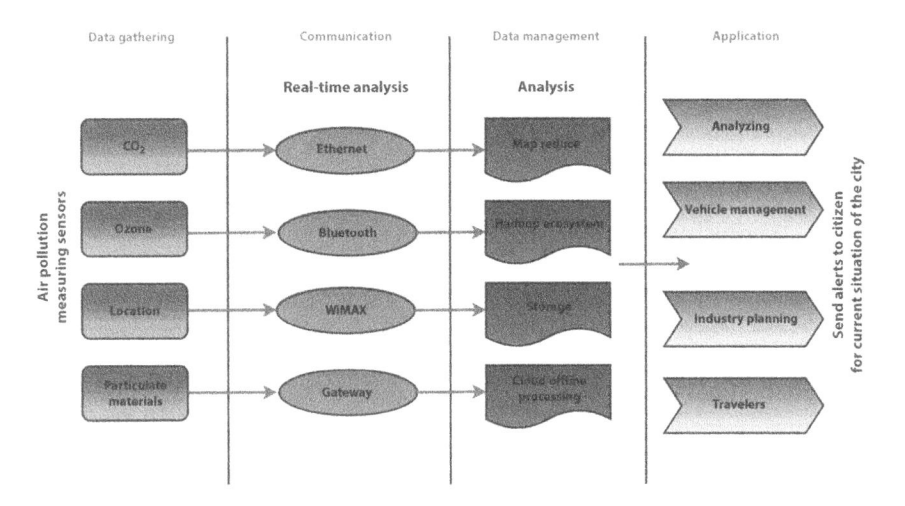

Figure 8.6 Structure of air pollution.

nitrogen dioxide, and particulate matter. Since there are numerous data sources, this is where the gathering and aggregate process happens. Here, pre-processing and filtering are carried out because the data may take on many formats. The process of finding and eliminating unnecessary data is known as pre-processing. Data gathered must be shifted to the layer from the accumulating layer after it via an intermediary communication layer. LTE, Wi-Fi, gateways, and other communications technologies comprise this layer. Every sensor's data are sent to the layer of data processing from this point on. This layer can be used with gateways that have real-time processing capabilities. Computing can be used to reduce fog latency. This is where decisions can be made in real time. The main layer for storing and analyzing data is data management. Several third-party solutions can be combined here because analysis requires instantaneous computing. A smart city's data collecting layer compiles information from a variety of networked devices, including sensors.

Figure 8.7 shows air pollution using sensors that are wireless. A wireless sensor network, which acts as an interface between sensors and the data they gather, is the central component of the system. This is a splendid illustration of a smart city that makes use of an SEM system to maintain the comfort and safety of its citizens. Among the crucial aspects of enabling the extended sustainability and viability of the global market is environmental preservation. Environments free from hazards and pollution are crucial for both public health and national growth. Accordingly, experts make use of information to perform thorough research and support executive.

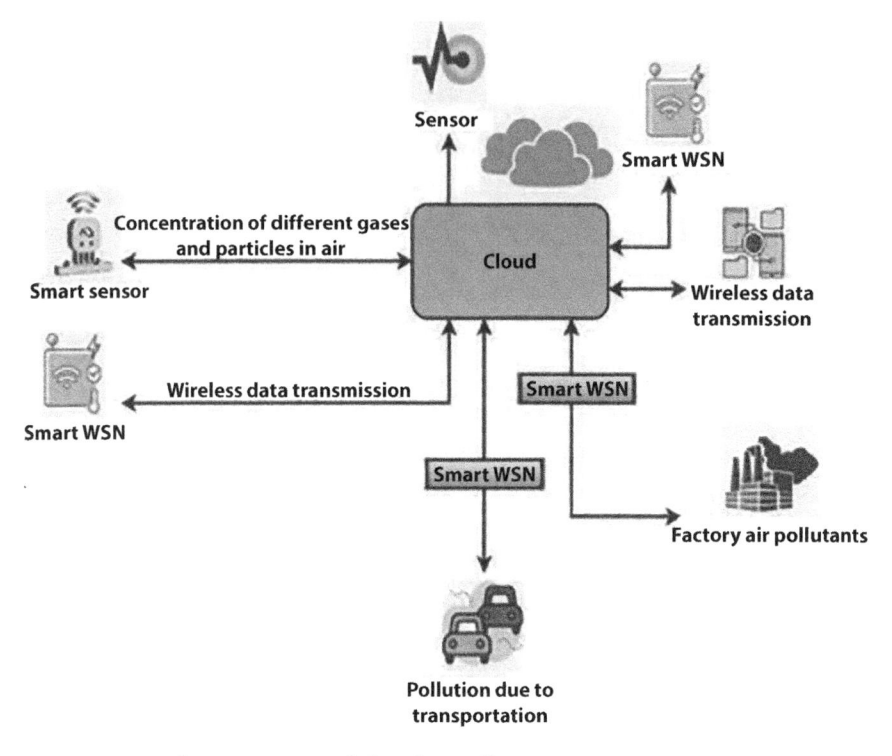

Figure 8.7 Wireless sensor network-based air pollution.

Data can be used by specialists and researchers, including local governors, to spot long-term patterns and support scientific investigations.

Smart city concepts are replacing conventional techniques in the design and planning of urban areas. When designing smart cities, wireless networks help in tracking the amount of automobile emissions in the area. Environmental protection is a global issue because a healthy environment is vital for prosperous economic growth, sustainable agriculture, and a wholesome community. Monitoring environment has been made easier by wireless networks and sensors. The conventional approaches to city planning and design are gradually being replaced by new technologies. Smart city planners can keep an eye on the quantity of car emissions in their city through the use of wireless networks. Because a healthy environment is necessary for economic growth, sustainable agriculture, and a wholesome community, environmental conservation is a major priority on a global scale.

8.9 Conclusion

When monitoring air quality in smart cities, mobility and other factors, including the mobile sensor's altitude, would result in a tolerable measurement error. This can be made up for by expanding the monitoring area at very little expense. As this chapter explains, for smart cities, air pollution monitoring devices have been constructed using the WSN network. Together with air detection sensors and a microcontroller, the system is created with a WSN module, database, and online monitoring. Air pollution monitoring stations are too expensive to be feasible. When installing monitoring stations is not practical, economical sensors that can track vital aspects influencing air effluence, like traffic and weather, can be used. As a result, any application that requires an accurate instantaneous assessment of the atmosphere through minimal infrastructure costs can use the suggested approach, the recommended method, for estimating the level of air quality in any municipal street or neighborhood. This chapter has created a elegant manifold sensor detection system for tracking air pollution in real time utilizing the WSN network given the significance of air pollution in intelligent urban areas. Low-cost sensors that can monitor significant factors influencing air pollution can be used as a stand-in for air pollution monitoring stations in locations where they are too expensive to be practical. The weather and traffic are two of these variables.

References

1. Karl, H. and Willig, A., *Protocols and Architectures for Wireless Sensor Networks*, John Wiley and Sons Ltd, The Atrium, Southern Gate, Chichester, West Sussex, England, 2005.
2. Culler, D., Estrin, D., Srivastava, M., *Overview of Sensor Networks*, IEEE Computer, USA, 2004.
3. Martinez, K., Hart, J.K., Ong, R., Environmental sensor networks. *IEEE Comput. J.*, 37, 8, 50–56, 2004.
4. Mainwaring, A., Culler, D., Polastre, J., Szewczyk, R., Anderson, J., Wireless sensor networks for habitat monitoring, in: *Proceedings of the 1st ACM International workshop on Wireless sensor networks and applications*, Atlanta, Georgia, USA, pp. 88–97, 2002.
5. Akyildiz, I.F., Pompili, D., Melodia, T., Underwater acoustic sensor networks: research challenges. *Ad Hoc Netw.*, 3, 3, 257–279, 2005.
6. Brook RD, F.B., Cascio, W., Hong, Y., Howard, G., Lipsett, M., Luepker, R., Mittleman, M., Samet, J., Smith, S.C., Jr, Tager, I., Air pollution andcardiovascular disease: a statement for healthcare professionals from the Expert Panel

on Population and Prevention Science of theAmerican Heart Association. *Circulation*, 109, 21, 2655–2671, 2004.
7. Ali, H., Soe, J.K., Weller, S.R., A real-time ambient air quality monitoring wireless sensor network for schools in smart cities, in: *the Proceedings of the IEEE First International Smart Cities Conference (ISC2'15)*, pp. 25–28, 2015.
8. Sarath, K. and Guttikundaab, R., Health impacts of particulate pollution in a megacity—Delhi, India. *Environ. Dev.*, 6, 8–20, 2013.
9. Kaiwen, C., Kumar, A., Xavier, N., Panda, S.K., An Intelligent Home Appliance Control-based on WSN for Smart Buildings, in: *the Proceedings of the IEEE International Conference on Sustainable Energy Technologies (ICSET)*, Hanoi, Vietnam, pp. 282–287, 2016.
10. Dutta, J., Air Sense: Opportunistic crowd-sensing based air quality monitoring system for smart city, in: *the Proceedings of the IEEE SENSORS*, Orlando, FL, USA, 2016.
11. Tham, K.W., Kumar, A., Kalluri, B., Panda, S., A Wireless Sensor-Actuator Network for Enhancing IEQ, in: *In the Proceedings of the 15th Conference of the International Society of Indoor Air Quality & Climate (ISIAQ)*, Philadelphia, PA, USA, 2018.
12. Wang, W., Yuan, Y., Ling, Z., The Research and Implement of Air Quality Monitoring System Based on ZigBee, in: *the Proceedings ofthe 7th International Conference on Wireless Communications, Networking and Mobile Computing (WiCom)*, Wuhan, China, 2011.
13. Kim Oanh, N.T., Pongkiatkul, P., Cruz, M.T., Dung, N.T., Phillip, L., Zhuang, G., Lestari, P., Chapter 3: Monitoring and source apportionment for particulate matter pollution in six Asian cities, in: *Integrated Air Quality Management: Asian Case Studies.Taylor and Francis Group*, CRC Press, 2012, ISBN 9781439862254.
14. Ma, Y., Richards, M., Ghanem, M., Guo, Y., Hassard, J., Air Pollution Monitoring and Mining Based on Sensor Grid in London. *Sensors*, 8, 6, 3601–3623, 2008.
15. Hassard, G., Ghanem, M., Guo, Y., Hassard, J., Osmond, M., Richards, M., Sensor Grids For Air Pollution Monitoring, in: *the Proceedings of 3rd UK e-Science All Hands Meeting*, 2004.
16. Prof. Ghewari, M.U., Mahamuni, T., Kadam, P., Pawar, A., Vehicular Pollution Monitoring using IoT. *Int. Res. J. Eng. Technol.*, 5, 2, 1734–1739, 2018.
17. Liu, J.-H., Chen, Y.-F., Lin, T.-S., Lai, D.-W., Wen, T.-H., Sun, C.-H., Juang, J.-Y., Jiang, J.-A., Developed urban air quality monitoring system based on wireless sensor networks, in: *Sensing technology, 2011 fifth international conference on IEEE*, pp. 549–554, 2011.
18. Surannavar, K., Tatwanagi, M., Nadaf, S.P., Hunshal, P.B., Patil, D., Vehicular pollution monitoring system and detection of vehicles causing global warming. *Int. J. Eng. Sci. Comput.*, 7, 6, 12611–12614, 2017.

19. Zhang, Y., Wang, Y., Gao, M., Ma, Q., Zhao, J., Zhang, R., Wang, Q., Huang, L., A Predictive Data Feature Exploration-Based Air Quality Prediction Approach. *IEEE Access*, 7, 30732–30743, 2019.
20. Xia, L. and Shao, Y., Modelling of traffic flow and air pollution emission with application to hongkong island. *Environ. Modell. Softw.*, 20, 9, 1175–1188, 2005.
21. Sharmila Devi, S., Kaviarasu, N., Kiruthikram, P.P., Praveen Kumar, K., Air Pollution Prevention Using Wireless Sensors, in: *IEEE International Conference on Advancements in Electrical, Electronics, Communication, Computing and Automation*, pp. 1–5, 2023.
22. Arroyo, P., Lozano, J., Suárez, J., II, Herrero, J.L., Carmona, P., Wireless Sensor Network for Air Quality Monitoring and Control, in: *Chemical Engineering Transactions*, pp. 217–222, 2016.
23. Kumar, A., Kar, P., Warrier, R., Kajale, A., Panda, S.K., Implementation of Smart LED Lighting and Efficient Data Management System for Buildings. *Energy Procedia*, 143, 173–178, 2017.
24. Marcello, A., Gómez Maureira, D.O., Teernstra, L., *ThingSpeak —an API and Web Service for the Internet of Things*, LIACS, Leiden University, Netherlands, 2011.
25. Zhao, M., Kumar, A., Ristaniemi, T., Chong, P.H., Machine-to-Machine Communication and Research Challenges: A Survey. *Wirel. Pers. Commun.*, 97, 3, 3569–3585, 2017.
26. Arun Kumar, H.Y.S., Wong, K.J., Chong, P.H.J., Location-Based Routing Protocols for Wireless Sensor Networks: A Survey. *Wirel. Sens. Netw.*, 9, 1, 25–72, 2017.
27. Wang, S.-L., Chen, Y.L., Kuo, A.M.-H., Chen, H.-M., Shiu, Y.S., Design and evaluation of a cloud-based Mobile Health Information Recommendation system on wireless sensor networks. *Comput. Electr. Eng.*, 49, 221–235, 2016. DOI: 10.1016/j.compeleceng.2015.07.017.
28. Brunelli, D. and Rossi, M., Enhancing lifetime of WSN for natural gas leakages detection. *Microelectron. J.*, 45, 12, 1665–1670, 2014.

9

Interconnected Healthcare 5.0 Ecosystems: Enhancing Patient Care Using Sensor Networks

Ashwini A.[1*], Kavitha V.[2] and Balasubramaniam S[3]

[1]Department of Electronics and Communication Engineering, Vel Tech Rangarajan Dr. Sagunthala R&D Institute of Science and Technology, Chennai, Tamil Nadu, India
[2]Department of Computer Science and Engineering, University College of Engineering, Kancheepuram, Tamil Nadu, India
[3]School of Computer Science and Engineering, Kerala University of Digital Sciences, Innovation and Technology, Thiruvananthapuram, Kerala, India

Abstract

The culmination of major changes to the healthcare system is Healthcare 5.0. Healthcare 5.0 is the most recent stage of significant systemic changes. The use of cutting-edge technology, particularly sensor networks, is changing healthcare in the present day of interconnected health ecosystems. This chapter explores the role that sensor networking play toward improving patient care in such interconnected Health 5.0 habitats. Since sensor networks provide the foundation for immediate monitoring of patients and diagnosis, it demonstrates how essential they are in today's era. The section concentrates on how developments in monitoring patients remotely were made available by networks of sensors, and they have removed the limitations of normal medical limits. For the advantage of patients and their families, it works by combining sensor networks that are wireless that makes it possible for continuous surveillance outside medical walls. Additionally, it examines how smart sensor networks are changing diagnostics and preventive medicine showing how data-driven decisions from such networks may help active health programs, specific therapy procedures, and rapid identification of diseases. The problems of obtaining complete health information and the combination of numerous streams of data are recognized by sensors and data collection methodologies. The security

---

*Corresponding author: a.aswiniur@gmail.com

Rajesh Kumar Dhanaraj, Malathy Sathyamoorthy, Balasubramaniam S and Seifedine Kadry (eds.) Networked Sensing Systems, (225–246) © 2025 Scrivener Publishing LLC

of patients in interconnected ecosystems is emphasized, with specific focus paid to patient security, protection, and ethical considerations about ongoing data use and surveillance. Examined are the structures, capacity, and accessibility concerns related to integrating wireless sensors for a range of client lifestyles, with a focus on measures that narrow access gaps and give patients fair distribution of the latest medical equipment. The final section additionally takes a look at the way data collected by sensors aid clinicians make medical choices by offering them the most current information on well-planned procedures as well as treatment plans. It highlights the latest advances in sensor that is being tested on network connectivity and makes forecasts concerning their possible consequences for the transition to Healthcare 6.0. It examines how sensor-enabled, connected Healthcare 5.0 environments have changed the field and ends with ideas for additional advances in caring for patients. This comprehensive study underlines the vital function that that sensors have in linked medical infrastructure and gives insight into the ways they can improve medical care for patients, foster creative thinking, and influence the method of delivery of health services in the years to come.

Keywords: Artificial Intelligence, electronic health records, Healthcare 5.0, Healthcare 6.0, Internet of Things, patient care real-time systems, privacy, sensor networks

9.1 Introduction to Healthcare 5.0

The major change in the medical sector can be explained by an array of causes, such as advances in technology, shifts toward public demands, and changes to the provision of health care methods. Healthcare 5.0, an era of integrated medical systems, which encourage patient-centered treatment, customized medicine, and easy integration of modern innovations, is at the forefront of this transition. Doctors and nurses handled most of their decisions about healthcare under Healthcare 1.0, and they tend to be conventional in approach. Handwritten documents, restricted availability of health information, and a focus on reactive cases rather than proactive prevention were features of this modern era. With the advent of Health 2.0, which increased the availability of medical knowledge and simplified medical procedures, electronic medical records, or EHRs, came into existence [1]. Clients had the opportunity to take an increased role in handling their health through the internet and patient portals, which showed an evolution favoring the empowerment of patients and individuality.

Healthcare 3.0 has witnessed an increase in the use of electronic medical devices, medical informatics, and e-medicine, which has led to the creation informed by data healthcare delivery. To enhance healthcare decisions and increase results on multiple fronts, population control, medical studies, and the merging of enormous volumes of data have gained importance. The merging of Internet of Things (IoT) devices, data mining (ML) algorithms, and machine learning (AI) in the transition to Healthcare 4.0 has brought in an entirely novel phase of digital health systems. This era was highlighted by a focus on monitoring patients remotely, tailored and anticipatory therapy, and easy integration of medical systems to facilitate data sharing throughout many care destinations. Healthcare 5.0 introduces an entirely novel model for the medical industry. An integrated strategy for patient care known as "Healthcare 5.0" places a strong emphasis on each patient's unique needs and preferences. It manages this by integrating cutting-edge innovations to offer prepared, collaborative, and customized healthcare experiences, such as gadgets, data networks, personalized medical care, and correct pharmaceuticals. The concept of interconnection refers to a seamless combination of medical users, technology, and data flows to create a unified network with the goal of improving quality of life. The core pillars of Healthcare 5.0 are enhancing outcomes for patients as well as creating affordable delivery of healthcare tactics.

Patient centricity is accepted in the modern era, with healthcare institutions tailoring their operations to each person's specific requirements and preferences. Patients are actively involved in their own care journeys assisted by healthcare experts that use cutting-edge technologies to give individualized interventions and preemptive preventive measures. Patients are given autonomy by real-time data insights. There are countless opportunities as Healthcare 5.0 is moved closer together [2]. A complete transformation in healthcare can be made in a new era of creativity, cooperation, and independence for every person involved by utilizing the power of linked healthcare ecosystems and adopting a patient-focused approach to care fulfillment.

9.1.1 Evolution from Healthcare 4.0 to Healthcare 5.0

The shift from Healthcare 4.0 to Healthcare 5.0 signifies a noteworthy progression in the healthcare sector propelled by technological breakthroughs, evolving patient demands, and the requirement for more proactive and customized care. Healthcare 4.0, sometimes referred to as the "era of connected health," brought Internet of Things (IoT) equipment, computational intelligence (AI), and information processing into clinical practice, laying the groundwork for the industry's digital transformation. The goal of Healthcare 4.0 was to use technology to increase productivity, improve patient outcomes, and simplify the delivery of healthcare. Key elements of

Healthcare 4.0 included the ability of hospitals to enable efficient transfer of information across multiple healthcare environments, predictive modeling for controlling illnesses, and monitoring patients from afar. The future of healthcare mostly focused on responsive care approaches, which involve reacting to existing health disorders with procedures rather than targeting proactive medical management. Technology developments made healthcare services more accessible and enabled better decision making, but a change toward a patient-centric approach was still required.

The evolution of Healthcare 5.0 is shown in Figure 9.1. Healthcare 5.0 expands on the framework established by Healthcare 4.0 by giving patients' unique needs and preferences top priority and enabling them to actively participate in their own well-being and health [3]. Technology is employed in Healthcare 5.0 not just to detect and treat illnesses but also to stop them in their tracks. The convergence of wearable technology, precision medicines, genomic medicine, and advanced sensor networks offer instantaneous insights and enable individualized therapy catered to each person's own health profile, thus making this feasible.

Moreover, Healthcare 5.0 highlights the significance of providing care that is holistic and addresses not just the physical but also the psychological, personal, and social dimensions of health. Care plans are extensive, coordinated, and in line with the patient's objectives and aspirations when healthcare practitioners collaborate together, and patients as well as their companions are more involved. The movement from Healthcare 4.0 to Healthcare 5.0, in general, signifies a move toward a more humane approach to health services, where technology is employed as a tool to

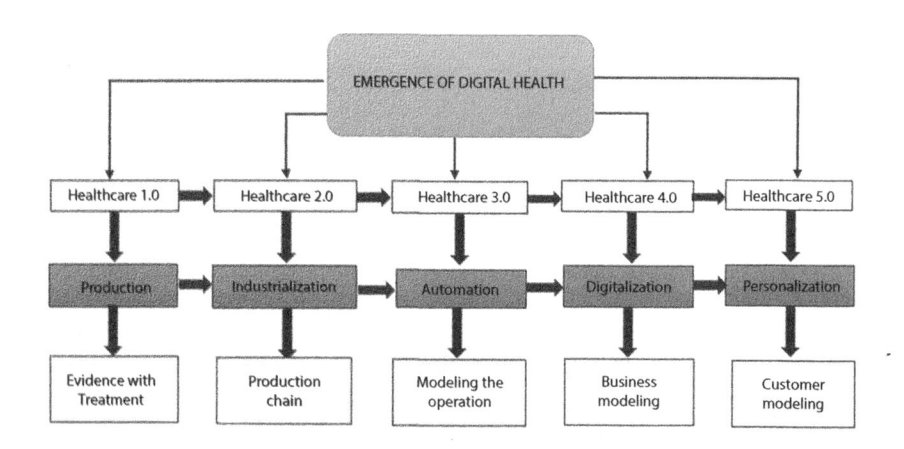

Figure 9.1 Evolution of Healthcare 5.0.

improve patient outcomes, foster wellness throughout life, and improve patient experience. Healthcare facilities can better serve the changing demands of individuals and communities by embracing this change, which will ultimately result in a population that is happier and healthier.

9.2 Real-Time Monitoring Using Sensor Networks

To provide complete and patient-centered care, a network of integrated healthcare systems acts as an essential framework that makes it easier for people to collaborate, share data, and integrate different healthcare components [4]. This framework includes a number of important components as shown in Figure 9.2.

Interoperability Standards
Interoperability standards are essential to interconnected healthcare systems because they guarantee that various healthcare IT gadgets, systems, and applications may successfully communicate with one another [5]. These standards control interfaces, protocols for communication, and data formats allowing smooth data integration and transfer between various systems.

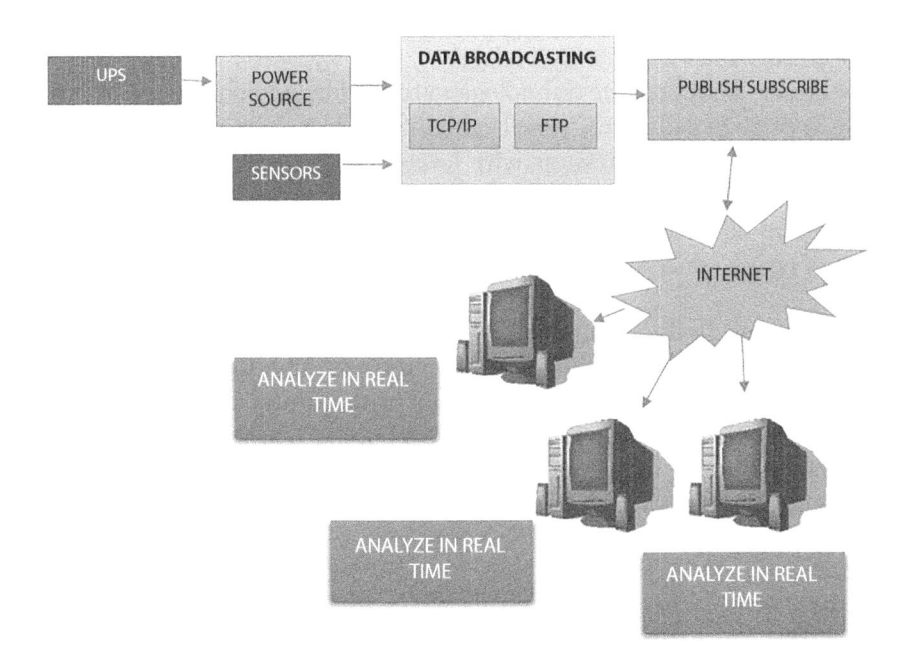

Figure 9.2 Real-time monitoring of sensor networks.

Health Information Exchange (HIE)
Health data communication platforms facilitate the digital transmission of individual medical data across healthcare practitioners, medical centers, pharmacies, and other participants. They are the backbone of networked healthcare systems. Regardless of the context or location of care, this platforms allow medical staff to access complete patient records instantly.

Electronic Health Records (EHRs)
EHRs are essential to the functioning of networked healthcare systems because they store and digitize patient health data in an electronic format. Throughout the care progression, authorized healthcare practitioners can access and update detailed medical histories, treatment plans, prescription drugs, laboratory findings, and other clinical data via electronic health records (EHRs).

Tele-Health and Remote Monitoring
Virtual healthcare delivery is made possible by tele-health and remote monitoring technologies, which let patients get medical attention, consultation, and monitoring from a distance. Virtual consultations, home-based care services, and remote patient–provider interactions are made possible by interconnected healthcare systems that make use of tele-health platforms, video conferencing tools, and remote monitoring equipment.

IoT Devices and Wearable Technology
IoT devices and wearable technologies are used by interconnected healthcare systems to gather real-time health information from patients, including indicators of wellness, activity levels, sleep habits, and medication adherence [6]. Through wireless communication, these devices connect to healthcare IT systems to provide personalized therapies, early identification of health concerns, and ongoing monitoring.

Analytics and Decision Support
Healthcare data is analyzed using technology and tools to find useful insights, identify patterns, and help with clinical decision-making. Healthcare systems that are interconnected utilize machine learning algorithms, predictive modeling approaches, and data analytics to optimize treatment regimens, improve care coordination, and improve patient outcomes.

Patient Engagement and Empowerment
By giving patients access to their health details, instructional materials, and self-management tools, interconnected healthcare systems place a

high priority on patient empowerment and involvement. Patients may take an active role in their own care, connect with physicians, and make educated decisions about their well-being and health thanks to patient portals, smartphone apps, and customized health dashboards.

Security and Privacy
For networked healthcare systems to safeguard sensitive patient data against cyber threats, illegal access, and breaches, security and privacy safeguards are essential components. Encryption strategies, access controls, authentication methods, and strong security measures protect patient data throughout the healthcare ecosystem guaranteeing its availability, confidentiality, and integrity.

In general, the network of linked healthcare systems fosters a unified and cooperative setting in which healthcare providers can easily cooperate to provide excellent, patient-centered care [7]. Interconnected healthcare systems strive to increase patient empowerment, care coordination, and health outcomes by utilizing technology, communication, communication of data, and patient participation.

9.3 Advancements in Remote Patient Monitoring

The state of healthcare has changed dramatically as a result of improvements in remote patient monitoring, which allow medical professionals to keep an eye on patients' health from a distance, gather data in real time, and act quickly when needed. The following are significant developments in virtual patient tracking, which is described in Figure 9.3.

Wireless Sensor Technology
Wearable technology and medical sensors that can continually monitor vital signs, including blood pressure, pulse, levels of glucose in the blood, oxygen saturation, and temperature, have been made possible by the advancement of wireless sensor technology. By transmitting data instantly to healthcare providers, these gadgets eliminate the need for tangible links and enable monitoring in real time.

Mobile Health (mHealth) Applications
Patients can now use smartphones and tablets to track their medical condition from the convenience of their own homes because of the growth of mobile health applications [8]. These apps frequently work with wearable technology to track different health data, remind users to take their

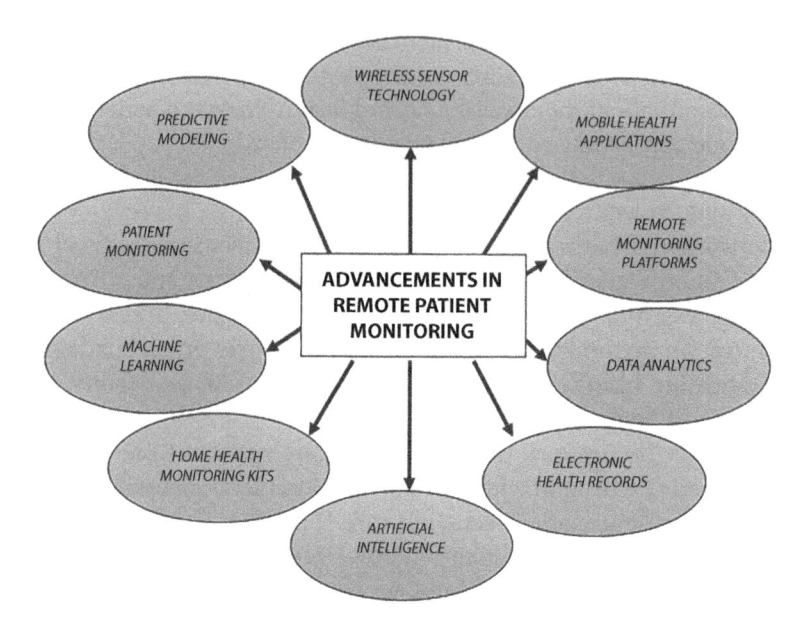

Figure 9.3 Advancements in remote patient monitoring systems.

medications, provide instructional materials, and help patients communicate with medical professionals.

Remote Monitoring Platforms
To make the process of remotely monitoring patients more efficient, telehealth systems and remote monitoring platforms have been developed. With the help of these systems, medical professionals may monitor and remotely access patient data, establish abnormal value thresholds, get notifications for important occurrences, and have secure video conferences or secure messaging conversations with patients.

Data Analytics and Predictive Modeling
By enabling healthcare personnel to examine vast amounts of patient data and spot trends, patterns, and anomalies, developments in data analytics and modeling for prediction have improved remote patient monitoring. By forecasting possible illnesses or drops, undesirable events can be effectively avoided by the use of algorithms for prediction.

Electronic Health Record Integration
Electronic health records, or EHRs, and systems for global surveillance of patients are becoming increasingly connected to ensure seamless data

interchange and connection throughout the clinical databases of healthcare professionals and distant surveillance equipment. This link allows improved medical continuation throughout multiple locations and enables an exhaustive review of a patient's medical data.

Artificial Intelligence (AI) and Machine Learning
AI and computational approaches to deep learning are being utilized to more accurately and effectively assess and analyze medical information that is collected remotely. Algorithms, such as these, are able to detect small variations in information about patients, detect correlations that indicate amplification or worsening of conditions, and supply individualized care prescriptions based on different health characteristics.

Home Health Monitoring Kits
Home health record packages have been developed to provide people with sophisticated monitoring tools in which they may employ in their own residences. These kits may include sizes, glucose monitors, pulse monitors, sensors for medical use, and connected devices [9]. This could enable individuals to follow multiple healthcare indexes and send data with physicians virtually.

Remote Patient Monitoring in Chronic Disease Management
Remote patient monitoring has significantly altered the way that persistent medical conditions like being overweight, heart attack, high blood pressure, and breathing problems disease are maintained. By continuously tracking key health factors, healthcare providers can see early signs of flare ups or challenges, adjust regimens consequently, and distribute medications appropriately to lower the rate of hospitalization and enhance the outcomes.

The recent developments in remote patient tracking have an opportunity to drastically alter the way care is provided by raising participation among patients, increasing accessibility to care, reducing expense, and eventually boosting patient the results and the level of existence.

9.3.1 Challenges in Healthcare 4.0

The incorporation of technological innovations, such as data analytics, IoT, and AI or Healthcare 4.0, has made a significant impact on the field of healthcare. But it also presents an array of challenges that must be fixed, including the following:

Data Security and Privacy Concerns
With a rise in the digital age of healthcare data, keeping data security and confidential for patients becomes paramount [10]. Healthcare companies have difficulties protecting private patient data from illegal access, cyber-attacks, and data breaches.

Interoperability Issues
Healthcare 4.0 requires the utilization of multiple technology and infrastructure, often from various vendors, that result in connectivity challenges. Ensuring smooth exchange of information and communication between different systems remains an essential obstacle for medical organizations obstructing care efficiency and coordination.

Limited Access to Technology
Differences in access to digital and technological competence among both patients and medical professionals may worsen healthcare inequalities. Some individuals have possession of smartphones, broadband connectivity, or the skills necessary to use electronic health tools properly leading to gaps in medical services and consequences.

Integration with Existing Workflows
It can be difficult and disruptive to integrate new digital technology into current healthcare workflows. Usability and adoption rates of digital tools may be impacted by healthcare providers' inability to integrate these tools in their daily procedures, workflow inefficiencies, and resistance to change.

Quality and Reliability of Health Data
Digital health technology can produce health data that varies greatly in terms of quality and dependability. Clinical decision making and patient care may be impacted by healthcare practitioners' difficulties in verifying the dependability and correctness of data gathered from wearable technology, surveillance systems as a whole, or patient-generated health data.

Regulatory and Compliance Requirements
When deploying digital health technology, healthcare institutions have to navigate complicated regulations and compliance standards [11]. Digital health projects become more complex and expensive when complying with standards like the General Data Protection Regulation (GDPR) and the Health Insurance Portability and Accountability Act (HIPAA).

Healthcare Professional Training and Education
To apply digital health technology and data-driven decisions in clinical practice, healthcare practitioners need to have the right training and education. Nevertheless, a lack of proper preparation for the intricacies of Healthcare 4.0 in numerous healthcare training programs results in a skills gap and reluctance to the use of new technologies.

Financial Sustainability and ROI
It is frequently necessary to make large upfront investments in infrastructure, equipment, and training to use digital health innovations [12]. It may be difficult for healthcare organizations to show the return on investment (ROI) of digital health efforts, especially in the near run. As a result, stakeholders may be reluctant to provide money.

To address these issues and ensure that everyone has equal use to digital health tools and that healthcare outcomes are improved, healthcare organizations, legislators, modern technology vendors, and other stakeholders must work together to develop solutions that promote data protection, interoperability, ease of use, and compliance with regulations.

9.4 Early Disease Detection Through Sensor Networks

A number of essential elements or building pieces are needed for early disease detection via sensor networks to efficiently track, examine, and comprehend health data. Figure 9.4 shows the overall block diagram of early disease detection through the wireless sensor networks. Among these blocks are the following:

Sensor Devices
For the purpose of early disease diagnosis, high-quality sensor systems that can measure pertinent physiological parameters like blood pressure, pulse, glucose levels, and saturation in oxygen, humidity, and activity levels are crucial. These sensors can be used independently or combined with medical equipment and wearable technology.

Wireless Communication
Data may be transferred from gadgets that sense to gathering and organizing systems with wireless communication protocols like Bluetooth, WiFi, Zigbee, or cellular networks [13]. Timely transmission of health data

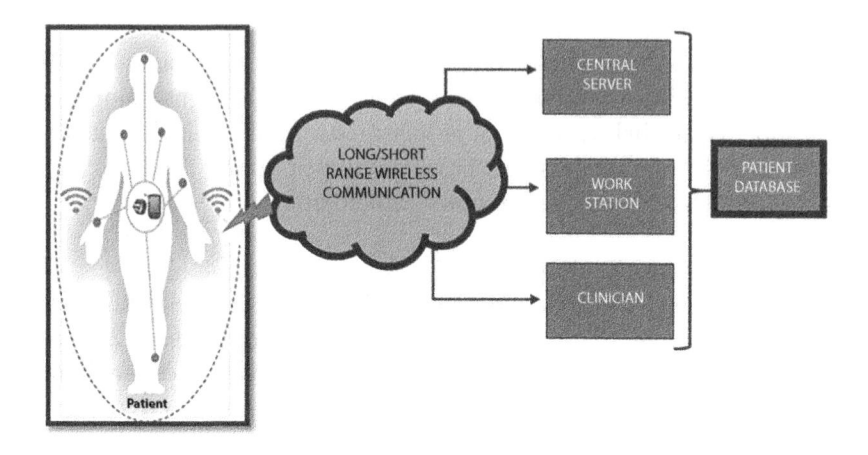

Figure 9.4 Early detection through wireless body sensor networks.

and real-time monitoring are ensured by means of dependable and secure wireless communication.

Data Collection and Aggregation
Health data are gathered from sensor devices by collection and aggregation infrastructure, which then integrate it into a cloud-based platform or central repository. These systems that gather, store, and arrange medical information for further study could be gateways, data collecting units, or cloud servers.

Data Processing and Analysis
Algorithms for processing and analyzing data examine unprocessed sensor data to find trends, patterns, and anomalies connected to early disease signs. To evaluate enormous quantities of medical data and produce useful insights, advanced analytics approaches, like AI, machine learning, and modeling with predictions, may be used.

Early Warning Systems
When potential health problems are identified, early warning systems use statistical algorithms to identify deviations from typical health trends and send out alerts or notifications. These systems can recognize warning signs of disease development or progression using threshold-based notification analysis of patterns or algorithmic prediction.

Decision Support Tools
Based on the examination of sensor data, technology that supports decisions offer healthcare providers practical insights and recommendations [14]. Healthcare providers can make better choices regarding treating patients and methods of intervention with the use of these technological devices. Such solutions include customized healthcare dashboards, risk assessment algorithms, and medical decision-making technologies.

Integration With Electronic Digital Records (EDRs)
Connectivity with digital medical files enables the transfer of health data throughout networks of sensors and the clinical systems of medical providers. This integration encourages continuation of care and informed taking decisions by guaranteeing that healthcare providers have access to information from sensors in a setting of the patient's health state.

Privacy and Security Measures
Sensor network-gathered private medical data must be protected with privacy and security measures. Data anonymization, access control, authentication, and encryption techniques assist in maintaining patient privacy and prevent unauthorized data access and breaches.

Regulatory Compliance
Sensor networks make sure the gathering, retention, and utilization of medical information complies with moral and legal norms by abiding to legal constraints, such as the act on health insurance. Patients and healthcare professionals can continue to have faith in disease detection technologies by adhering to regulatory frameworks [15].

Healthcare organizations can efficiently utilize wireless sensor networks to keep track of patients' health status, identify possible medical problems early, and take fast action to enhance the health of patients and their quality of life by incorporating these blocks into early illness detection systems.

9.5 Leveraging Multisensor Data for Comprehensive Health Insights

Multiple modes utilizing multisensor data for holistic health insights entail combining information from several sensors to provide a more complete picture of a person's activity and overall health. The following are the main elements of using multisensor data to obtain thorough health insights, which are shown in Figure 9.5.

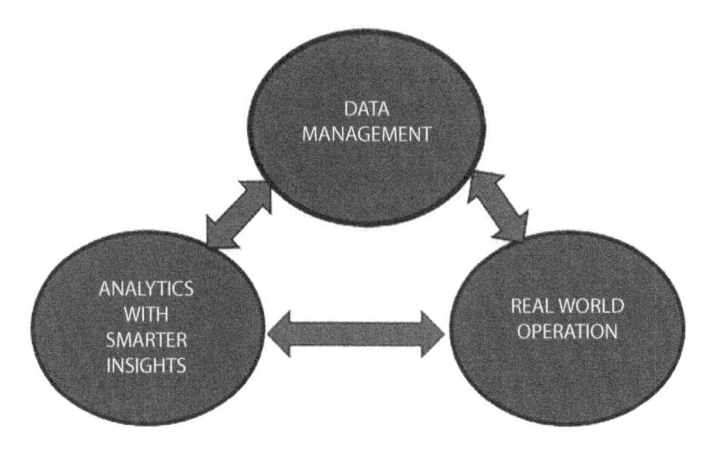

Figure 9.5 Analytics with real-world insights.

Data Fusion and Integration
To create a single dataset, multisensor data fusion entails merging data from several sensors, such as cellphones, wearable technology, medical sensors, and environmental sensors [16]. Data integration strategies make sure that, in preparation for analysis, information from various sensors is synchronized and coordinated.

Rich Health Data Sources
A wide range of health-related data is available from multisensor data sources, such as physiological indicators (e.g., blood pressure, pulse, oxygen in saturation), intensity of activity, sleep schedules, atmospheric variables (e.g., humidity, temperature, and the level of quality of air), and behavioral indicators (e.g., engaging in physical activity, food choices, habits).

Comprehensive Health Monitoring
Healthcare providers can track several facets of a patient's health in real-time by utilizing multisensor data. With this all-encompassing approach, vital signs, levels of exertion, levels of sleep, and environmental exposures may be continuously monitored giving a more comprehensive picture of the person's general well-being.

Early Detection of Health Issues
Through the identification of trends, patterns, and anomalies in medical information, multisensor data analysis facilitates the early detection of possible health disorders. Variations in physiologic markers, which means amounts of movement, or surroundings can all serve as early indicators of

possible health hazards or the beginning of disease. Subsequently, these changes may prompt such therapies and preventive actions.

Personalized Health Insights
Based on a person's characteristics, interests, and health targets, multisensor analysis of data enables the collection of personalized health data [17]. By analyzing several data streams, medical professionals can uncover unique suggestions for medical treatments, lifestyle modifications, or preventive actions based on specific patterns of health.

Predictive Analytics and Machine Learning
Advanced analytic techniques, such as preventative forecasting and algorithms powered by AI, can assess the likelihood of medical conditions, categorize individuals based on their health status, and predict future health outcomes by leveraging multisensor data analysis. These predictions enable specific treatment therapies and preventive interventions to optimize medical outcomes.

Clinical Decision Support Systems
Multisensor data analysis supports clinical decision making by providing healthcare providers with helpful information and recommendations [18]. Algorithms, which promote clinical choice, can integrate multisensor data with EHRs, or electronic health records, to assist healthcare providers in making intelligent choices about the treatment of patients and their care regimens.

Research and Population Health Management
By identifying risk factors, health trends, and epidemiological patterns at the population level, multisensor data analysis aids in the management of population health. Research endeavors benefit from this. By analyzing multisensor data from large samples, researchers can learn about the prevalence of diseases, differences in health, and the influence of the outside world on health results.

In a nutshell, leveraging multisensor data to derive complete health information offers a powerful way to monitor, assess, and understand various aspects of an individual's well-being [19–22]. By merging data from many sensors, healthcare professionals can get an improved understanding of the health of their patients' state, spot problems with health early, and tailor medical treatments to improve their health.

9.6 Security Measures for Protected Health Information

Protecting protected health information is essential for upholding patient privacy following regulations, such as the Health Care Portability and Ethics Act, and averting data breaches and unauthorized access. Some essential safety measures to protect Protected Health Information (PHI) are as follows:

> ➤ Encrypting sensitive health data protects it from unauthorized access and monitoring both during transit and storage. Robust methods of encryption and protocols should be used to protect data both at rest, as while kept on databases or other devices, and in transit, as during digital communication.
> ➤ By putting access controls in place, businesses can be sure that only people with permission can access PHI. To restrict access to PHI by considering users' positions, duties, and privileges, access control mechanisms, such as multi-factor authentication and user authentication, like passwords, biometrics, and role-based access control, should be used.
> ➤ Reducing the amount of PHI that is gathered, used, and retained can help lower the chance of exposure and unauthorized access [23]. When it comes to patient care, therapy, payment, or healthcare operations, only gather and hold the bare minimum of PHI that is required. If it is no more necessary, securely destroy PHI.
> ➤ In the case of a security attack or data breach, maintaining audit trails and recording access to PHI make it easier to monitor and track user activities, identify suspicious activity or illegal access, and support reaction and forensic investigation. Information about user access, data updates, and system operations pertaining to PHI should all be documented in audit logs.
> ➤ Physical security measures guard against unauthorized access to PHI-containing buildings, servers, gadgets, and storage media. To stop theft, tampering, or unauthorized access to PHI-bearing assets, safekeeping facilities, alarm systems, and monitoring systems must be put in place.
> ➤ Implementing procedures for disaster recovery and consistently backing of PHI can help lessen the effects of system

failures, natural catastrophes, and data breaches [24]. PHI backup copies should be kept in a secure location, encrypted, and routinely checked to guarantee data availability and integrity in the case of a data loss occurrence.
- ➤ Employee awareness of security standards, compliance with HIPAA, and the value of protecting PHI are increased, and the likelihood of negligence or carelessness resulting in data breaches is decreased. Professionals should receive training on how to safeguard private medical data securely, spot fraudulent efforts, and report questionable activity right away.
- ➤ Ensuring that third-party vendors or service providers, such as cloud providers or software vendors, abide by the security and privacy guidelines specified in business associate contracts if they have access to PHI [25–27]. To protect PHI communicated with outside parties, conduct due diligence evaluations, keep an eye on provider regulation, and enforce contractual duties.
- ➤ Create reaction strategies and procedures to deal with security events, data breaches, and unapproved PHI disclosures. Create procedures for looking into events, stopping breaches, reducing risks, and informing impacted parties, law enforcement, and other stakeholders in accordance with breach disclosure regulations.

Healthcare companies can reduce the risk of data breaches, uphold patient confidence, and prove compliance with legal requirements for protecting sensitive health information by putting these security measures into place and taking a comprehensive approach to PHI protection.

9.7 Overcoming Infrastructure and Connectivity Barriers

Solving network and connectivity obstacles is essential in the context of Healthcare 5.0 to fully utilize new technology in the provision of patient-centered care. This means utilizing cutting-edge approaches that go beyond conventional constraints to guarantee that a variety of populations have fair access to healthcare treatments [28]. The goal of Healthcare 5.0 is to improve accessibility for marginalized communities and expand the reach of medical care beyond geographical limits by implementing e-health platforms, health apps for smartphones, and networked healthcare ecosystems. Additionally, to enable the smooth transfer of health data and enable remote consulting, monitoring, and tele-health interventions, investment in a strong telecommunications infrastructure is required. This infrastructure includes mobile networks and high-speed internet connectivity. Accepting cutting-edge technology, like satellite-based communication systems, improves connectivity even more in isolated and rural locations where there may not be enough terrestrial infrastructure [29].

Moreover, to advance infrastructure development projects and close the digital divide, cooperative relationships between governments, healthcare institutions, technology companies, and community organizations are essential [30, 31]. Healthcare 5.0 may promote inclusivity, increase the effectiveness of care delivery, and ultimately enhance health outcomes for all people, regardless of socioeconomic level or geographic location, by giving infrastructure and connectivity improvements first priority.

9.8 Improving Treatment Plans Through Sensor-Generated Insights

Utilizing real-time sensor data to optimize patient care, refining therapies through sensor-generated insights, is a revolutionary paradigm in healthcare delivery. Sensor-generated insights give medical professionals an in-depth understanding of their patients' health status by continually tracking vital signs, levels of activity, and other health variables [32]. This allows for targeted and timely actions. With the use of these insights, medical professionals can customize treatment programs to meet the unique needs, preferences, and therapy responses of each patient. The comparison of various services in the healthcare industry describing its advantages and challenges are shown in Table 9.1.

Wearable technology and remote monitoring systems, for instance, can monitor medication compliance, identify early warning indicators of problems, and evaluate the efficacy of treatments enabling proactive modifications to treatment plans [33, 34]. Furthermore, sensor-generated data help healthcare professionals make data-driven decisions by allowing them to see patterns, trends, and correlations that could guide specific therapies and enhance patient outcomes. Healthcare professionals can improve patient involvement, optimize treatment regimens, and provide more

Table 9.1 Comparison of various services in the healthcare industry.

Technologies	Cloud computing	Blockchain	IoT	AI
Advantages	High efficiency, improved storage, increased flexibility	Higher security, higher reliability, decentralized, higher latency	Low latency is required, lightweight algorithms, Heterogeneity	High efficiency, higher compatibility, high availability
Challenges	Secured, portability, reliability, interoperability	Security issues, high bandwidth, resource consumption, poor scalability	Low data storage, low computational capacity, privacy challenges	High communication, less protection of data, low efficiency, high computation costs

individualized, efficient care that is catered to the specific needs of each patient by utilizing sensor-generated information [35–37].

9.9 Conclusion

In summary, interconnected Healthcare 5.0 ecosystems, enabled by the easy integration of sensor networks to improve patient care, offer a revolutionary paradigm shift in the delivery of healthcare [38]. Healthcare 5.0 enables continuous evaluation, individualized actions, and preventive healthcare measures by leveraging cutting-edge technologies including wearables, data analytics, and remote monitoring systems. Healthcare professionals can provide timely, patient-centered treatment by using sensor-generated insights to obtain a comprehensive picture of patients' health state. By providing care outside of hospitals and clinics, this integrated approach to healthcare transcends traditional boundaries and gives patients the tools they need to take charge of their own health. In addition, Healthcare 5.0 encourages cooperation between healthcare providers, which stimulates innovation, enhances care coordination, and eventually improves patient outcomes. To move forward with the vision of completely personalized, proactive, and revolutionary healthcare for people around the world globally, the developers are embracing the potential of Communicating Healthcare 5.0 ecosystems powered by networks of sensors and data-driven insights.

References

1. Bida, M.N., Mosito, S.M., Miya, T.V., Demetriou, D., Blenman, K.RM., Dlamini, Z., Transformation of the Healthcare Ecosystem in the Era of Society 5.0, in: *Society 5.0 and Next Generation Healthcare: Patient-Focused and Technology-Assisted Precision Therapies*, pp. 223–248, 2023.
2. Natarajan, R., Lokesh, G.H., Flammini, F., Premkumar, A., Venkatesan, V.K., Gupta, S.K., A Novel Framework on Security and Energy Enhancement Based on Internet of Medical Things for Healthcare 5.0. *Infrastructures*, 8, 2, 22, 2023.
3. Sharma, A., Singh, P., Kesarwani, S., Singh, A.P., Internet of Things and Sensor Networks in Industry 5.0. *Emerging Technol. Digital Manuf. Smart Fact.*, 67, 67–78, 2023.
4. Gardasevic, G., Katzis, K., Bajic, D., Berbakov, L., Emerging wireless sensor networks and Internet of Things technologies—Foundations of smart healthcare. *Sensors*, 20, 13, 3619, 2020.
5. Anzola, C., Wilder, E., Moreno, M.A.M., Enhanced living environments (ELE): a paradigm based on integration of industry 4.0 and society 5.0 contexts with ambient assisted living (AAL), in: *Gerontechnology III: Contributions to the Third International Workshop on Gerontechnology, IWoG 2020*, Springer International Publishing, pp. 121–132, 2021.
6. Chander, B., Pal, S., De, D., Buyya, R., Artificial intelligence-based internet of things for industry 5.0. *Artif. Intell.-Based Internet Things Syst.*, 12, 3–45, 2022.
7. Mishra, P. and Singh, G., Internet of medical things healthcare for sustainable smart cities: current status and future prospects. *Appl. Sci.*, 13, 15, 8869, 2023.
8. Dlamini, Z., Miya, T.V., Hull, R., Molefi, T., Khanyile, R., Ferreira de Vasconcellos, J., Society 5.0: Realizing Next-Generation Healthcare, in: *Society 5.0 and Next Generation Healthcare: Patient-Focused and Technology-Assisted Precision Therapies*, pp. 1–30, 2023.
9. Ashwini, A,.S., Quadruple spherical tank systems with automatic level control applications using fuzzy deep neural sliding mode FOPID controller. *J. Eng. Res.*, 2023. Preprint, https://doi.org/10.1016/j.jer.2023.09.022.
10. Sadeghi, M. and Mahmoudi, A., Synergy between blockchain technology and internet of medical things in healthcare: A way to sustainable society. *Inf. Sci.*, 660, 120049, 2024.
11. Ashwini, A., Purushothaman, K.E., Rosi, A., Vaishnavi, T., Artificial Intelligence based real-time automatic detection and classification of skin lesion in dermoscopic samples using DenseNet-169 architecture. *J. Intell. Fuzzy Syst.*, 4, 6943–6958, 2023. Preprint.
12. Mbunge, E. and Muchemwa, B., Towards emotive sensory Web in virtual health care: Trends, technologies, challenges and ethical issues. *Sens. Int.*, 3, 100–134, 2022.

13. Ozdemir, V. and Hekim, N., Birth of industry 5.0: Making sense of big data with artificial intelligence, "the internet of things" and next-generation technology policy. *OMICS: J. Integr. Biol.*, 22, 1, 65–76, 2018.
14. Mishra, K.N. and Pandey, S.C., Enhancing the Concert of M-health Technologies in Smart Societies Using Cloud-IoT-Based Distributive Networks, in: *Cloud-IoT Technologies in Society 5.0*, pp. 133–161, 2023.
15. Osama, M., Ateya, A.A., Sayed, M.S., Hammad, M., Pławiak, P., Abd El-Latif, A.A., Elsayed, R.A., Internet of medical things and healthcare 4.0: Trends, requirements, challenges, and research directions. *Sensors*, 23, 17, 7435, 2023.
16. Chibuike, M.C., Sara, G.S., Adele, B., Overcoming Challenges for Improved Patient-Centric Care: A Scoping Review of Platform Ecosystems in Healthcare, in: *IEEE Access*, 2024.
17. Ashwini, A. and Kavitha, V., Automatic Skin Tumor Detection Using Online Tiger Claw Region Based Segmentation–A Novel Comparative Technique. *IETE J. Res.*, 69, 1–9, 2021.
18. Ciasullo, M.V., Orciuoli, F., Douglas, A., Palumbo, R., Putting Health 4.0 at the service of Society 5.0: Exploratory insights from a pilot study. *Socio-Economic Plann. Sci.*, 80, 101–163, 2022.
19. Tallat, R., Hawbani, A., Wang, X., Al-Dubai, A., Zhao, L., Liu, Z., Min, G., Zomaya, A.Y., Alsamhi, S.H., Navigating Industry 5.0: A Survey of Key Enabling Technologies, Trends, Challenges, and Opportunities. *IEEE Commun. Surv. Tutor.*, 26, 2, 1080–1126, 2023.
20. Maddikunta, P.K.R., Pham, Q.-V., Prabadevi, B., Deepa, N., Dev, K., Gadekallu, T.R., Ruby, R., Liyanage, M., Industry 5.0: A survey on enabling technologies and potential applications. *J. Ind. Inf. Integr.*, 26, 8163–8189, 100257, 2022.
21. Bhat, J.R. and Salman, A., Alqahtani, 6G ecosystem: Current status and future perspective. *IEEE Access*, 9, 43134–43167, 2021.
22. Garai, A., Péntek, I., Adamko, A., Revolutionizing healthcare with IoT and cognitive, cloud-based telemedicine. *Acta Polytech. Hung.*, 16, 2, 163–181, 2019.
23. Ashwini, A. and Murugan, S., Automatic Skin Tumour Segmentation Using Prioritized Patch Based Region–A Novel Comparative Technique. *IETE J. Res.*, 66, 1–12, 2020.
24. Adel, A., Future of industry 5.0 in society: Human-centric solutions, challenges and prospective research areas. *J. Cloud Comput.*, 11, 1, 1–15, 2022.
25. Ferlin, A.A. and Rosi, V., Iot based object perception algorithm for urban scrutiny system in digital city, in: *2023 International Conference on Circuit Power and Computing Technologies (ICCPCT)*, pp. 1788–1792, 2023.
26. Gomathi, L., Mishra, A.K., Tyagi, A.K., Industry 5.0 for Healthcare 5.0: Opportunities, Challenges and Future Research Possibilities, in: *2023 7th International Conference on Trends in Electronics and Informatics (ICOEI)*, pp. 204–213, IEEE, 2023.

27. Ashwini, A., Purushothaman, K.E., Banu Priya, P., Jenath, M., Prasanna, Automatic Traffic Sign Board Detection from Camera Images Using Deep learning and Binarization Search Algorithm, in: *2023 International Conference in recent advances in Electrical, Electronics, Ubiquitous Communication and Computational Intelligence (RAEEUCCI)*, IEEE, 2023.
28. Javaid, M., Haleem, A., Singh, R.P., Suman, R., 5G technology for healthcare: Features, serviceable pillars, and applications, in: *Intelligent Pharmacy*, 2023.
29. Amr, A., Unlocking the future: fostering human–machine collaboration and driving intelligent automation through industry 5.0 in smart cities. *Smart Cities*, 6, 5, 2742–2782, 2023.
30. Sathyamoorthy, M., Dhanaraj, R.K., Vanitha, C.N., Krishnasamy, L., Augmented Reality Based Medical Education, in: *2023 Intelligent Computing and Control for Engineering and Business Systems (ICCEBS)*, Chennai, India, pp. 1–6, 2023, doi: 10.1109/ICCEBS58601.2023.10449124.
31. Pateraki, M., Fysarakis, K., Sakkalis, V., Spanoudakis, G., Varlamis, I., Maniadakis, M., Lourakis, M., Biosensors and Internet of Things in smart healthcare applications: Challenges and opportunities, in: *Wearable and Implantable Medical Devices*, pp. 25–53, 2020.
32. Balakrishnan, A., Kadiyala, R., Dhiman, G., Ashok, G., Kautish, S., Yadav, K., Prasad, M.N., A personalized eccentric cyber-physical system Architecture for smart healthcare, in: *Secur. Commun. Netw.*, pp. 1–36, 2021.
33. Li, C., Wang, J., Wang, S., Zhang, Y., A review of IoT applications in healthcare, in: *Neurocomputing*, p. 127017, 2023.
34. Gollagi, S.G. and Balasubramaniam, S., Hybrid model with optimization tactics for software defect prediction. *Int. J. Model. Simul. Sci. Comput.*, 14, 02, 2350031, 2023.
35. Krishnasamy, L., Dhanaraj, R.K., Gupta, M., Rai, P., Sruthi, K., T, G., Detection of diabetic Retinopathy using Retinal Fundus Images, in: *2022 4th International Conference on Advances in Computing, Communication Control and Networking (ICAC3N)*, Greater Noida, India, pp. 449–455, 2022, doi: 10.1109/ICAC3N56670.2022.10074340.
36. Muthumeenakshi, R., Singh, C., Sapkale, P.V., Mukhedkar, M.M., An Efficient and Secure Authentication Approach in VANET using Location and Signature-Based Services. *Adhoc Sens. Wirel. Netw.*, 53, 2022.
37. Subhadra Sarngadharan, A., Narasimhamurthy, R., Sankaramoorthy, B., Singh, S.P., Singh, C., Hybrid optimization model for design and optimization of microstrip patch antenna. *Trans. Emerging Telecommun. Technol.*, 33, 12, e4640, 2022.
38. Balasubramaniam, S., Prasanth, A., Kumar, K.S., Kavitha, V., Medical Image Analysis Based on Deep Learning Approach for Early Diagnosis of Diseases, in: *Deep Learning for Smart Healthcare*, Auerbach Publications, pp. 54–75, 2024.

10
Farming 4.0: Cultivating the Future with Internet of Things Empowered on Smart Agriculture Solutions

Ashwini A.[1]*, S.R. Sriram[2], J. Manoj Prabhakar[3] and Seifedine Kadry[4]

[1]*Department of Electronics and Communication Engineering, Vel Tech Rangarajan Dr. Sagunthala R&D Institute of Science and Technology, Chennai, Tamil Nadu, India*
[2]*Department of ECE, College of Engineering Guindy, Anna University, Chennai, Tamil Nadu, India*
[3]*Department of CSE, Dhaanish Ahmed Institute of Technology, Coimbatore, Tamil Nadu, India*
[4]*Department of Applied Data Science, Noroff University College, Kristiansand, Norway*

Abstract

Agriculture is one of the major sectors that have been considered to be essential to ensuring food security. The world's population is growing, and there are many natural factors that make nourishing billions of people challenging. The application of Internet of Things (IoT) in agriculture is revolutionizing the field and creating opportunities for accurate monitoring and data-driven farming. The Internet of Things with the sensors and unmanned aircraft, which helps in tracking the farming lands based on the phenomenon of humidity, crop performance, livestock, and temperature are termed as Farming 4.0. This research chapter holds the smart agriculture concept that highlights the usage of Internet of Things in its evolutionary concept. This nominally increases the food output, as the year 2050 is predicted to have food shortage due to growing population, traditional farming methods, and outdated skills of farmers in field pattern. This fine goal has led to the enlarged connectivity between the digital scale of the marketed items using the internet. The loss reduction with increase in yield values is determined by the process of gathering data continuously with a precise level of monitoring. By considering

*Corresponding author: a.aswiniur@gmail.com

the periodic data with the current trends, the farmers predict the yield with the disease outbreaks enhancing the consumer preferences with accurate data-driven methods. Making decisions based on strategic values helps in choosing the best strategic plan. Managing the strategies in remote locations using the IoT helps in providing the data with real-time access creating web alerts responding to disease outbreak with a high range of accuracy at nominal consumer preferences. This is made possible by connecting the sensors to the land. This helps in monitoring the storage conditions optimizing the tracking of shipments on agricultural supply chain values. This ensures producing high-quality food with waste reduction enhancing the efficiency of the supply chain process. The data gathered using the IoT based on artificial intelligence and various training procedures help the supporting system with decisions packed, thus highlighting suggestions and various insights. The agro-based sensor devices used for evaluation are connected to farming, service types, and the readiness of actual technological levels chosen for practicing. The results are framed with the process of investigating the technologies at the digital level such as machine learning embedded on the robotic systems and the Internet of Things. Thus, embracing security with IoT protocols helps to protect the assets on the farmland fostering long-term technological viability. Thus, the future of agriculture lies in leveraging the global IoT, which helps in achieving improved choices and accurate level of monitoring with suitable practices.

Keywords: Decision support system, Farming 4.0, Internet of Things, livestock production, machine learning, smart agriculture, sensor devices

10.1 Introduction to Smart Agriculture and IoT Integration

The need to produce food more efficiently and sustainably, as well as technological advances, has driven the transformation of agriculture over the past few decades. The concept of "smart agriculture" emerged as a paradigm shift, using cutting-edge technology to improve agricultural practices and increase production while reducing resource use and environmental impacts. The Internet of Things is at the heart of this transformation transforming the way we farm and revolutionizing precision farming and the way we make decisions based on data.

Initially, the agricultural practice was entirely based on the concept of manual labor. The most innovative form of today's modern world is smart agriculture, which acts as a promising approach in monitoring and controlling various aspects of agricultural operations using connectivity in data analysis [1]. Enhancing the use of agriculture components with the Internet of Things helps in observing and gaining real-time data on water level in soil, conditions of the crops, and the behavior of carious farm animals, thus taking vital decisions in fertilization, and pest and water controlling patterns. It helps in improving efficiency, thus decreasing risk and increasing the sustainable aspects by integrating with the various IoT devices.

10.1.1 Evolution of Agriculture: From Traditional to Smart Farming

The field of agriculture has always evolved with a consistent phase over millions of years with prominent advancement and survival rate. All the traditional farming methods require an in-depth foundation, which proved to be labor intensive using all types of hands-on knowledge of all rudimentary tools that are handled over time period. This method was very successful, but their scope was confined such that they are more susceptible to various shifts in environmental ranges. Thus, it results in producing erratic results with more wastage of money. As there was predominant increase in population with the rise in technological advances, agricultural business is equipped with immense transformation [2]. The recent development of "Smart Farming," with the World Wide Web of Things, has brought various changes in the agricultural environment patterns.

Smart Farming also holds various traditional methods of farming practices. This helps in tracking and controlling all the aspects of agricultural outputs depending on digitization. Thus, agricultural producers get all the present data about the environment, soil water stages, animal behavior, and crop yields through the integration of Internet of Things sensors. Producers can improve utility, manufacturing, and responsibility for the environment by providing data on pest reduction techniques, applying fertilizer plans, feeding animals, and moisture plans.

Figure 10.1 shows the Smart Farming role in developing a sustainable environment. A change in point of view that acknowledges the mutual dependence of agricultural undertakings and the need for creativity to solve contemporary issues have been key to the shift from normal to smart farming. An era of potential where skills and imagination combine to build an era for agriculture, is more adaptable, feasible, and lucrative, as it negotiates the intricate nature of the global farming sector, is heralded by the change from specialization to innovation.

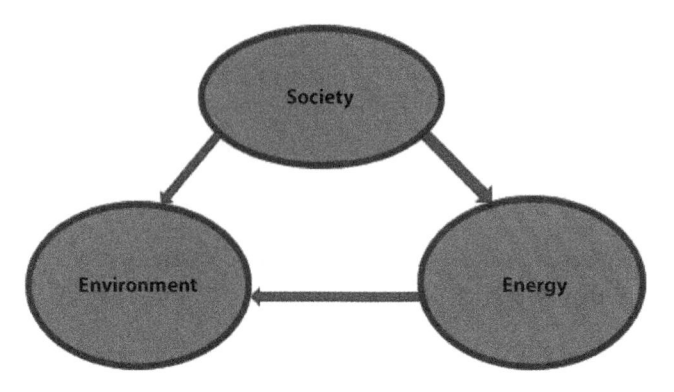

Figure 10.1 Smart farming in a sustainable environment.

10.2 IoT Sensor Networks in Farming

IoT sensor networks play a critical role for modern agriculture by enabling the transformation of traditional farming techniques into based on data, precision-driven businesses [3]. The aforementioned networks are intricate systems of carefully chosen, interconnected sensors that monitor, assess, and make choices in real time about land, stock services, and agricultural assets. A group of sensors intended to record different environmental data, agriculture metrics, and health of animal evaluations make up the heart of the Internet of Things network of sensors. In this regard, moisture in soil checks give a farmer useful knowledge about how moistened the agricultural areas are assisting them in planning irrigation and saving water. In a comparable way, sensors for humidity and temperature offer important data to assess the environmental conditions in livestock pens or greenhouses, and assuring optimum conditions for livestock development or maintenance.

An IoT sensor connection to a network is equally crucial because it enables sharing of data, and simple interaction among remote sensors and centralized control systems is possible [4]. Such networks may span large agricultural areas enabling landowners to access critical agricultural data swiftly and securely despite geographical constraints. They accomplish so by employing wireless communications, like as Zigbee wireless and Lora Internet acquaintances, or cellular networks for that matter. The Internet of Things networks of sensors depend on data analytics, which transforms processed data generated by sensors into information and insights that could be put to use. Agriculture can routinely detect variations and spot patterns of information gathered by sensing streams through complicated computational algorithms and mathematical simulations.

A new era in the practice of accurate farming is signaled by the use of the Internet of Things (IoT) connected sensors, which allow for the use of information-driven findings in place of gut feeling when making farming decisions. Farmers may be able to achieve previously unachievable levels of conservation, effectiveness, and yield by utilizing sensors, connection, and analytics. This will usher in an age of clever methods of farming that are prepared to take on the challenges that are presented by the changing environment of farming.

10.2.1 IoT Sensors and Their Applications

Various Internet of Things devices are used to monitor modern agricultural elements of the farming process enabling various livestock procedures. This helps in increasing the yields during the farming operations as they are helpful in various forms of applications [5]. Figure 10.2 shows the role of various IoT sensors in agricultural industry.

LIGHT SENSOR Automated Lightning system Energy saving	SOIL AND WATER SENSOR Maintaining the soil quality Monitoring the water levels	SOLENOID VALVES Remote operations Response time is very fast Low power consumption
TEMPERATURE AND HUMIDITY SENSOR Rugged construction Wireless connection More battery life time		SOIL MOISTURE SENSOR Provides real time values Water proof device Ultra low power consumption
GPS TRACKER Remote data access Locating live data Asset management	FERTILIZER PUMP Enables easy spraying of fertilizers Ease of operation	FLOW METER Ultra low rate of power consumption Provides high accuracy

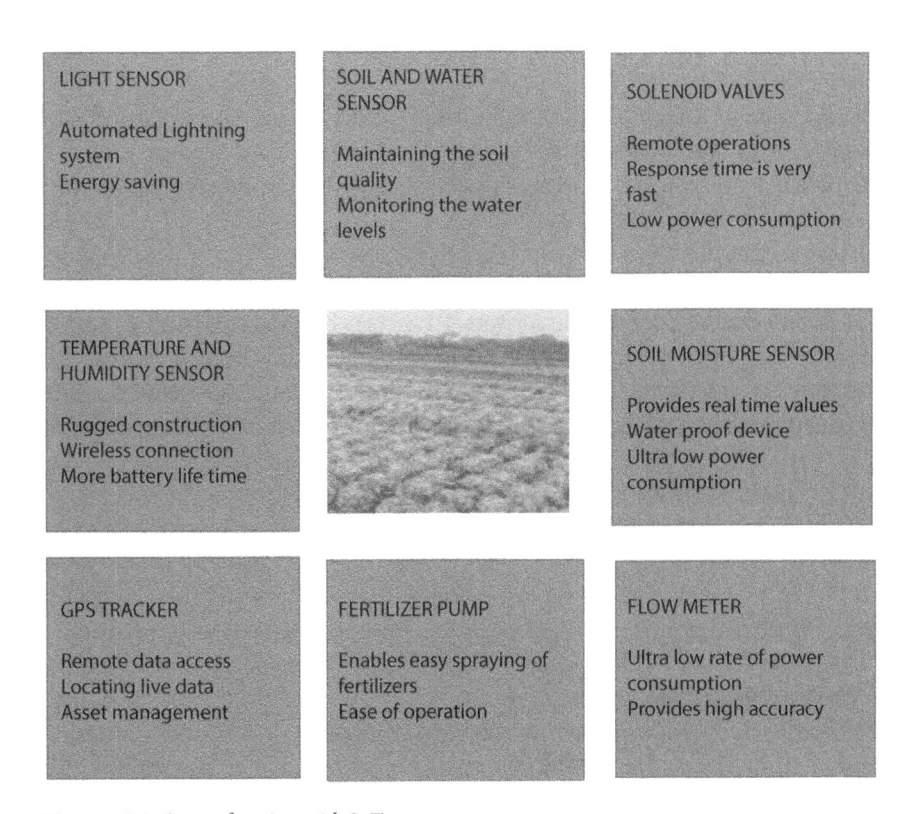

Figure 10.2 Smart farming with IoT sensors.

Soil Moisture Sensors
The amount of water content in the soil helps in providing the schedule for irrigation. This helps the farmers in efficiently saving the water, both during the flood and drought conditions, thus optimizing the irrigation period maintaining the moisture level of the soil.

Weather Sensors
Weather sensors helps in monitoring all the meteorological characteristics like humidity, temperature, wind speed, and precipitation [6]. This solely helps in making correct decisions on harvesting patterns and control of pests on various environmental conditions.

Crop Health Sensors
Crop health sensors keeps track of the plant physiological state. It includes pigment meters, leaf dryness sensors, and spectrum detectors. This enables farmers to identify disease outbreaks, with stress values taking prompt remedial action, changing fertilizer application, and the use of specialized pest control methods.

Environmental Sensors
Environmental sensors help in monitoring various parameters, like carbon dioxide, light intensity, and air quality. Temperature data sensors help the farmers in increasing the growth of plants safeguarding and growing crops [7].

Livestock Monitoring Sensors
Various forms of animal behavior, general crop conditions, and productivity, are obtained using monitoring sensors. RFID tags helps in confirming animal identity. Temperature meters help in identifying cattle's heat-related stress. Farmers can effectively improve breeding results, getting well-refined management methods, thus detecting and consistently monitoring livestock data.

Water Quality Sensors
The pH, chemical composition, and oxygen level of water are analyzed more effectively by the usage of water quality sensors. Farmers can be more confident in generating irrigation patterns with agricultural development, thus eliminating nutrient imbalances and contaminants of water quality.

Precision Agriculture Sensors
Numerous sensors are used in scouting, mapping of fertilizer contents, and yield of crops. This helps in making decisions, which maximizes the resource usage with lower cost of production boosting farm profitability.

10.3 Smart Pest and Disease Control in Crop Production

Pests and diseases are serious challenges to crop production because they negatively affect agricultural livelihoods, diminish yields, and jeopardize food security. To effectively lower risks, decrease chemical inputs, and protect crop health, successful pest and disease control strategies make use of cutting-edge technology, including Internet of Things devices, data analysis, and precise application approaches. Using a network of sensors, IoT-enabled rat monitoring devices find and count insects in real time. These sensors have the ability to identify insects using a variety of techniques, including pheromone traps, sound gauges, and picture recognition software [8]. Farmers that regularly monitor insect numbers are better able to pinpoint pest hotspots, evaluate population trends, and, where needed, use focused control techniques. Operators may boost crop resilience, lessen their dependency on chemicals, and encourage sustainable farming methods that guarantee long-term yields and safeguard the natural environment in the agricultural industry by utilizing creative disease and pest control techniques [9].

10.3.1 Meticulous Fertilization and Nutrition Control

Accurate fertilizer and nitrogen management are critical to modern agriculture's goal of increasing crop output while reducing its negative environmental effects. This method precisely monitors soil nutrient levels, ensures effective crop nutrient absorption, and modifies fertilizer applications using cutting-edge technology [10]. These technologies include the Internet of Things sensors, data analysis, and variable rate treatment equipment. The following is a summary of the main elements and techniques of fertilization and targeted nutrition management:

Soil Nutrient Monitoring
Farmers use the Internet of Things, soil nutrient detectors, to continually monitor vital minerals such as vitamins, potassium, phosphorus, and nitrogen (N). Farmers are able to determine which sections of their soil require replenishment and evaluate the nutritional content of their soil with accuracy using real-time data from these sensors on soil characteristics such as pH levels and nutrient concentrations.

Nutrient Mapping and Soil Sampling

When combined with GPS-enabled soil-sampling equipment, high-quality soil mapping technology makes it simpler to create precise nutrient maps that show the geographic variations in soil fertility within fields. Farmers may create site-specific fertilizer treatments that are suitable for various crop zones by utilizing maps, like this one, to pinpoint crop zones with varying nutrient requirements.

Variable Rate Fertilization

Based on crop specifications, production objectives, and soil nutritional levels, farmers may apply fertilizers at varying rates throughout fields by combining a variety of rate application techniques with satellite navigation guidance and precise farming software [11]. By adjusting the rate of fertilizer use in real time according to soil and prescription maps, farmers can maximize nutrient dispersion, avoid overfertilization, and boost crop uptake efficiency.

Nutrient Management Planning

Farmers can construct comprehensive nitrogen management strategies (NMPs) that align inputs, like fertilizer, with agronomic, economical, and environmental goals with the aid of nutrient software for management and based on data decision support tools. These plans encompass factors, including crop rotation, nutrient cycling, soil health, and legal and regulatory compliance, to guarantee sustainable long-term nutrient management practices.

Remote Monitoring and Control

With the assistance of IoT-enabled remote surveillance solutions, farmers may obtain data on soil nutrients, documentation of applied fertilizer, and current insights into fertilizer fluctuations in the field remotely. By regularly monitoring soil and crop reactions, farmers may correct nutrient surpluses or shortages, optimize nutrient management strategies, and make timely adjustments to fertilization procedures.

Nutrient Use Efficiency and Environmental Stewardship

Fertilizer application is ensured by precise handling of nutrients and enrichment procedures, which maximize crop requirements while minimizing nutrient waste to the environment. Farmers can lower greenhouse gas emissions, leaching, and nutrient runoff by implementing targeted fertilization strategies [12]. This will assist in upholding regulatory compliance and environmental sustainability. Farmers may increase crop yields, encourage sustainable agricultural practices, and enhance resource efficiency by utilizing precision fertilization and fertilizer management strategies. Farmers can optimize fertilizer investment returns, sustainably raise crop yields, and successfully manage the problems of feeding more people worldwide by using this data-driven method.

10.3.2 Accurate Irrigation Techniques and Water Administration

In light of the growing demand for irrigation and the unpredictable nature of climate change, intelligent irrigation systems have become indispensable instruments for boosting crop productivity and improving water management in the agricultural sector. These systems employ cutting-edge technologies like Internet of Things sensors, machine learning, and data analytics to provide crops the exact quantity of water at the correct time and place. The main elements and tactics of water management and precision irrigation systems are examined in the section that follows:

Soil Moisture Sensors
Soil moisture sensors are installed in fields to monitor the soil's moisture content in real time. Because such Internet of Things-enabled gadgets provide data on soil moisture levels at different depths and locations, farmers can assess the water condition of harvests and make informed decisions about when and how much to water.

Weather-Based Irrigation Scheduling
Algorithms for irrigation scheduling incorporate meteorological information such as humidity, wind speeds, temperature, and precipitation projections. Precision irrigation systems constantly modify irrigation schedules to account for shifting environmental conditions and maximize water use efficiency by integrating real-time weather data [13].

Drip and Micro-Irrigation Technologies
Water is precisely delivered to the root zone of plants by drip system and micro-irrigation systems, which use an interconnected system of emitters, tube, and valves. Particularly in water-stressed regions, these methods maximize crop yields while minimizing water losses from vaporization, runoff, and deep percolation.

Variable Rate Irrigation (VRI)
With VRI systems, farmers may adjust irrigation rates and patterning within fields according to topographical factors, crop drinking requirements, and soil moisture maps [14–17]. VRI systems minimize over-irrigation, decrease waterlogging or salt problems, and maximize water use efficiency by modifying water distribution based on geographical variability in soil characteristics and crop growth stages.

Remote Monitoring and Control
Farmers may remotely monitor irrigation systems, obtain accurate information on soil moisture levels, and modify irrigation settings from any location thanks to IoT-enabled remote monitoring tools. Farmers are able to optimize water distribution, reduce crop stress or water wastage, and rapidly address concerns by receiving warning for unexpected conditions or system faults.

Data Analytics and Decision Support Tools
To provide insights and management recommendations for irrigation, data analytics algorithms examine past irrigation data, trends in soil moisture, and crop consumption of water patterns. With the use of decision support systems, farmers can make well-informed decisions to optimize crop output and water efficiency by receiving actionable information on the best times, lengths, and frequencies for irrigation [18].

Water Recycling and Reuse
Through the optimization of water application rates and the reduction of losses, precision irrigation systems promote the widespread utilization of water reuse as well as recycling methods. Reusing treated wastewater or agricultural drainage system effluent for irrigation helps to promote sustainable water management techniques by lowering dependency on freshwater sources.

Farmers can reduce the risks caused by drought and climate variability, conserve water resources, and sustainably increase the growth of crops to meet the nutritional requirements of growing world populations by incorporating exactness drip irrigation and techniques for managing water into agricultural practices [19]. With the employment of these technologies, farmers may maximize water efficiency, reduce their negative effects on the environment, and develop financial strength in the face of shifting agricultural landscapes.

10.4 Automation and Robotics in Agriculture

Modern agriculture is being significantly changed by automation and robotics, which are transforming conventional farming methods and changing the agricultural landscape. These technologies offer new opportunities to improve the sustainability, efficiency, and productivity of agricultural operations [20, 21]. They include a wide variety of robotic systems, AI-driven solutions, and self-driving cars. Figure 10.3 looks at the key components and applications of automation and robotics in agriculture:

Robotic Farm Machinery
Unmanned cultivators, tractors, and planters equipped with GPS, devices, and AI algorithms provide precise and efficient field operations. With minimal human help, these autonomous machines can do tasks, like cultivating, reaping, spraying, and sowing, reducing labor costs and improving operational efficiency.

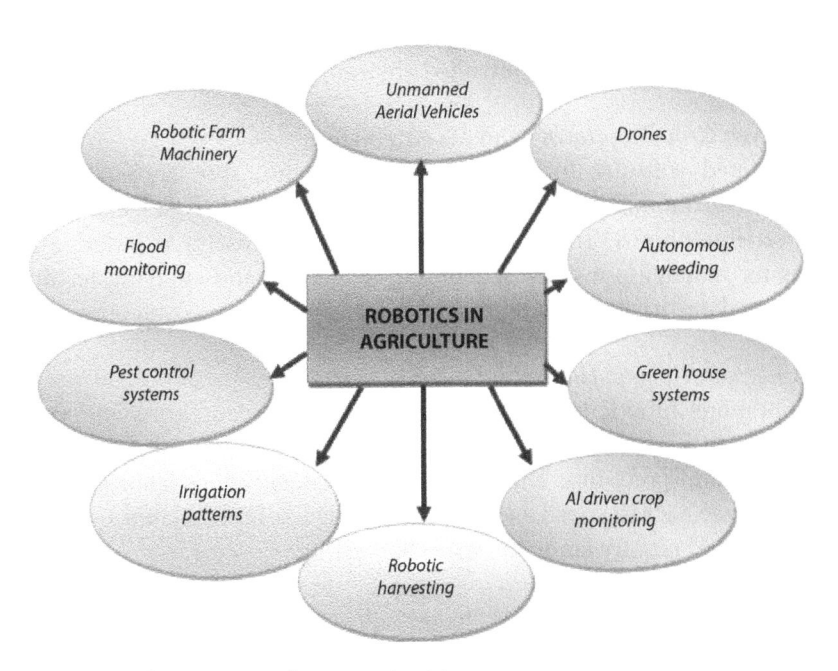

Figure 10.3 Robotics in agriculture Farming 4.0.

Unmanned Aerial Vehicles (UAVs) and Drones
Using suspended imagery along with information from UAVs and drones equipped with sensors, multispectral sensors, and LiDAR technology, farmers may map fields, identify pests, and monitor crops. Precision agriculture methods and timely decision making are aided by the rapid, inexpensive, and high-quality analysis capabilities that these unmanned drones offer.

Autonomous Weeding and Pest Control
Deep learning techniques and neural networks can be used to program robots to recognize and remove weeds and pests from agricultural regions on their own. These solutions reduce environmental contamination, decrease the need for chemical herb and pesticides, and promote environmentally friendly plant and insect management techniques [22, 23].

Automated Greenhouse Systems
Automated greenhouse systems incorporate sensors, switches that are needed, and control systems to optimize environmental factors such as moisture level, temperature, light level, and CO_2 concentration. These technologies enable accurate oversight over growth conditions, enhance crop quality and yield, and reduce resource inputs, all of which support year-round production in controlled environments.

AI-Driven Crop Monitoring and Management
AI-powered software tools monitor crop health, spot anomalies, and enhance agronomic practices by analyzing data from IoT gadgets, drones, and satellite photos. These platforms give farmers useful knowledge and ideas for managing insects, scheduling irrigation, and applying fertilizer enabling data-driven decision making and improved crop outcomes.

Autonomous Robotic Harvesting
Autonomous robotic systems equipped with automatic arms and vision systems can harvest produce, fruits, vegetables, and specialty crops with efficiency and precision. These robotic harvesters reduce labor shortages and increase harvest speed and precision using less human labor, especially for crops that require labor-intensive harvesting.

Smart Farming System Integration
The management solutions of farmlands helps in integrating the digital platforms combining the automated robotic systems. This helps in providing the resource allocations of optimized agricultural working patterns, which helps in increasing farm production with profitable increase.

The operation effectiveness can be increased by the farmers using the automated robotic systems embedded with the farming technologies. This opens the way for providing more productive, technologically robust, and sophisticated production of crops satisfying the feeding patterns of growing population.

10.4.1 Agricultural Operations Using Automatic Systems

Vehicles that are autonomous have brought about an important shift in farming processes. They will remake conventional methods of farming and improve efficiency, sustainability, and earnings across a range of activities and interests [24]. These vehicles' modern electronic devices, GPS tracking systems, and usage of AI techniques let themselves do operations, including soil production of cells, grasping, delivery by spraying, and blossom, with little help from humans. The following is an outline of the key parts and potential uses of self-driving cars in farming settings:

Autonomous Tractors
Intelligent machines are equipped with detectors for detection of obstacles, GPS aid, and inertial navigating that allow for a wide range of operations such as farming, planting, and nurturing. Such cars may function with incredible accuracy, following predetermined directions, covering a greater distance by employing less energy, and they decrease user discomfort.

Robotic Harvesters
With mechanized and artificial intelligence systems, unmanned robotic harvesters are capable of gathering veggies, fruits, and specialized crops. These devices, which comprehend ripened produce, modify harvesting approaches, and reduce crop harm, might improve both efficiency and yield while reducing cost of labor and reliance upon people.

Unmanned Aerial Vehicles (UAVs) for Crop Monitoring
AI systems with the autonomous robotic harvesters help to effectively pick rotten vegetables between the specialized crops [25]. These tools help maintain the effective agricultural yield reducing the use of labor and their dependency. This method helps in safeguarding the crops against damage due to pests.

Autonomous Sprayers and Spreaders
Autonomous sprays and spreading devices help in administering pesticides and nutrients related to plant growth, which help in providing the

tracking embedded with the GPS system of navigation panels. Thus, either by increasing or decreasing the amounts with crop requirements, environmental factors, the navigation vehicles or drones can limit the chemical spraying using an AI-controlled mechanism, thus increasing the resource sustainability and decreasing various limitations.

Soil Sampling and Mapping
Soil sampling, with the topological mapping procedures, enables autonomous vehicles to collect oil samples, thus increasing soil fertility for greater analysis with the detailed mapping procedures. This helps in bringing out various classifications based on soil fertility enabling the farmers to decide which crop has to grow at which season. Moreover, it also helps in implementing prior practices depending on various crop zones [26].

Connectivity to Farm Management Systems
Tools used for farm management, along with the self-driving vehicles, help in data transfer with remote monitoring categories and in managing various operations. These technologies improve the productivity of agriculture and efficiency in operations by simplifying procedures and reducing unproductive time by making setup, organizing, and carrying out assignments simpler.

Environmental Benefits and Sustainability
Autonomous vehicles are beneficial to the environment as they consume less fuel, minimize compacting of land, and employ insecticides more efficiently [27]. Through highly accurate and efficient operations, these vehicles preserve natural resources, mitigate the adverse ecological effects of conventional cropping methods, and advance sustainable agriculture operations. By adopting autonomous cars into their operations, farmers may overcome personnel shortages, increase productivity during operations, and apply environmentally friendly agricultural techniques. These innovative findings open the door to an agricultural sector that is more affluent, self-sufficient, and environmentally conscious, capable of feeding the increasing world population while preserving nature and the planet's resources.

10.4.2 AI in Farm Automation

Artificial intelligence has created revolutionary patterns enabling the analysis of predictive systems with higher decision-making procedures. Using vast amounts of data collected from sensors, aerial motor vehicles, satellites, and agricultural equipment, machine learning and artificial intelligence algorithms can analyze trends, spot anomalies, and improve farming operations with hitherto unheard-of precision and efficiency [28]. Below is a summary of the key components and applications of neural networks and artificial intelligence in agricultural automation:

- AI and ML systems look at both historical and present data on climatic trends, soil health, fruit condition, and management practices to estimate harvest rates with an outstanding level of accuracy.
- Farmers are able to plan ahead for crop unpredictability, allocate resources as efficiently as possible, and make well-informed decisions about sowing, reaping, and marketing tactics to these predictive models.
- Drone or field camera digital photos are analyzed driven by AI image recognition algorithms to find early indicators of pest infestations, nutrient deficits, and crop diseases [29]. These algorithms enable quick action, targeted therapies, and preventive pest management tactics to minimize the loss of crops and maximize yields by detecting minute changes in plant shape, color, and texture.
- Machine learning algorithms trained on large datasets of weed species and crop kinds may recognize and categorize plants in real time based on visual signals captured by cameras or sensors mounted on agricultural equipment. By cutting chemical inputs and mitigating environmental effect, these algorithms discriminate between goods and weeds. This makes synchronized weed control approaches and herbicide software specifically designed for the purpose of suppressing weeds possible.
- Robotics equipped with engines detectors and image recognition driven by artificial intelligence (AI) may carry out a variety of autonomous operations including grafting, weeding, sprayed, and harvesting. These robotic systems use data mining (ML) algorithms to recognize objects, navigate difficult settings, and change their habits in real time. This helps in effective completion of job thus increasing the effectiveness with minimal labor charges.
- To improve water efficiency, a fertilization schedule is created. This helps in evaluating rainfall rates, climate, and water predictions in predicting the various soil data. This helps in

mitigating the process of over-irrigation, waste water management, and sustainable water resources giving alignment with changes to various environmental conditions [30].
- Artificial intelligence (AI) helps in analyzing data from sensors, such as tractor kinds, mixes, and drip irrigation systems, on the equipment to monitor various operations, and determining project maintenance needs. This helps in prolonging the operational lifetime by anticipating equipment breakdowns before they happen.
- Artificial intelligence (AI)-driven solutions increase farm sustainability and facilitate data-driven decision making more effectively.

10.5 Cloud Computing for Agricultural Data Management

The process and capacity to store agricultural data provides flexible, scalable, and reasonable cloud computing strategies bringing a revolutionary change in data management. Producers, investigators, and other agricultural interests can use cloud-based platforms to access powerful computing resources, advanced analytics tools, and collaborative spaces to obtain insights, enhance decision making, and promote innovation in agriculture. The key components and applications of the internet of things for agricultural data management is given in Table 10.1.

Table 10.1 Key components and applications.

Key components	Applications
Data storage and accessibility	Farmers use cloud-based storage solutions
Data integration and interoperability	Seamless data transfer
Big Data analytics and insights	Massive volumes of data related to farming are processed and analyzed by cloud-based analytics systems
Remote sensing and imagery processing	Crop health can be tracked using the cloud sensors

Data Storage and Accessibility
Farmers use cloud-based storage solutions to securely and centrally store agronomist models, meteorological records, data collected by sensors, satellite shots, and adjustable, centralized databases for a range of agricultural data. Owing to their remarkable dependability and ease of use, these platforms promote knowledge sharing and stakeholder participation by enabling consumers to observe and exchange information from any location with access to the web [31].

Data Integration and Interoperability
Because cloud computing makes it easier to integrate various data sources and formats, it facilitates seamless data transfer between farming systems and apps. Combining data from a variety of sources, including smart handheld devices, agricultural management uses, and meteorological APIs, could aid farms gain an in-depth comprehension of their operations, speed up techniques, or improve their choices.

Big Data Analytics and Insights
Massive volumes of data related to farming are processed and analyzed by cloud-based analytics systems that use advanced algorithms and deep neural networks techniques to find correlations and insightful data. These advancements enable farmers to accurately and successfully manage their crops with the goal of minimizing risks and increasing yields. Producers can keep track of current events, compare previous performance, and estimate projected outcomes.

Remote Sensing and Imagery Processing
Crop health can be tracked using the cloud sensors indicating spot deviations and accessing the environmental parameters tracking the spot deviations, crop health, using drone feeds. It provides various details with the bioenergy calculations, and soil pattern moisture levels where the watering crop scouting may all benefit by these data.

Precision Agriculture Applications
The creation of customized agricultural guidance systems that cater to unique wants is made feasible in the cloud. With the help of these tools, input quality, resource management, and overall farm production can be increased using prediction models, visual maps, and adjustable dashboards. Agricultural applications help farmers use cloud precision to make smart decisions, improving efficiency through different recommendations and solutions.

Scalability and Cost Efficiency
Scalability to agricultural businesses with the cloud systems adjusts their capacity meeting the upfront expenses and advanced equipment usage, thus reducing the overall cost of the agricultural modern information systems.

Security and Data Privacy
Cloud computing companies safeguard agricultural data from cyber threats, unauthorized access, and breaches by putting strong security controls and compliance standards in place. Cloud-based solutions protect users' trust and confidence by guaranteeing the security, honesty, and ready accessibility of agricultural data through the use of encrypting data-restricted access, and data governance regulations.

Farmers and other agricultural stakeholders can leverage knowledge gained from data, sophisticated statistical analysis, and communication tools to optimize farm operations, boost productivity, and promote sustainable agricultural growth using virtualization for agricultural data management [32]. These cloud-based technologies enable farmers to achieve success in a quickly changing agricultural landscape by helping them to adapt to shifting difficulties and grab new possibilities.

10.6 Big Data Analytics for Predictive Farming

The agricultural industry has undergone a transformation thanks to big data analytics, which has given farmers access to predictive knowledge and decision-making tools to maximize yields, reduce risks, and optimize crop production. Using sophisticated analytics, machine learning computer programs, and massive data processing, predictive farming forecasts crop performance, foresees environmental effects, and recommends proactive management practices. Below is a summary of the key components and applications of substantial information analytics for farming prediction:

- The compilation of information from a run of sources, counting authentic yields of crops, climate reports, soil properties, fawning photographs, and agronomic models, is the introductory arrangement in prescient cultivating. To construct enormous datasets that accurately delineate the complexity of cultivating settings and frameworks, the information sources are combined, cleansed, and standardized.

- Model estimates that assess edit yields within the future are made by utilizing relapse examination, determining of drift, gathering strategies, and other analytics-based huge information approaches to chronicle abdicate information. Machine learning calculations are able to precisely and dependably expect edit yields by looking at the relationships between surrender variables, such as soil properties, agrarian strategies, and climatic conditions.
- Large-scale meteorological information, such as precipitation, mugginess, temperature, and wind designs, may be prepared and examined using big information analytics to form prescient models of climate alteration and its change. These models offer assistance to ranchers who expect dangers and adjust their administration strategies by assessing the impacts of extraordinary climate occasions, such warm waves, surges, and storms, on the advancement of crops, expansion, and surrender potential.
- To make models for the long-run soil quality and richness, huge information expository instruments coordinate different soil information, such as soil surface, pH, carbon substance, and supplements [33]. These models give the ideal conditions for edit development and efficiency by moderating nitrogen crevices and lacks, optimizing fertilizer applications, and directing supplement choices through the assessment of the soil's characteristics and supplement dynamics.
- Analytics for enormous information bug and ailment flare-ups are anticipated utilizing methods counting topographical investigation, algorithmic learning, artificial intelligence, and satellite photography. These strategies are based on crop phenology, environmental factors, and insect life cycles. To pinpoint high-risk regions and locations for pest infestations, predictive models make use of historical information on pest occurrence, disease outbreaks, and crop sensitivity. This enables prompt intervention and the implementation of tailored management strategies.
- Big data analytics technology gives farmers access to multimedia presentations with forecasting data and crop management guidance, as well as decision assistance tools. To assist in making decisions on planting, hydro scheduling, pest

tracking, and harvest scheduling, these systems use real-time data flows, modeling for forecasting, and agronomic expertise. This enables farmers to make well-informed decisions and maximize the use of available resources.
- Response loops and iterative procedures are used by predictive agricultural systems to enhance their mathematical models and management techniques over time. Predictive farming approaches may be optimized by farmers through the update of management plans, optimization of prediction algorithms, adaptation to changing conditions with the integration of new data, measuring model performance, and evaluation of results.

10.7 Sustainable Practices with IoT in Agriculture

With networked sensors, analysis of data, and automated processes, the Web of Things (IoT) has made it easy for farmers to optimize resource use, minimize environmental impact, to enhance resilience as time goes on through a range of agricultural operations [34]. By monitoring crop condition, fertilizer levels, and subsurface levels in real time, farmers may effectively tailor their irrigation and fertilization methods to meet the specific needs of their agricultural produce with the assistance of Internet of Things (IoT) sensors. Also, IoT enables proactive methods of infection and insect control reducing the need for dangerous insecticides while also improving the well-being of ecosystems. By using IoT-enabled pest tracking tools with deep learning, farmers may identify infestations early, manage activities more closely, and adopt coordinated pest control techniques that combine predatory settings and biologic methods for management. Because it optimizes feed economic performance, decreases emission of greenhouse gases, and improves animal well-being, IoT is necessary for the farming and raising of livestock. Farmers can ensure that their farm animals are living in optimal circumstances, quickly identify the signs of stress or illness, and administer the appropriate care by establishing the livestock with Internet of Things (IoT)-capable equipment for behavioral evaluation, present sensing, and nutritional monitoring [35]. As IoT grows and permeates the agricultural business, its role in encouraging environmentally friendly behaviors will become increasingly important in ensuring the continued existence and prosperity of the sector.

10.8 The Future Landscape of Farming 4.0

The farming landscape of the future, sometimes known as "Farming 4.0," is a paradigm change toward a more data-driven, technologically sophisticated, and environmentally conscious agricultural industry. Farming 4.0, which builds on the achievements of earlier agricultural revolutions, uses state-of-the-art tools, creative thinking, and teamwork to meet the changing opportunities and challenges that face the world's food production. With the incorporation of cloud computing platforms, sensor networks, and Internet of Things (IoT) devices into agricultural operations, Farming 4.0 is defined by a broad digital transformation and connection. The utilization of networked systems facilitates continuous surveillance, gather data, and decision making, hence promoting increased productivity, resilience, and productivity in agricultural activities.

Farmers may boost crop yields, cut down on input waste, and maximize resource use using precision agriculture techniques with the help of intelligent machines (AI), artificially intelligent algorithms, and advanced data analytics. Massive amounts of data on soil conditions, weather patterns, growth of crops, and insect dynamics are analyzed to help farmers make well-informed decisions, adapt management strategies, and achieve more exact results in a sustainable manner. Farming 4.0 automates repetitive tasks, decreases labor-intensive employment, and increases operational efficiency in agriculture through the use of robotics and autonomous technologies. Drones equipped with artificial intelligence (AI) algorithms, autonomous autos, and automated harvesters enable precision in the planting interpretation, applications, harvesting, and monitoring. This increases agricultural operations' output while reducing the demand for human labor.

Sustainability, which prioritizes minimizing environmental impact, protecting natural resources, and promoting ecosystem health, is one of Farming 4.0's guiding concepts. The future viability of agricultural systems is ensured by combining technology-driven methods with sustainable practices, such as regenerative agriculture, organic farming, and agroforestry, which also enhance soil health, ecosystems, and carbon sequestration. Farming 4.0 promotes cooperation and knowledge sharing through digital platforms, ecosystem partnerships, and data exchanges. An even more flexible and linked agricultural ecosystem is promoted by agriculture, scholars, especially innovators in technology, and legislators working together to co-create novel approaches and share best practices to address common concerns about food availability, economic growth, and climate change adaptation. The key elements of Farming 4.0 are traceability throughout the nutritional supply chain, transparency, and consumer interaction [36]. Consumers may now obtain details on the origin of food, production processes, and sustainability certifications because of digital platforms, smart labels, and decentralized blockchain technology. Customers are better equipped to make judgments as a result, and consumer–producer trust is increased. In Farming 4.0, which uses technology-enabled solutions to lower risks, boost farm adaptability, and develop adaptive ability, adaptability and climate change adaptability are given first attention [37, 38].

10.9 Conclusion

Ultimately, Farming 4.0 brings in a new era in cultivating crops, one in which innovative agricultural technology, made possible by the rise of the Internet of Things, replaces obsolete methods of cultivation with data-driven, efficient, and ecologically friendly approaches. Using data analysis along with remote monitoring to improve livestock management, diminishing insect and infection events with rapid identification and aimed at behavior, and optimizing the use of mineral resources through irrigation and fertilization are just a few of the methods wherein Farming 4.0 uses the Internet of Things. Farming 4.0 increases market access for smallholder farmers, promotes openness and confidence around the food supply chain, and speeds up the widespread use of sustainable practices by utilizing digital technology, ecosystem alliances, and already-existing data exchanges [39, 40]. Fairness, toughness, and acceptability must be given top priority as Farming 4.0 is developed to guarantee that all farmers, despite their geographic location or size, may take use of the exciting potential of IoT-enabled farm solutions. Farming 4.0 has the potential to feed the increasing global population, encourage environmental stewardship, and build a more healthy and successful agricultural sector for future generations by promoting innovation, cooperation, and sustainability.

References

1. Javaid, M., Haleem, A., Singh, R.P., Suman, R., Enhancing smart farming through the applications of Agriculture 4.0 technologies. *Int. J. Intell. Netw.*, 3, 150–164, 2022.

2. Jararweh, Y., Fatima, S., Jarrah, M., AlZu'bi, S., Smart and sustainable agriculture: Fundamentals, enabling technologies, and future directions. *Comput. Electr. Eng.*, 110, 108799, 2023.
3. Sharma, A., Sharma, A., Tselykh, A., Bozhenyuk, A., Choudhury, T., Alomar, M.A., Sánchez-Chero, M., Artificial intelligence and internet of things oriented sustainable precision farming: Towards modern agriculture. *Open Life Sci.*, 18, 1, 20220713, 2023.
4. Ashish, G., Chakraborty, D., Law, A., Artificial intelligence in Internet of things. *CAAI Trans. Intell. Technol.*, 3, 4, 208–218, 2018.
5. Sathiya, S. and Antony, C., & Ghodke, P. K., Smart Agriculture: Emerging and Future Farming Technologies, in: *Recent Trends and Best Practices in Industry 4.0*, pp. 135–181, 2023.
6. Prakash, C., Singh, L.P., Gupta, A., Lohan, S.K., Advancements in smart farming: A comprehensive review of IoT, wireless communication, sensors, and hardware for agricultural automation. *Sens. Actuators, A*, 47, 114605, 2023, https://doi.org/10.1016/j.sna.2023.114605.
7. Bhangar, N.A. and Shahriyar, A.K., IoT and AI for Next-Generation Farming: Opportunities, Challenges, and Outlook. *Int. J. Sustain. Infrastruct. Cities Soc.*, 362, 8, 2, 4–26, 2023.
8. Polymeni, S., Plastras, S., Skoutas, D.N., Kormentzas, G., Skianis, C., The Impact of 6G-IoT Technologies on the Development of Agriculture 5.0: A Review. *Electronics*, 12, 12, 2651, 2023.
9. Khan, A., Hassan, M., Shahriyar, A.K., Optimizing onion crop management: A smart agriculture framework with IoT sensors and cloud technology. *Appl. Res. Artif. Intell. Cloud Comput.*, 6, 1, 49–67, 2023.
10. Rahman, M.B., Chakma, J.D., Momin, A., Islam, S., Uddin, M.A., Islam, M.A., Aryal, S., Smart Crop Cultivation System Using Automated Agriculture Monitoring Environment in the Context of Bangladesh Agriculture. *Sensors*, 23, 20, 8472, 2023.
11. Ashwini, A. and Sriram, Quadruple spherical tank systems with automatic level control applications using fuzzy deep neural sliding mode FOPID controller. *J. Eng. Res.*, 2023, Preprint, https://doi.org/10.1016/j.jer.2023.09.022.
12. Simo, A., Dzitac, S., Duțu, A., Pandelica, I., Smart Agriculture in the Digital Age: A Comprehensive IoT-Driven Greenhouse Monitoring System. *Int. J. Comput. Commun. Control*, 18, 6, 6147, 2023.
13. Ashwini, A., Purushothaman, K.E., Rosi, A., Vaishnavi, T., Artificial Intelligence based real-time automatic detection and classification of skin lesion in dermoscopic samples using DenseNet-169 architecture. *J. Intell. Fuzzy Syst.*, 4, 1–16, 2023, Preprint.
14. Pawar, A. and Deosarkar, S.B., IoT-based smart agriculture: an exhaustive study. *Wirel. Netw.*, 29, 1–14, 2023.
15. Rehman, K.U., Andleeb, S., Ashfaq, M., Akram, N., & Akram, M. W., Blockchain-enabled smart agriculture: Enhancing data-driven decision making and ensuring food security. *J. Clean. Prod.*, 427, 138900, 2023.

16. Balkrishna, A., Pathak, R., Kumar, S., Arya, V., Singh, S.K., A comprehensive analysis of the advances in Indian Digital Agricultural architecture. *Smart Agric. Technol.*, 5, 100318, 2023, https://doi.org/10.1016/j.atech.2023.100318.
17. Asha, K.N., Gahana, N., Gowda, S., Harsha, C.R., Chandana, D.M., Agriculture Automation System using Machine Learning and Internet of Things. *Int. J. Eng. Manage. Res.*, 13, 3, 163–167, 2023.
18. Kethineni, K. and Pradeepini, G., Intrusion detection in internet of things-based smart farming using hybrid deep learning framework. *Cluster Comput.*, 27, 1–14, 2023.
19. Ashwini, A. and Kavitha, V., Automatic Skin Tumor Detection Using Online Tiger Claw Region Based Segmentation–A Novel Comparative Technique. *IETE J. Res.*, 69, 1–9, 2021.
20. Chandraprabha, M. and Dhanaraj, R.K., Deep Learning Based Bagged CNN with Whale Optimization Algorithm to Forecast the Productivity of Rice Crop using Soil Nutrients, in: *2023 International Conference on Advances in Computation, Communication and Information Technology (ICAICCIT)*, pp. 707–711, IEEE, 2023, November.
21. Malathy, S., Vanitha, C.N., Kotteswari, S.V., S., P., M., E., Rainfall Prediction for Enhancing Crop-Yield based on Machine Learning Techniques, in: *2022 International Conference on Applied Artificial Intelligence and Computing (ICAAIC)*, pp. 437–442, Salem, India, 2022, 10.1109/ICAAIC53929.2022.9792793.
22. Adli, H.K., Remli, M.A., Wan Salihin Wong, K.N.S., Ismail, N.A., González-Briones, A., Corchado, J.M., Mohamad, M.S., Recent Advancements and Challenges of AIoT Application in Smart Agriculture: A Review. *Sensors*, 23, 7, 3752, 2023.
23. Huynh, H.X., Tran, L.N., Duong-Trung, N., Smart greenhouse construction and irrigation control system for optimal Brassica Juncea development. *PLoS One*, 18, 10, e0292971, 2023.
24. Sulastri, Smart Greenhouse Development: A Case Study in West Java, Indonesia, in: *Smart Agriculture for Developing Nations: Status, Perspectives and Challenges*, pp. 69–76, Springer Nature Singapore, Singapore, 2023.
25. Gangadevi, E., Rani, R.S., Dhanaraj, R.K., Nayyar, A., Spot-out fruit fly algorithm with simulated annealing optimized SVM for detecting tomato plant diseases. *Neural Comput. Appl.*, 36, 8, 4349–4375, 2024.
26. Alahmad, T., Neményi, M., Nyéki, A., Applying IoT Sensors and Big Data to Improve Precision Crop Production: A Review. *Agronomy*, 13, 10, 2603, 2023.
27. Ferlin, A.A. and Rosi, V., Iot based object perception algorithm for urban scrutiny system in digital city, in: *2023 International Conference on Circuit Power and Computing Technologies (ICCPCT)*, pp. 1788–1792, 2023.
28. Frikha, T., Ktari, J., Zalila, B., Ghorbel, O., Amor, N.B., Integrating blockchain and deep learning for intelligent greenhouse control and traceability. *Alexandria Eng. J.*, 79, 259–273, 2023.

29. Ashwini, A., Purushothaman, K.E., Priya, B., Prathaban, M., Jenath, P., Automatic Traffic Sign Board Detection from Camera Images Using Deep learning and Binarization Search Algorithm, in: *2023 International Conference in recent advances in Electrical, Electronics, Ubiquitous Communication and Computational Intelligence (RAEEUCCI)*, IEEE, 2023.
30. Herdiansyah, H., Antriyandarti, E., Rosyada, A., Arista, N., II, Soesilo, T.E.B., Ernawati, N., Evaluation of Conventional and Mechanization Methods towards Precision Agriculture in Indonesia. *Sustainability*, 15, 12, 9592, 2023.
31. Polkowski, Z., Mishra, S.K., Mishra, B.K., Borah, S., Mohanty, A., Impact of internet of things in social and agricultural domains in rural sector: a case study. *Int. J. Cloud Comput.*, 12, 1, 90–105, 2023.
32. Debangshi, U., Sadhukhan, A., Dutta, D., Roy, S., Application of Smart Farming Technologies in Sustainable Agriculture Development: A Comprehensive Review on Present Status and Future Advancements. *Int. J. Environ. Clim. Change*, 13, 11, 3689–3704, 2023.
33. Balasubramaniam, S. and Gollagi, S.G., Software defect prediction via optimal trained convolutional neural network. *Adv. Eng. Softw.*, 169, 103138, 2022.
34. Gebresenbet, G., Bosona, T., Patterson, D., Persson, H., Fischer, B., Mandaluniz, N., Nasirahmadi, A., A concept for application of integrated digital technologies to enhance future smart agricultural systems. *Smart Agric. Technol.*, 5, 100255, 2023, https://doi.org/10.1016/j.atech.2023.100255.
35. Balasubramaniam, S., Kumar, K.S., Kavitha, V., Prasanth, A., Sivakumar, T.A., Feature selection and dwarf mongoose optimization enabled deep learning for heart disease detection. *Comput. Intell. Neurosci.*, 2022, 2022, https://doi.org/10.1155/2022/2819378.
36. Zhang, H. and Zhang, L., A Reliable Data-Driven Control Method for Planting Temperature in Smart Agricultural Systems. *IEEE Access*, 11, 38182–38193, 2023.
37. Balasubramaniam, S., Vijesh Joe, C., Sivakumar, T.A., Prasanth, A., Satheesh Kumar, K., Kavitha, V., Dhanaraj, R.K., Optimization enabled deep learning-based DDoS attack detection in cloud computing. *Int. J. Intell. Syst.*, 2023, 2023, https://doi.org/10.1155/2023/2039217.
38. Gumbi, N., Gumbi, L., Twinomurinzi, H., Towards sustainable digital agriculture for smallholder farmers: A systematic literature review. *Sustainability*, 15, 16, 12530, 2023.
39. Yadav, N. and Sidana, N., Precision Agriculture Technologies: Analysing the Use of Advanced Technologies, Such as Drones, Sensors, and GPS, In Precision Agriculture for Optimizing Resource Management, Crop Monitoring, and Yield Prediction. *J. Adv. Zool.*, 44, 255, 2023.
40. Balasubramaniam, S. and Kavitha, V., Geometric data perturbation-based personal health record transactions in cloud computing. *Sci. World J.*, 2015, 2015, https://doi.org/10.1155/2015/927867.

11

Public Safety Management in Smart Society 5.0: A Blockchain-Based Approach

P.N. Senthil Prakash[1], S. Karthic[2]* and M. Saravanan[2]

[1]School of Computer Science and Engineering, Vellore Institute of Technology, Chennai, Tamil Nadu, India
[2]Department of Computer Science and Engineering, KPR Institute of Engineering and Technology, Coimbatore, Tamil Nadu, India

Abstract

Society 5.0 represents a framework to indicate representation of cyberspace and physical space, which is aimed to provide solutions to the societal challenges, promoting economic growth using technology. In the recent years, significant advancements across various technological domains have produced various benefits to the human community. For instance, traffic management systems is capable of dynamically adjusting the signal timings considering the real-time vehicle flow, which helps to eliminate traffic in smart cities. Another scenario is the innovations in autonomous vehicles including cars and other passenger vehicles. These are capable of providing more secure traveling experience reducing the need for human drivers. Similarly, smart healthcare systems efficiently provide essential treatments to patients and support doctors with detailed reports helping in further diagnosis and treatment more accurately. Similarly, the innovation in technology plays a significant role in industries, such as automotive, construction, and food, which provide improvements in terms of efficiency and productivity. This leads to streamlined process of the industries to meet the evolving demands of the market. For instance, consider the manufacturing sector; regulatory bodies are essential to ensure the production activities by conducting periodic audits to ensure compliance with safety, environmental, and quality standards. Despite such supervision, deviation incidents still occur causing risks to industry workers and the public residing near the industries. One of the areas where blockchain technology provides a promising solution is to improve the regulatory overhead and overcome industrial risks. Data are captured from industrial operations and stored in blocks

---

*Corresponding author: karthi131@gmail.com

and transmitted to regulatory bodies in real time utilizing blockchain technology. This ensures that data stored in the distributed ledger remains unaltered enhancing transparency and accountability of the industrial information. Smart contracts are a key feature of blockchain technology, which is utilized to enforce compliance with regulatory standards, since smart contracts execute automatically and perform the specified tasks when the specific criteria are met. From the huge amount of data generated, information, like quality of product, regulatory compliance, and environmental impact, can be obtained. Applying blockchain with these data can ensure regulatory compliance and quality of the product. The integration of blockchain into an industry process can provide significant improvement and contribute positively toward Society 5.0. The industries can utilize blockchain technology to improve environment practices being followed. This also enables the industries to have proper planning toward waste disposal, pollution emission, and resource consumption effectively. Thus, integrating blockchain into industrial processes contribute to sustainability and enhance the quality of life.

Keywords: Society 5.0, smart cities, blockchain, distributed ledger

11.1 Introduction

Society 5.0 represents a significant evolvement in our society by the seamless integration of digital and physical spheres with the aim of enhancing the well-being of humans. It is being developed based on the shortcomings of the existing methods. Society 5.0 initiates a transformative transition where technology plays a central role in driving innovation focused on human needs [1]. Society 5.0 focuses to address the global challenges by involving latest technologies such as artificial intelligence, big data analytics, and the Internet of Things (IoT). Society 5.0 seeks to address global challenges while promoting inclusive growth and environmental sustainability [2]. This ensures that the development of technologies effectively contributes to the economic growth and societal well-being. Through a thorough exploration of its core principles and guiding philosophies, it aims to unravel the intricate fabric of Society 5.0 and its significant implications for the future direction of humanity. Society 5.0 gives priority to environmental sustainability to protect and preserve the natural resources and the ecosystem. The focus is to create a better ecosystem and living environment utilizing the latest technologies.

Due to the technological advancements in technologies like AI, IOT, and big data, the society is undergoing a digital transformation. The advancement in these technologies can contribute to efficiency improvement and connectivity across various sectors. Society 5.0 deals with integration of technology into real-time environment. For example, smart cities can use the technologies for energy conservation, parking space allotment, and enhance public services. Thus, the integration of technologies ensures that the transformation of industries, like healthcare, manufacturing, and agriculture, can improve productivity and meet consumer needs [3]. This indicates that it is necessary to utilize the cutting-edge technologies to improve the well-being of the public.

Blockchain technology is considered as a novel approach in handling data to revolutionize storage, management, and transfer of data in a manner that is both secure and transparent. It functions as a decentralized ledger system, where transactions are recorded and verified across a network of interconnected computers, known as nodes, in a chronological and immutable fashion [4]. During the transactions, it is broadcasted and verified by other nodes using consensus mechanism. This structure ensures that each block of data is securely linked to the next block forming an unbroken chain of information.

Blockchain is employed in a variety of applications to ensure security and integrity of data. Blockchain has gained popularity in recent years. Figure 11.1 presents the statistics of a Google search of the terms blockchain and smart city for the past few years. It is seen that blockchain has gained momentum from the year 2017 onward.

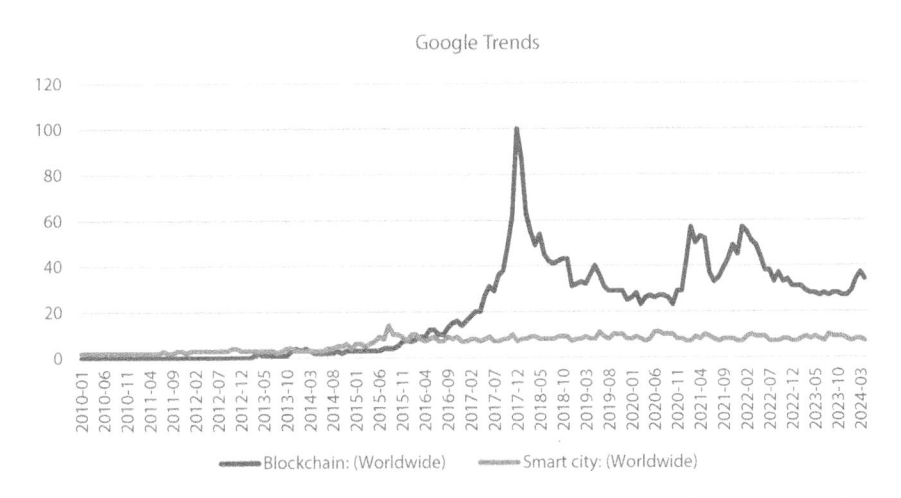

Figure 11.1 Google worldwide trends for the search term "blockchain" and "smart city" from January 2010 to March 2024.

The important aspect of blockchain technology lies in its transparency as the ledger is distributed across multiple nodes such that all participants within the network can have access to an identical copy of the information. This fosters an environment for trust and accountability. The immutable nature of blockchain ensures that once a transaction is successfully completed, it remains tamper-proof and resistant to alteration ensuring security and reliability [5].

Blockchain technology evolved from the term bitcoin. It utilizes the decentralized ledger to record all the data and transactions. From the evolution, blockchain has gained attention across various industries due to the potential to address various real-time challenges [6]. Blockchain reduces the manual operational cost thereby reducing the operational expenses in the industries.

Blockchain possesses a potential to reshape how digital information is stored and managed. It provides providing a decentralized, secure, and transparent alternative to traditional centralized systems [7]. Usually, the data are distributed across the network of nodes to avoid a single point of failure and ensure trust. As technology is growing toward digitalization and has security challenges, blockchain is considered as an alternative secure mechanism in sectors like healthcare, finance, identity management, and so on. This has made blockchain an essential technology to reshape the future and business. Figure 11.2 presents the essential features of blockchain in a variety of applications.

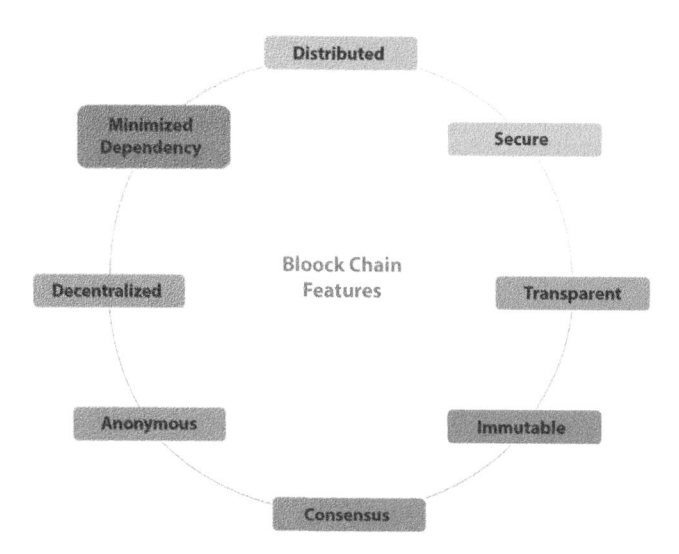

Figure 11.2 Features of blockchain.

Society 5.0 refers to the convergence of various technologies like blockchain, artificial intelligence, IOT, and big data in real-time scenarios. It represents a human centric technology development that helps in improving the welfare of the public. The integration of various technologies possesses various challenges and security concerns including data privacy and ethical use of technology. Hence, integration of these technologies requires attention and measures to ensure the safety and security of individuals [8].

Due to the availability of the internet, almost all the devices are interconnected devices and digital platforms are being accessed for a variety of applications, Society 5.0 becomes vulnerable to cyberattacks, data breaches, and identity theft. Deficiencies in cybersecurity infrastructure may result in the compromise of sensitive data, financial harm, and the disruption of vital services in real-time applications [9].

The blockchain technology was originally developed from the concept of bitcoin and cryptocurrency. It has now emerged as a framework providing its application in various sectors contributing to Society 5.0. Blockchain has evolved into a versatile tool with extensive implications across diverse sectors [10]. The decentralized nature of blockchain eliminates a single point of failure and makes it difficult to tamper the data present in the blocks. In Society 5.0, blockchain can help address the challenges of public safety management integrating with other technologies.

The crucial part of blockchain is consensus mechanism that contribute to the network security and efficiency. Consider the block chain ledger as a book that records all the transactions. The consensus algorithm functions as the guardian of the book ensuring no unauthorized access to the data. If any unauthorized access to the data happens, the consensus algorithm detects it and takes necessary actions to prevent data update [11]. This ensures the integrity of the data and makes difficult for the unauthorized user in making changes to the data. The consensus algorithm is also involved in ensuring the smooth functioning of the network. These algorithms maintain consistency across the entire blockchain. This consistency acts an agreement to the participant about the validity and the order of the transactions. This agreement ensures all the users have access to the same view of a ledger. The consensus mechanism also ensures the fault tolerance of the distributed ledger. Even when some parts of the network fail or act incorrectly, the network continues to function properly. They also prevent double spending of same cryptocurrencies. Different consensus mechanisms are developed for various application with the blockchain systems [12]. These methods are developed to address various requirements, and they vary in the way they achieve agreement among the participants. Some common consensus mechanisms include Proof of Work (PoW), Proof of

Stake (PoS), Delegated Proof of Stake (DPoS), Byzantine Fault Tolerance (BFT), Practical Byzantine Fault Tolerance (PBFT), Proof of Authority (POA), Proof of Identity (POI), Proof of Burn (POB), etc. Each consensus mechanism has its own strategy in validating the given transactions and adding them to the distributed ledger. Different consensus algorithms provide different levels of security mechanisms [13]. The selection of appropriate consensus mechanisms depends on the application and has impact on the performance of the blockchain network [14].

11.2 Security Challenges in Society 5.0

Society 5.0 is the combination of various technologies such as AI, IOT, blockchain, and big data. AI is used for automated decision making, and IOT offers connectivity to devices. Blockchain ensures secure transactions, and big data facilitates analysis of vast amounts of data. The integration of such technologies offers various benefits leading to improved productivity and quality of life. The rapid digitalization of technologies is exposed to security challenges like cyber threats and malware attacks [15]. The interconnected nature in Society 5.0 introduces vulnerabilities where cyberattacks can compromise sensitive data. As technology keeps evolving, new types of threats keep emerging [16].

The rapid improvements in technologies like AI, machine learning, quantum computing, and biotechnology presents novel cybersecurity challenges. AI-driven systems are vulnerable to adversarial attacks, while malevolent entities tamper with algorithms to yield erroneous or biased outcomes [17]. Quantum computing faces a significant risk to cryptographic protocols potentially obsoleting current encryption techniques and jeopardizing data integrity. The advancements in biotechnology, such as gene editing and synthetic biology, raise apprehensions regarding the potential exploitation of biological data and substances for malicious intents [18].

In Society 5.0, various systems and devices interact seamlessly through IoT networks enabling smooth communication and data exchange. However, this integration also expands the potential targets for cybercriminals to exploit the data from the network. Vulnerabilities within one interconnected system can spread across the entire network leading to multiple failures and widespread disruptions. Examples of interconnected systems at risk include smart homes, smart cities, critical infrastructure, and industrial control systems.

The seamless integration of cutting-edge technologies and widespread connectivity gives rise to a multitude of privacy risks affecting individuals, businesses, and societal applications. The extensive utilization of IoT devices, intelligent sensors, and digital platforms leads to the ongoing accumulation of extensive personal data [19]. As data keeps growing exponentially, security challenges also keeps increasing. These data frequently encompass details such as location information, medical data, and behavioral patterns of the users. The perpetual surveillance facilitated by these technologies elicits apprehensions regarding privacy violations and the risk of unauthorized intrusion into personal data. There is an exponential growth in the deployment of biometric identification systems encompassing technologies such as facial recognition, fingerprint scanning, and iris recognition [20, 21]. If biometric data are once compromised, they lack the ability to be protected from alteration or substitution rendering individuals susceptible to identity theft and illicit surveillance [22].

The usage of AI algorithms with blockchain introduces the concern of algorithmic. It represents systematic inaccuracies or inequities in the results generated by AI algorithms. This can arise due to various sources like incorrect training flows in algorithm design and defects in data collection process. These biases have a significant impact on social applications like employment, financial services, and law enforcement [23]. Mitigating algorithmic biases necessitates meticulous examination of training data, algorithmic architecture, and decision-making methodologies to uphold principles of fairness, transparency, and accountability [24].

The exponential progress within Society 5.0 presents intricate challenges to regulatory frameworks and privacy legislation. Existing regulations often lag behind the rapid evolution of emerging technologies resulting in deficiencies in safeguarding privacy and enforcing compliance [25]. As data transcends geographical borders and legal jurisdictions, achieving regulatory compliance becomes markedly intricate that necessitates global cooperation and the harmonization of privacy standards [26].

11.3 Blockchain in Society 5.0

In the increasing population, the number of industries is also growing to meet the demand. The industries include manufacturing industries, food processing industries, chemical industries, textile industries, refinery industries, etc. These industries are monitored by regulatory bodies to ensure that these industries are following regulations to ensure the safety of the employees, people around the industry, and the common public utilizing the products produced by those industries [27]. For instance, the Food Safety and Standards Authority of India (FSSAI) is a regulatory body monitoring all the food processing industries, and the Central Pollution Control Board (CPCB) is a statutory body regulating all the industries to ensure the quality of air and water. Presently, the regulatory body team visits the industry periodically to conduct reviews. Mostly, the reviews are not conducted regularly due to the increase in the number of industries and shortage of manpower in the regulatory organizations. Due to these reasons, the audits are not conducted regularly, and in turn, the industries are failing to ensure standards to be followed [28] even though industries have internal audit teams whose responsibility is to conduct audits and document it. But most of the time, the audit documents are just fabricated, thereby violating the standards. Blockchain technology, which is a distributed and immutable ledger, can be adopted in regulating the standards to be followed by the industries, thereby ensuring the safety of the people. In industries, smart sensors can be installed in its premises to collect data at regular intervals.

The captured data can be stored in the blockchain, which is an immutable ledger, i.e., the recorded data are tamper proof. Regulatory team members can see the complete history of the captured data, and if there are any violations, then appropriate measures can be taken [29]. Blockchain technology has the concept of a smart contract, which refers to a self-executing code. Smart contracts can be used to execute immediate steps to mitigate the violations. There are mainly three types of blockchains, namely, public blockchain, private blockchain, and consortium blockchain. The regulatory body can choose the appropriate blockchain based on their need. In marching toward Society 5.0, it would be better if all the regulatory bodies use a common public blockchain so all the industries can be monitored.

Token refers to a digital assert that resides on the blockchain itself and can be transferred from one account to another account through smart contracts. Basically, there are two types of tokens, namely, fungible tokens and non-fungible tokens. Fungible tokens have common characteristics, i.e., one token can be exchanged with another token as they have the same value. But non-fungible tokens (NFTs) have unique characteristics, and hence, one NFT cannot be replaced with another NFT as they both have different values. For this regulatory body scenario, fungible tokens will be the right choice. Any industry to operate in the region should have sufficient tokens in their wallet; only then will they be allowed to start their production. During the initial stage itself, the industries will be asked to buy tokens by paying some amount as a security deposit. Then, once the production starts, the smart sensors will start collecting data about the air quality, water quality, etc., based on the type of industry. The smart contract deployed in the blockchain will monitor the collected data, and if any abnormal instance is encountered, then appropriate measures, like transferring certain tokens from a particular industry back to the regulatory body account, are taken. If more violations are recorded, then a greater number of tokens will be transferred, and at a particular stage, the industries will run out of sufficient tokens, thereby forcing them to stop their production. Further, to ensure sustainable development, incentives can be provided for the industries who are maintaining the standards for a long period. Incentives can be issued to industries in terms of tokens, and industries can be rated based on the tokens they hold. Such mechanism will not only regulate the standards but also encourage the industries to conduct periodical trails to ensure their standards. A decentralized application (Dapp) can be developed upon the blockchain, and using this application, the industry members and the regulatory members can see the captured data recorded in the blockchain in terms of transactions [30]. Dapp will also have a user interface that allows the common public to access the data so that they will also be aware of the quality levels in the industry.

11.3.1 Blockchain for Refinery Industry

Oil refinery industries and chemical industries are the major threats for people residing in nearby locations. During the refinery process, refinery sludges or some gas leakages, which may include sulfur dioxide, nitrogen oxide, etc., may be hazardous to the nearby residents. Hence, the blockchain-based application will record all the data captured through the smart Internet of Things (IoT) devices in the immutable ledger [31]. Since the recorded data are tamper proof, it can be used to take any legal actions against the violating industry. A complete history of data collected by the sensors is available in the ledger, and it may also provide further scope of research. Figure 11.3 presents the overview of blockchain and its application in the chemical and refinery industry.

11.3.2 Blockchain in Identity Management

Blockchain technology can be also adopted in identity management. Presently, majority of the financial institutions are maintaining their customers' data (KYC) in a separate database requiring customers to repeatedly provide their details whenever they shift to a different financial institution. Blockchain technology can address this issue by maintaining

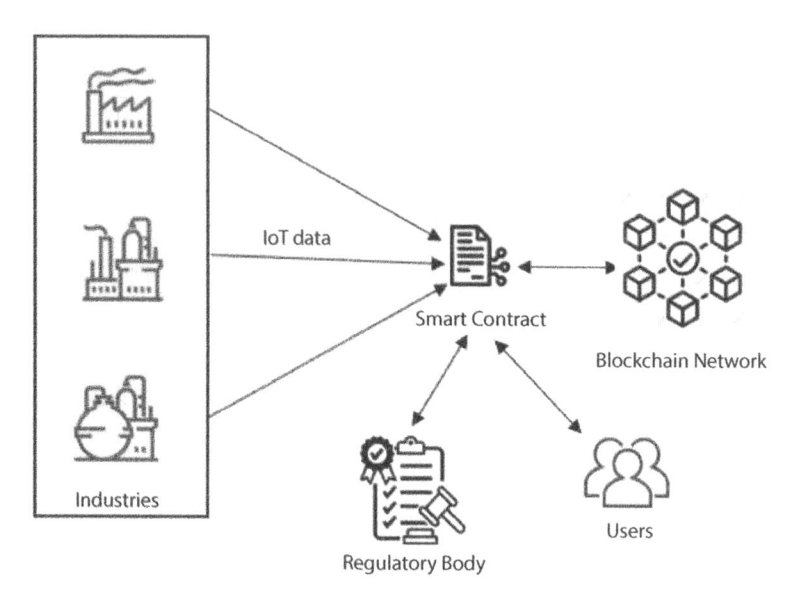

Figure 11.3 Role of blockchain to mitigate the pollution caused by industries.

customers' KYC details in a consortium chain that could be accessible by all the financial institutions [32]. Customers have to provide their details only for the first time, which are then stored in the chain. Thereafter, whenever the customer wants to provide their details, they can just share their crypto address. Any financial institution can access the customers' data using their crypto address. Thus, it reduces the burden of customers in resubmitting their details every time whenever the customers shift to a different financial institution. Whenever a financial institution accesses a particular customer's KYC, then this access transaction will also be stored in the chain [33]. The process of blockchain in performing identity management is presented in Figure 11.4.

Thus, customers will be aware of the financial institutions accessing their data, thereby increasing the transparency. Customers have the authority to determine which entities are permitted to access their KYC data. Tokenomics can be also applied to improve the effective participation of the customers and financial institutions [34]. Customers with a good credit score for a certain period can be awarded with tokens, and the financial institutions can categorize the customers based on the number of tokens they have in their wallet. Even financial institutions can be awarded with tokens based on their customer feedback. These tokens can be used to avail additional services provided by the blockchain-based Dapp.

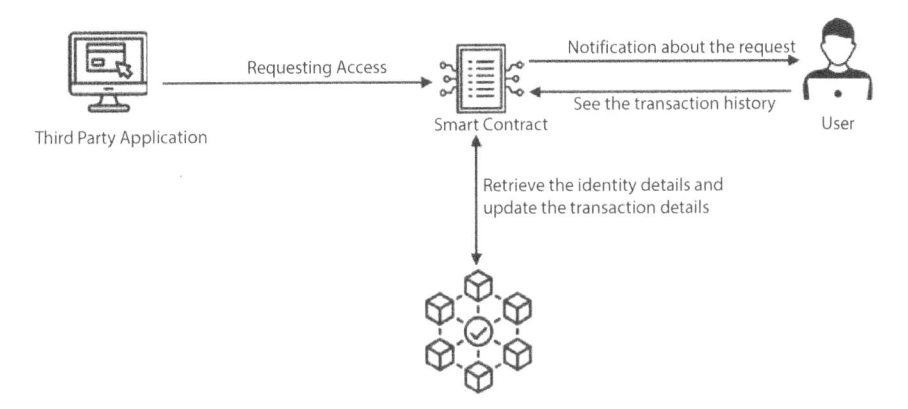

Figure 11.4 Overview of identity management using blockchain.

11.3.3 Blockchain and Its Impact in Healthcare

Healthcare is another challenging domain, which can ensure the safety of the people. The healthcare sector comprises mainly three entities, namely, Providers, Payers, and Patients [35]. Providers are the healthcare service providers, i.e., hospitals. Payers are the ones who pay for the service provided by the hospital, and the insurance companies will be payers in most of the cases. Usually, a patient's health records are stored by the hospitals in their databases. Patients have to spend money to access their health conditions and those data are stored in the hospital database itself. Getting second opinion or treatment at a different hospital can be a burden for patients because of both additional costs and the delay caused due to repeated tests. Another factor is that the patients do not have any control or access to their health records stored in the hospitals, and the hospitals may share the data with other organizations for their business benefits [36]. The application of blockchain in the healthcare sector is illustrated in Figure 11.5. All the stakeholders of the healthcare sector get benefited with the utilization of blockchain technology in healthcare.

Adoption of blockchain technology will address these issues by allowing all the hospitals to store data in the blockchain. Since the same blockchain is used, the patient's data stored by one hospital can also be accessed by another hospital through the patient's crypto address. Any hospital can access the health records only by knowing the crypto address; hence, the patient has the control of their data, and they can decide with whom they can share their data. Blockchain's immutability property ensures that the patient's data cannot be edited or deleted, thereby ensuring the stored

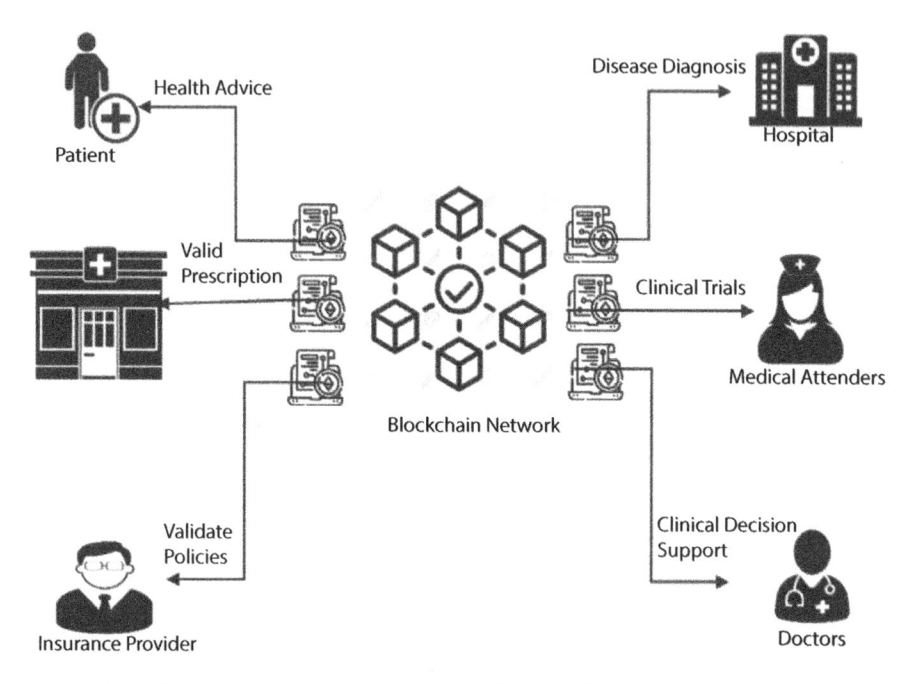

Figure 11.5 Blockchain to facilitate the 3 P's, i.e., Patients, Payers, and Providers.

health records are tamper proof [37]. Interplanetary File System (IPFS) is a distributed file storage protocol, which can be integrated with blockchain to store large health records including images and other formats. Tokens can be used in this blockchain ecosystem to improve the security and reliability of the application.

Hospitals can be awarded with tokens based on the patients' feedback, and they can also be penalized if they breach any regulations like sharing patient's data with any third-party organizations without their consent. Hence, the blockchain-based solution enables the patients to be the owner of their medical records, thereby securing their health information [38–40]. To extend further, the patient can share needed medical records with insurance providers for availing insurance. In the present scenario, the patient or the hospitals will be sharing the medical records with the insurance providers either through any legacy applications or through e-mail [41]. Sharing the data in this manner is not secure, and hence, blockchain-based application can ensure the data are shared in a secure way. The insurance provider can access the health record of its customer only through the customer's crypto address, and they cannot share further without the customer's consent. Similar to hospitals, the insurance providers can also be penalized in terms of tokens if they share the customers health record with others without the customers' consent.

11.3.4 Blockchain for Supply Chain Management

Supply chain is another domain where the adoption of blockchain technology brings many advantages. Ensuring the availability of high-quality food is an important aspect in fostering healthy individuals advancing toward Society 5.0. Few retailers already started exploring blockchain technology to provide complete information of the food items tracing their journey from processing plants to the store [42]. Details regarding vegetables and fruits, such as the farm of cultivation, cultivation date, their package specs, and their journey particulars, can be stored in the blockchain. A customer can access the details of the items just by giving a unique code and thereby ensuring the quality of the items they purchase. Since the data stored in the blockchain are tamper proof, the farm details, packaging specs and journey details cannot be modified or deleted. This mechanism not only ensures that quality food is provided to the customer but also helps in tracing if adulteration happens in the ecosystem [43]. Such details will enable the regulatory bodies to appropriate actions. Tokenomics can be adopted in this ecosystem to ensure effective participation from all entities. Farmers can be awarded with tokens for the quality of food they provide, and similarly, the supply chain organizations and the stores can also be awarded with tokens for the quality of service they provide [44]. If any entity is found to be violating the standards and regulations, then such entity can be penalized in terms of tokens. Regulatory bodies can decide on the quality of service provided by the entities through the number of tokens each entity has in its wallet. The process of enabling traceability in supply chain management is depicted in Figure 11.6.

11.3.5 Blockchain in Asset Management

Blockchain can offer numerous benefits in the real estate domain through its decentralized decision-making strategy [45]. The blockchain-based decentralized application (Dapp) can facilitate converting real-world assets to digital tokens, and it can be made available in the blockchain-based real estate market. The sellers can tokenize their physical asset by attributing its location details, price, and other facilities through the Dapp. The tokenized assets will be available in the market place where the potential buyers can see the assets along with its details. If the buyers are interested with the

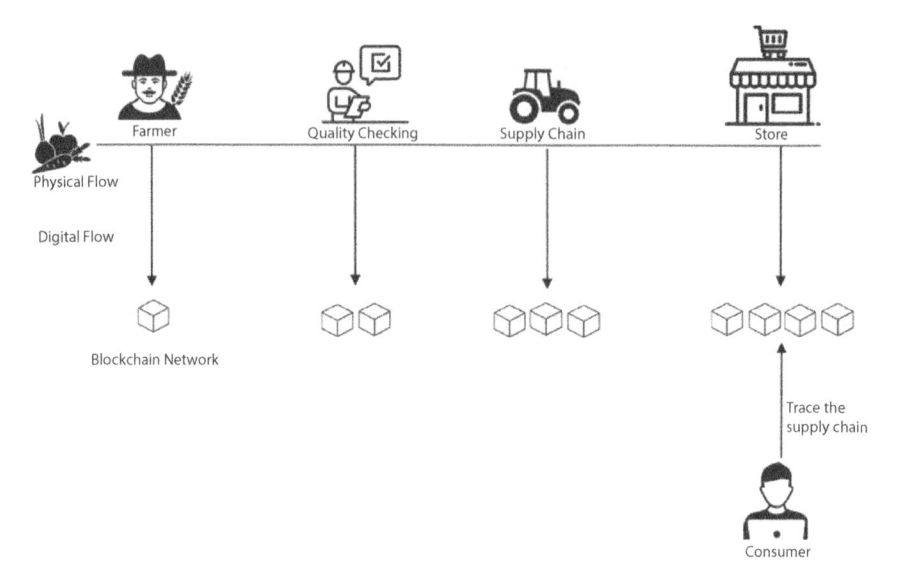

Figure 11.6 Blockchain application enabling traceability in the supply chain.

asset, they can buy the property through cryptocurrency itself without the involvement of a middleman. Smart contracts can be defined for the transaction, which can further ensure the availability of sufficient balance in the buyer's digital wallet [46]. The contract will also ensure that the quoted asset price will be transferred from the buyer's wallet to the seller's wallet, and after successful execution, the ownership details will be also updated in the ledger. Since the transactions are recorded in the immutable distributed ledger, the information stored is tamper proof. Thereby, the distributed blockchain-based application ensures the transactions are secure, transparent, traceable, and immutable [47]. For storing the images and other geographical details, the Inter Planetary File System (IPFS) can be used as the blockchain allows to store transactions with less memory capacity. IPFS is a distributed file storage protocol, which can be easily integrated with blockchain-based decentralized applications. Real estate Dapp, when employed in smart cities, can eliminate long queues waiting for land registration in the registration offices and can prevent fraudulent land registrations, thereby ensuring secure transactions for buying and selling of physical assets [48]. Tokenomics can be also applied in this use case to incentivize the genuine buyers and sellers. Blockchain is widely used in asset management task. Figure 11.7 illustrates the process involved in managing assets using blockchain technology.

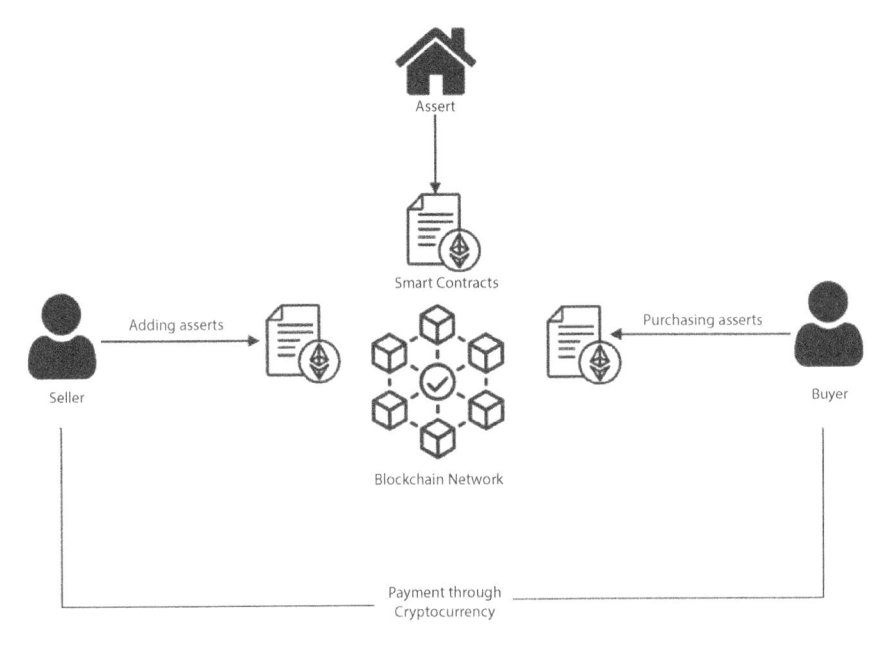

Figure 11.7 Tokenizing physical assets using blockchain technology.

11.3.6 Blockchain in Copyright Management

Blockchain application can be in copyright management applications. The content created by the user can be tokenized in the form of non-fungible tokens (NFTs) and then can be added in the ledgers [49]. Each non-fungible token will have unique properties including its value in terms of cryptocurrency, owner details, and other specifications. NFTs can be used in any application, but their properties remain immutable in the blockchain. For instance, the user may create a music album and can be tokenized as NFT in the blockchain [50]. This music album can be streamed through some web 3.0 applications where the end user can listen to the music but the ownership still remains with the album creator. In certain instances, the owner can sell the copyrights to some organizations, but the ownership will still remain with the creator [51]. Entire transactions starting from NFT creation to transferring the ownership details are stored in the distributed ledger. Further, the entire flow of the applications can be implemented through smart contracts. Any interaction with the NFTs is done only through the defined smart contracts, thereby ensuring the ownership and copyright of the digital contents [52].

Hence, adoption of blockchain technology helps to ensure safety of the people. Blockchain facilitates the people to live a healthier life, and it also ensures sustainable growth. In Society 5.0, every individual object in the society will be built as digital twins in cyberspace restructured in terms of systems, business design, urban and regional development, etc., and then reflected in physical space to transform society [53].

High-Performance Computing (HPC) data centers use a cluster of powerful processors and GPUs to execute given instructions and solve complex problems at extremely high speed. Such powerful computing resources are used to perform analytics in big data, scientific researches, genomic sequencing, medical researches, etc. HPC used in data centers can offer the potential to optimize consensus algorithms in blockchain networks [54]. HPC excels in handling complex computations and massive data processing tasks making it well suited to improve the efficiency and performance of consensus mechanisms. By adopting the computational power of HPC systems, blockchain networks can process transactions more rapidly and achieve faster consensus enhancing overall scalability. Such improvement in the performance ensures that blockchain can be used for critical applications handling millions of transactions per second. Moreover, robust infrastructure and advanced cooling systems of HPC data centers can contribute to the energy efficiency of consensus algorithms mitigating concerns related to high energy consumption associated with certain consensus mechanisms like PoW.

The integration of HPC capabilities with blockchain network contributes to the reduction of computational bottlenecks that occurred due to high volumes of transactions and complex operations. The integration enhances the efficiency of consensus mechanism and the effective adaptation of blockchain technology in a variety of applications [55]. Consensus mechanisms, like PoW and PoS, require significant resources to validate the transactions. The larger data-handling capacity HPC ensures faster response and scalability of transactions.

The widespread adoption of blockchain and cryptocurrencies have brought forth exciting opportunities for various applications. However, the usage of blockchain and cryptocurrencies has raised concerns about energy usage. The consensus mechanism, like PoW, utilize energy for validating the transactions. The PoW method uses miners to solve complex problems. This leads to high energy usage in the blockchain networks. As the network grows and becomes larger, more miners are required, and the puzzles become harder leading to even greater energy consumption. This creates a scenario where the blockchain network, utilizing a PoW consensus mechanism, consumes a larger amount of energy for their transactions.

The energy consumption nature of PoW has raised concerns about their impact on the sustainability of blockchain networks. As the energy required for these networks are generated from non-renewable energy sources, it has become a greater concern of environmental sustainability.

HPC data centers are facilities accompanied with powerful computational resources that are able to tackle complex computational tasks effectively. As mentioned, the blockchain method using a PoW consensus mechanism, in particular, suffers from energy utilization concerns. Apart from computing capabilities, HPC possesses energy efficient infrastructure. HPCs employ techniques, like liquid cooling, smart power, and energy-efficient components to reduce energy consumption with maximized performance. By leveraging blockchain with HPC, the energy-intensive process of blockchain mining can be optimized. This optimization aims to reduce the overall energy consumption required for mining activities.

11.4 Conclusion

Blockchain technology has garnered widespread interest from individuals and organizations across diverse sectors. Its transformative potential lies in a range of features, including decentralization, anonymity, persistence, and auditability. These features collectively redefine established industry norms and pave the way for innovative solutions in various domains. Using blockchain in healthcare systems ensures the privacy and data security of patient health records. It provides a flexibility for the users to finalize who can access their medical data. Blockchain technology revolutionizes the identity management tasks to perform in a few seconds where the existing methods take larger time for the same. With blockchain-based identity management schemes, the users get a higher level of security and privacy compared to traditional methods. Block eliminates the need for intermediaries in verification processes. This restricts access to sensitive information only to the verifier. Also, the blockchain ensures robust protection against fraud and identity theft in identity management. User data can be securely encrypted and stored within their identity wallet apps thereby safeguarding confidentiality and integrity. Integrating blockchain technology into supply chain management provides greatest levels of transparency, traceability, and operational efficiency. Using the mutable ledger blockchain ensures data integrity, security, and trust among stakeholders. Automated smart contracts expedite transaction settlements, while blockchain's auditability aids in regulatory compliance and adherence to sustainability standards thereby reinforcing accountability across the supply chain network.

Incorporating blockchain into Society 5.0 can improve the well-being of humans by providing improvements in healthcare, manufacturing, finance, and other sectors. This also provides a greater impact in improving the security challenges of the society due to technical advancements. By harnessing the computational power of HPC, blockchain networks can efficiently process large volumes of data. This capability is particularly valuable for enhancing AI algorithms for real-time decision-making capabilities. AI algorithms rely on data to make informed decisions, and the ability to process data rapidly and efficiently enables to make decisions in real time. This leads to more responsive and effective outcomes. The synergy between HPC and blockchain technology has the potential to revolutionize data-sharing systems. Combining the computational power of HPC with blockchain data-sharing systems can achieve enhanced security, privacy, and efficiency. The decentralized and tamper-resistant nature of blockchain ensures the integrity and confidentiality of shared data, while HPC enables rapid and efficient processing of large datasets. The combination of HPC's computational power and blockchain opens up new possibilities for innovation and efficiency paving the way for transformative changes across various applications. As the number of users increases in blockchain, the time in locating the blocks will take longer. The design of novel algorithms and methods are further required to enhance the performance in Society 5.0

References

1. Paulavičius, R., Grigaitis, S., Igumenov, A., Filatovas, E., A Decade of Blockchain: Review of the Current Status, Challenges, and Future Directions. *Informatica*, 30, 729–748, 2019. https://doi.org/10.15388/Informatica.2019.227.
2. Werth, J., Berenjestanaki, M.H., Barzegar, H.R., Ioini, N.E., Pahl, C., A Review of Blockchain Platforms Based on the Scalability, Security and Decentralization Trilemma, in: *International Conference on Enterprise Information Systems*, 2013, https://doi.org/10.5220/0011837200003467.
3. Tyagi, A.K., Dananjayan, S., Agarwal, D., Ahmed, H.F., Blockchain—Internet of Things Applications: Opportunities and Challenges for Industry 4.0 and Society 5.0. *Sens. (Basel, Switzerland)*, 23, 2, 20232023. https://doi.org/10.3390/s23020947.
4. Villarreal, E.R., García-Alonso, J.M., Moguel, E., Alegría, J.A., Blockchain for Healthcare Management Systems: A Survey on Interoperability and Security. *IEEE Access*, 11, 5629–5652, 2023. https://doi.org/10.1109/ACCESS.2023.3236505.

5. Zheng, Z., Xie, S., Dai, H., Chen, X., Wang, H., Blockchain challenges and opportunities: a survey. *Int. J. Web Grid Serv.*, 14, 4, 352–375, 2018. https://doi.org/10.1504/IJWGS.2018.095647.
6. Zheng, Z., Xie, S., Dai, H., Chen, X., Wang, H., An Overview of Blockchain Technology: Architecture, Consensus, and Future Trends, in: *2017 IEEE International Congress on Big Data (BigData Congress)*, pp. 557–564, 2017, https://doi.org/10.1109/BIGDATACONGRESS.2017.85.
7. Leng, J., Zhou, M., Zhao, L.J., Huang, Y., Bian, Y., Blockchain Security: A Survey of Techniques and Research Directions. *IEEE Trans. Serv. Comput.*, 15, 2490–2510, 2022. https://doi.org/10.1109/10.1109/TSC.2020.3038641.
8. Mourtzis, D., Angelopoulos, J.D., Panopoulos, N., A Literature Review of the Challenges and Opportunities of the Transition from Industry 4.0 to Society 5.0. *Energies*, 15, 7, 2022. https://doi.org/10.3390/en15176276.
9. Zharfan, M. and Harmain, H., Changing role of millennial accountants in the information revolution era (Industry 4.0) and challenges in the society generation scope (Society 5.0). *Enrichment :J. Manage.*, 13, 1, 376–384, 2023. https://doi.org/10.35335/enrichment.v13i1.1222.
10. Minoli, D. and Occhiogrosso, B., Blockchain mechanisms for IoT security. *Internet Things*, 1, 1–13, 2018. https://doi.org/10.1016/j.iot.2018.05.002.
11. Xu, J., Wang, C., Jia, X., A Survey of Blockchain Consensus Protocols. *ACM Comput. Surv.*, 55, 1 – 35, 2023. https://doi.org/10.1145/3579845.
12. Guru, A., Mohanta, B.K., Mohapatra, H., Al-Turjman, F.M., Altrjman, C., Yadav, A., A Survey on Consensus Protocols and Attacks on Blockchain Technology. *Appl. Sci.*, 2023. https://doi.org/10.3390/app13042604.
13. Xiong, H., Chen, M., Wu, C., Zhao, Y., Yi, W., Research on Progress of Blockchain Consensus Algorithm: A Review on Recent Progress of Blockchain Consensus Algorithms. *Future Internet*, 14, 47, 2022. https://doi.org/10.3390/fi14020047.
14. Gramoli, V. and Tang, Q., The Future of Blockchain Consensus. *Commun. ACM*, 66, 79 – 80, 2023. https://doi.org/10.1145/3589225.
15. D., D.S., Smart Contract Based Industrial Data Preservation on Block Chain. *J. Ubiquitous Comput. Commun. Technol.*, 2020. https://doi.org/10.36548/jucct.2020.1.005.
16. Mahajan, H.B. and Reddy, K.V., Secure gene profile data processing using lightweight cryptography and blockchain. *Cluster Comput.*, 2023. https://doi.org/10.1007/s10586-023-04123-6.
17. Hussain, A.A. and Al-Turjman, F., Artificial intelligence and blockchain: A review. *Trans. Emerging Telecommun. Technol.*, 32, 9, e4268, 2021. https://doi.org/10.1002/ett.4268.
18. Li, X., Jiang, P., Chen, T., Luo, X., Wen, Q., A survey on the security of blockchain systems. *Future Gener. Comput. Syst.*, 107, 841–853, 2020. https://doi.org/10.1016/j.future.2017.08.020.
19. Rizwan, A., Karras, D.A., Kumar, J.S., Sánchez-Chero, M.J., Mogollón Taboada, M.M., Altamirano, G.C., An Internet of Things (IoT) Based Block

Chain Technology to Enhance the Quality of Supply Chain Management (SCM). *Math. Probl. Eng.*, 2022, 2022. https://doi.org/10.1155/2022/9679050.
20. Haddouti, S.E. and Ech-Cherif El Kettani, M.D., Analysis of Identity Management Systems Using Blockchain Technology, in: *2019 International Conference on Advanced Communication Technologies and Networking (CommNet)*, Rabat, Morocco, pp. 1–7, 2019, https://doi.org/10.1109/COMMNET.2019.8742375.
21. Salem, S.H., Hassan, A.Y., Moustafa, M.S., Hassan, M.N., Blockchain-based biometric identity management. *Cluster Comput.*, 2023. https://doi.org/10.1007/s10586-023-04180-x.
22. Rathee, T. and Singh, P., A systematic literature mapping on secure identity management using blockchain technology. *J. King Saud Univ. Comput. Inf. Sci.*, 34, 8, 5782–5796, 2021. https://doi.org/10.1016/j.jksuci.2021.03.005.
23. Ning, L. and Yuan, Y., How blockchain impacts the supply chain finance platform business model reconfiguration. *Int. J. Logist. Res. Appl.*, 26, 9, 1081–1101, 2023. https://doi.org/10.1080/13675567.2021.2017419.
24. Ren, Y., Ma, C., Chen, X., Lei, Y., Wang, Y., Sustainable finance and blockchain: A systematic review and research agenda. *Res. Int. Bus. Finance*, 64, 2023. https://doi.org/10.1016/j.ribaf.2022.101871.
25. Zhang, D., Wahab, N.H., Kadir, K.A., Aldhaqm, A., Nasir, H.M., Wong, K.Y., Research on Blockchain: Privacy Protection of Cryptography Blockchain-Based Applications, in: *2023 3rd International Conference on Emerging Smart Technologies and Applications (eSmarTA)*, pp. 1–6, 2023, https://doi.org/10.1109/eSmarTA59349.2023.10293507.
26. Lai, Z., Comparison of current blockchain privacy protection technologies and prospects for future trends, in: *International Conference on Green Communication, Network, and Internet of Things*, 2023, https://doi.org/10.1117/12.2667239.
27. Buterin, V., Illum, J., Nadler, M., Schär, F., Soleimani, A., Blockchain Privacy and Regulatory Compliance: Towards a Practical Equilibrium. *SSRN Electron. J.*, 2023. http://dx.doi.org/10.2139/ssrn.4563364.
28. Makani, S., Pittala, R., Alsayed, E., Aloqaily, M., Jararweh, Y., A survey of blockchain applications in sustainable and smart cities. *Cluster Comput.*, 25, 3915–3936, 2022. https://doi.org/10.1007/s10586-022-03625-z.
29. Balzano, W., Lapegna, M., Stranieri, S., Vitale, F., Competitive-blockchain-based parking system with fairness constraints. *Soft Comput.*, 26, 4151–4162, 2021. https://doi.org/10.1007/s00500-022-06888-1.
30. Maya, P.C. and Salam, P.A., Implementation of a blockchain based DApp for P2P electricity trading, in: *2023 5th International Conference on Energy, Power and Environment: Towards Flexible Green Energy Technologies (ICEPE)*, pp. 1–6, 2023, https://doi.org/10.1109/ICEPE57949.2023.10201530.
31. Umran, S.M., Lu, S., Abduljabbar, Z.A., Lu, Z., Feng, B., Zheng, L., Secure and Privacy-preserving Data-sharing Framework based on Blockchain Technology for Al-Najaf/Iraq Oil Refinery, in: *2022 IEEE Smartworld,*

Ubiquitous Intelligence & Computing, Scalable Computing & Communications, Digital Twin, Privacy Computing, Metaverse, Autonomous & Trusted Vehicles (SmartWorld/UIC/ScalCom/DigitalTwin/PriComp/Meta), pp. 2284–2292, 2022, https://doi.org/10.1109/SmartWorld-UIC-ATC-ScalCom-DigitalTwin-PriComp-Metaverse56740.2022.00325.

32. Venkatraman, S. and Parvin, S., Developing an IoT Identity Management System Using Blockchain. *Systems*, 10, 39, 2022. https://doi.org/10.3390/systems10020039.

33. Yawalkar, P.M., Paithankar, D., Pabale, A.R., Kolhe, R.V., William, P., Integrated identity and auditing management using blockchain mechanism. *Meas.: Sens.*, 2023, 2023. https://doi.org/10.1016/j.measen.2023.100732.

34. Rathee, T. and Singh, P., Secure data sharing using Merkle hash digest based blockchain identity management. *Peer-to-Peer Netw. Appl.*, 202114, 3851 – 3864. https://doi.org/10.1007/s12083-021-01212-4.

35. Haleem, A., Javaid, M., Singh, R.P., Suman, R., Rab, S., Blockchain technology applications in healthcare: An overview. *Int. J. Intell. Netw.*, 2, 130–139, 2021. https://doi.org/10.1016/j.ijin.2021.09.005.

36. Mohey Eldin, A., Hossny, E.K., Wassif, K.T., Omara, F.A., Federated blockchain system (FBS) for the healthcare industry. *Sci. Rep.*, 13, 2023. https://doi.org/10.1038/s41598-023-29813-4.

37. Dewangan, N.K. and Chandrakar, P., Patient-Centric Token-Based Healthcare Blockchain Implementation Using Secure Internet of Medical Things. *IEEE Trans. Comput. Social Syst.*, 10, 3109–3119, 2023. https://doi.org/10.1109/TCSS.2022.3194872.

38. Pachouri, V., Pandey, S., Gehlot, A., Negi, P., Kathuria, A., Pandey, R., Hospital 4.0: Blockchain Helping in Hospitality Services, in: *2023 3rd International Conference on Pervasive Computing and Social Networking (ICPCSN)*, pp. 140–1143, 2023, https://doi.org/10.1109/ICPCSN58827.2023.00193.

39. Treiblmaier, H., Rejeb, A., Strebinger, A., Blockchain as a Driver for Smart City Development: Application Fields and a Comprehensive Research Agenda. *Smart Cities*, 3, 3, 853–872, 2020. https://doi.org/10.3390/smartcities3030044.

40. Qahtan, S., Sharif, K.Y., Zaidan, A.A., Alsattar, H.A., Albahri, O.S., Zaidan, B.B., Zulzalil, H.B., Osman, M.H., AlAmoodi, A.H., Mohammed, R.T., Novel Multi Security and Privacy Benchmarking Framework for Blockchain-Based IoT Healthcare Industry 4.0 Systems. *IEEE Trans. Ind. Inf.*, 8, 6415–6423, 2022. https://doi.org/10.1109/TII.2022.3143619.

41. Bodemer, O., Transforming the Insurance Industry with Blockchain and Smart Contracts: Enhancing Efficiency, Transparency, and Trust. *Eng. OA*, 1, 2, 105–110, 2023. https://doi.org/10.33140/eoa.01.02.08.

42. Dutta, P., Choi, T., Somani, S., Butala, R., Blockchain technology in supply chain operations: Applications, challenges and research opportunities. Transportation Research. *Part E, Logist. Transp. Rev.*, 142, 102067–102067, 2020. https://doi.org/10.1016/j.tre.2020.102067.

43. Singh, R., Khan, S., Dsilva, J., Centobelli, P., Blockchain Integrated IoT for Food Supply Chain: A Grey Based Delphi-DEMATEL Approach. *Appl. Sci.*, 2023. https://doi.org/10.3390/app13021079.
44. Chandan, A., John, M., Potdar, V., Achieving UN SDGs in Food Supply Chain Using Blockchain Technology. *Sustainability*, 2023. https://doi.org/10.3390/su15032109.
45. Peter, A., Kumar, A.A., Rajeev, A., Baiju, B., Chooralil, V.S., Real Estate Management System using Blockchain, in: *2023 International Conference on Innovations in Engineering and Technology (ICIET)*, pp. 1–4, 2023, https://doi.org/10.1109/ICIET57285.2023.10220623.
46. Jeong, S. and Ahn, B., Implementation of real estate contract system using zero knowledge proof algorithm based blockchain. *J. Supercomputing*, 77, 11881 – 11893, 2021. https://doi.org/10.1007/s11227-021-03728-1.
47. Yu, Y., Xu, Y., Yuan, J., Wu, C., Liu, X., Su, M., Keycrux: A New Design of Distributed and Convenient Blockchain Digital Wallet, in: *2022 IEEE Smartworld, Ubiquitous Intelligence & Computing, Scalable Computing & Communications, Digital Twin, Privacy Computing, Metaverse, Autonomous & Trusted Vehicles (SmartWorld/UIC/ScalCom/DigitalTwin/PriComp/Meta)*, pp. 2444–2451, 2022, https://doi.org/10.1109/SmartWorld-UIC-ATC-ScalCom-DigitalTwin-PriComp-Metaverse56740.2022.00342.
48. Peddibhotla, U., Chandran, S.C., Kumar, P., Kumar, R., *2024 16th Int. Conf. COMmunication Syst. & NETworkS (COMSNETS)*, 427–429, 2024. AU: Please provide article title and volume number.
49. Chen, X., Yang, A., Weng, J., Tong, Y., Huang, C., Li, T., A Blockchain-Based Copyright Protection Scheme With Proactive Defense. *IEEE Trans. Serv. Comput.*, 16, 2316–2329, 2023. https://doi.org/10.1109/TSC.2023.3246476.
50. Ciriello, R., Torbensen, A.C., Hansen, M.R., Müller-Bloch, C., Blockchain-based digital rights management systems: Design principles for the music industry. *Electron. Mark.*, 33, 1–21, 2023.
51. Galphat, Y., Gole, O., Baradkar, P., Gawali, S., Sahane, Y., Blockchain based Music Streaming Platform using NFTs, in: *2023 International Conference on Sustainable Computing and Smart Systems (ICSCSS)*, pp. 1598–1603, 2023, https://doi.org/10.1109/ICSCSS57650.2023.10169304.
52. Centorrino, G., Naciti, V., Rupo, D., A new era of the music industry? Blockchain and value co-creation: the Bitsong case study. *Eur. J. Innov. Manage.*, 2022, 2022. https://doi.org/10.1108/ejim-07-2022-0362.
53. Deguchi, A., Hirai, C., Matsuoka, H., Nakano, T., Oshima, K., Tai, M., Tani, S., What Is Society 5.0?, 2020, https://doi.org/10.1007/978-981-15-2989-4_1.
54. Yu, S., Lv, K., Shao, Z., Guo, Y., Zou, J., Zhang, B., A High Performance Blockchain Platform for Intelligent Devices, in: *2018 1st IEEE International Conference on Hot Information-Centric Networking (HotICN)*, pp. 260–261, 2018, https://doi.org/10.1109/HOTICN.2018.8606017.

55. Liang, X., Zhao, Y., Zhang, D., Wu, J.-F., Zhao, Y., SBHPS: A High Performance Consensus Algorithm For Blockchain, in: *2021 International Conference on High Performance Big Data and Intelligent Systems (HPBD&IS)*, pp. 6–11, 2021, https://doi.org/10.1109/hpbdis53214.2021.9658348.

12

Virtualization of Smart Society 5.0 Using Artificial Intelligence and Virtual Reality

Sakthivel Sankaran[1]*, M. Arun[2] and R. Kottaimalai[3]

[1]Department of Biomedical Engineering, Kalasalingam Academy of Research and Education, Krishnankoil, Virudhunagar, Tamil Nadu, India
[2]Department of Computer Applications, Kalasalingam Academy of Research and Education, Krishnankoil, Virudhunagar, Tamil Nadu, India
[3]Department of Electronics and Communication Engineering, Kalasalingam Academy of Research and Education, Krishnankoil, Virudhunagar, Tamil Nadu, India

Abstract

Considering the framework of Smart Society 5.0, the present chapter examines the revolutionary possibilities of combining artificial intelligence (AI) and virtual reality (VR). Smart Society 5.0 aims to create a world where technology seamlessly integrates with everyday life, enhancing both our overall well-being and societal progress, as the world becomes increasingly interconnected. It explores the complementary nature of AI and VR and looks at how they work together to create a simulated smart society. The plot starts out by explaining the core ideas of Smart Society 5.0 and highlighting the necessity of machine learning that can adjust to the various demands of individuals and groups. It then delves into how AI is facilitating the complex interactions between information connection, and self-governing choice making in this dynamic social environment. The entire scope of Smart Society 5.0 can be realized through the use of AI, which is emerging as an essential driver for customized offerings and forecasting.

The section that follows subsequent to the comprehensive field of virtual reality (VR) examines the ways in which VR systems aid in the production of realistic, virtual settings. It investigates the possibility of improved interactions between humans and machines, distant cooperation, and hands-on training by smoothly incorporating VR within the framework of Smart Society 5.0. Because virtual

---

*Corresponding author: sakthivelsankaran1992@gmail.com

Rajesh Kumar Dhanaraj, Malathy Sathyamoorthy, Balasubramaniam S and Seifedine Kadry (eds.) Networked Sensing Systems, (297–322) © 2025 Scrivener Publishing LLC

reality is realistic, people may interact with their environment in a fresh manner and develop a stronger bond among the actual and simulated worlds. The section also looks at research findings and real-life scenarios in which combining AI with VR proved to be beneficial in fields including planning for cities, medical services, and teaching. Furthermore, it discusses difficulties, moral concerns, and possible directions for further study in this rapidly evolving discipline. The chapter concludes by highlighting the revolutionary potential of combining VR and AI in the development of Smart Society 5.0. Everyone may establish a simulated smart society, which responds with the constantly evolving demands of its citizens by utilizing these advancements in concert providing a preview of the linked, highly intelligent, and comprehensive sociocultural realities that are still ahead.

Keywords: Smart society 5.0, artificial intelligence (AI), virtual reality (VR), synergy, transformative potential

12.1 Introduction to Smart Society 5.0

12.1.1 Smart Society 5.0 and Its Key Characteristics

According to Japanese specialists in a smart society, the advanced social model known as "5.0" must pervade all aspects of the Internet of Things (IoT) and handle social and economic concerns. This elevates society to new heights by leveraging artificial intelligence (AI) and the benefits of cyberspace to provide the greatest solutions to societal and economic challenges. The IoT, in which technical advancements in the manufacturing sector are highly recognized across all societal segments, is substantially responsible for the intellectual Smart Society 5.0. The integration of wireless networks, the IoT, and cyberspace into the physical world will transform society into a highly intelligent and sophisticated society. In addition to Society 5.0, a variety of next-generation infrastructure agents are coordinated [1].

The Japanese Cabinet has introduced a new program called "Smart Society 5.0," which integrates AI, cyberspace, and other technological advancements to create an intelligent and comfortable society. In Society 5.0, human dispute resolution and economic problem solving are centralized. Providing sustainability across the social, political, economic, and environmental spheres is the central idea of the Smart Society. This concept encompasses every evolutionary stage that has occurred, including hunting, agriculture, industry, information, and the current integration of AI and the internet. The three primary areas of focus for Smart Society 5.0 are productivity, mobility, and health [2].

With its goal of facilitating universal access to resources and products, City 5.0 is a socio-material system that treats its residents as customers. Society 5.0, as defined in this article, envisions a livable city that is (re) modeled through digitization to provide public services and amenities, thereby removing barriers for its civilians. The ultimate aim is to enhance the livability of the city by avoiding or preventing restrictions in society that limit access to available services and resources for all citizens [3].

The human-centric Smart City 5.0 paradigm focuses on human desires, interests, and emphasizes how individuals can benefit from using products and services. This is achieved through the digitalization of public goods and services, which eliminates any limitations or issues. These limitations include factors, such as livability, constraints, governance, management, and restrictions, that hinder the use of products and resources. The approach revolves around the intelligent utilization of created commodities and services enabling people or citizens to derive benefits from the use of public goods and services [4].

The idea behind Society 5.0, originating in Japan, advocates for the application of technology to improve both environmental sustainability and human well-being. Society 5.0 builds upon earlier stages of technological and societal development represented by Society 1.0, 2.0, 3.0, and 4.0. In this latest stage, societal concerns, such as inequalities, hunger, and climate change, are addressed through technologies like AI, IoT, data mining, and robotics. The intended outcome is an equitable, intelligent, and networked society with the intention of paving the way for an environmentally friendly and compassionate era [8].

12.1.2 Evolution from Previous Smart Society Models

The concepts of livability and the removal of social constraints are pivotal in the transition from Industry 4.0 to Smart City 5.0. City 5.0 advances the notion of smart consumption emphasizing the experience and time during production, whereas the preceding model, Industry 4.0, focuses on smart production representing production time. AI, robotics, the IoT, people, and knowledge are all integrated in City 5.0. While Industry 4.0's primary objective is achieving frictionless manufacturing, Smart City 5.0 is more oriented toward frictionless consumption. Comparatively, we observe that City 5.0 holds more tangible and significant implications than Industry 4.0. Furthermore, while Industry 4.0 aimed at removing constraints, Smart City 5.0 introduces a modified theme: eliminating restrictions during consumption. The evolution of society is described in Figure 12.1 [3].

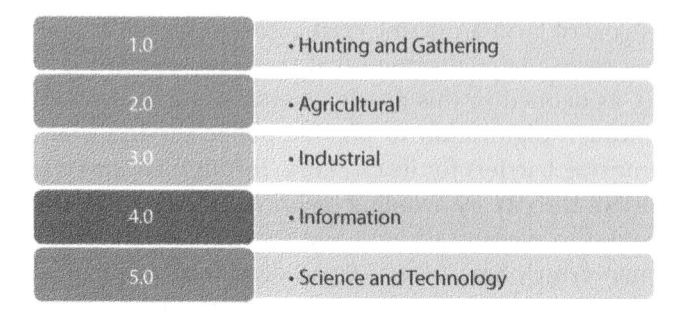

1.0	• Hunting and Gathering
2.0	• Agricultural
3.0	• Industrial
4.0	• Information
5.0	• Science and Technology

Figure 12.1 Evolution of society.

The principles underlying Industry 4.0, Health 4.0, Banking 4.0, and similar concepts revolve around the development of low-cost or cost-effective technologies. The idea of City 5.0, a citizen-focused approach, emphasizes the prudent utilization of these products and technologies in accordance with the needs and welfare of citizens. Prioritizing the application of technology to address social issues characterizes the Smart Society 2.0 paradigm. Additionally, the forthcoming model, 4.0, is entirely dedicated to the intelligent manufacturing of various commodities and items.

Although designated as "human centric," the focus of smart City 5.0 revolves around people, their demands, and interests in city administration concerning how residents can benefit from employing technology and purchasing the commodities and products created. This paradigm enables people or customers to utilize the most cutting-edge technologies already in use without any restrictions [4].

The transition from Smart Cities 1.0 and 2.0 to Smart Cities 5.0 is characterized by the necessity of infrastructure, technology, and human resources for regional development. Realizing the potential in every sector and region is crucial for the creation of Smart City 5.0. Three fundamental elements commonly required for regional development are infrastructure, technology, and human resources. This builds upon the earlier paper Smart City 3.0, which utilized the PDCA method to become global. In this context, the formula $E = KMC^2$ is applied, where K represents knowledge, C1 and C2 denote computer and communication technologies, respectively, and E stands for energy or a company's worth. Further formulas were derived from this equation. The use of mathematical methods to determine the stage of the incoming new edition is how this research differs from the previous model [5].

The key idea here is the integration of AI into a smart society, particularly in relation to waste management, energy consumption, and traffic control in this scenario. The primary objective is a complete transformation of urban society. Unlike other smart society models, this one aims to protect urban residents' livelihoods, increase productivity, and optimize services. The study's findings have the potential to focus the attention of politicians, technologists, and urban planners on improving and advancing smart societies [6].

Industry 4.0 focuses solely on industrial production, whereas Society 5.0 addresses all facets of society. The primary goals of "Smart Society 5.0" are to decrease harmful effects on society and accelerate the speed of social progress in Japan. Industry 4.0 focuses on the expansion of a specific area, but Society 5.0 sees all of society as its operational area. Additionally, whereas Industry 5.0 concentrates on fusing the real and virtual worlds, Society 5.0 aims to promote economic growth and social management to address problems. Moreover, Civilization 5.0 is designed to serve as a framework for sustainable human development [7].

12.2 Foundations of Virtual Reality

12.2.1 Brief History and Development of Virtual Reality

Utilizing head-mounted displays (HMDs), like the Google Cardboard and Xiaomi headsets, suggests that consumer-grade virtual reality (VR) technology is still in its initial phase. The usage of mobile VR devices, such as those powered by cell phones, indicates a period between 2019 and 2020 when smartphones began to drive VR experiences. References to high-immersion VR technology for education point to developments in this sphere, most likely occurring in the years 2022–2023. The employment of separate VR visors and controllers with regular gaming PCs points to a period around 2021 when these configurations were being utilized for specific projects. Research utilizing IVR devices with powerful computers, head-mounted displays, and controllers likely took place in 2020 demonstrating the integration of VR into classroom environments. Clusters of desktop PCs with 360° cameras and analytics software suggest a period around 2021 when desktop VR systems were being employed for teaching. An exploration of such devices for immersive VR experiences around 2021 is revealed through the usage of a spherical video-based VR system. Although the year of the full VR classroom setup is not mentioned, it suggests a more advanced level of VR integration into educational environments, most likely in the mid- to late 2020s [9].

Jaron Lanier coined the term "virtual reality" (VR) in 1987 marking the first attempt to employ VR technologies. Cinematographer Heilig invented Sensorama, a 1957 device that could recreate various environments including noises, scents, and sensual elements like motions and breeze. However, since the technology was new, the majority of entrepreneurs at the time were unable to make sensible use of it. Afterward, in 1968, Sutherland created a prototype of a head-mounted gadget, and in 1970, Massachusetts Institute of Technology scientists created the initial interactive map. Virtual reality was largely utilized in games during the 2000s leading to the creation of a plethora of devices, including head motion devices, specifically for gaming. These devices enabled users to not only view but also engage with a virtual world [10].

12.2.2 Key Components and Technologies in VR

Virtual reality is portrayed in Figure 12.2. Mandal (2013) divides technology into three groups: desktop VR systems that are non-immersive, semi-immersive, and immersive. The most basic form of virtual reality, with a lower degree of immersion, is the non-immersive system. These systems only require a computer screen; no other hardware is necessary. Conversely, enhanced desktop VR offers semi-immersive VR systems. For example, such devices can provide motion and head tracking, thereby

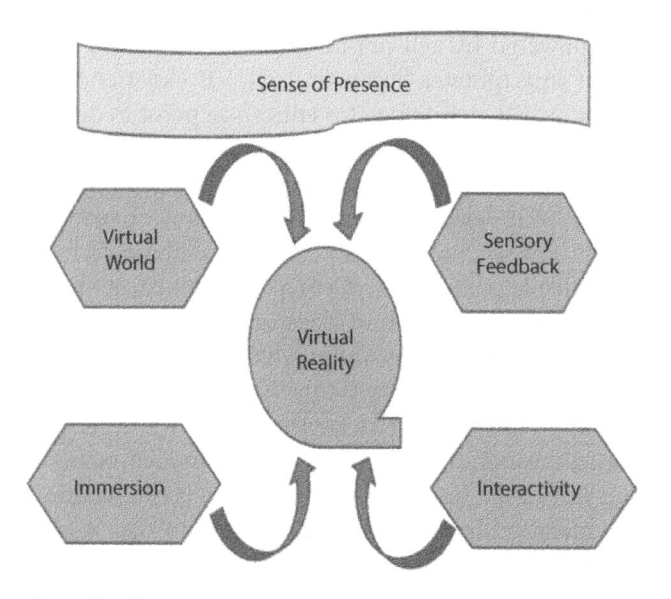

Figure 12.2 Virtual reality.

offering the user a greater "sense of being there" and enhancing immersion. At the extremely least, the most immersive systems enable viewers to completely submerge inside a 3D world with HMDs. Immersion, interactivity, and presence are the main elements of VR. Immersion is a psychological condition in which an individual engages in a setting that offers a constant flow of sensations and stimuli. Interactive refers to a user's virtual interaction with the surroundings enabling them to observe the outcomes of their actions [11].

Two types of virtual reality (VR) exist based on technological characteristics: immersive VR, which uses devices like head-mounted displays (HMDs) or mobile audiences to totally isolate users from the real world, and non-immersive VR, which displays simulations on a PC or television, notably in desktop VR programs and Cave Automatic Virtual Environment (CAVE) systems. Virtual reality (VR) is an artificially created representation of a 3D space that people are able to engage upon. Typically, expert electronics including virtual reality headphones are used for this. The goal of the system is to fully engage customers and give them a sense of existence giving the impression that they are actually there in the simulated setting. VR offers a complete and engaging encounter, which goes beyond the limitations imposed by conventional 2D model, and is used in a wide range of industries, including entertainment activities, medical, learning models, and pleasure. An enhanced VR equipment known as a Cave Automatic Virtual Environment (CAVE) is defined through the arrangement of several huge displays in an isolated space or area. If paired with specialist tracking methods, such stereoscopic picture panels create a complete three-dimensional experience for viewers.

The confined area in which users are able to completely lose themselves in the digital realm is referred to as a cave. People may interact as well as explore the simulated setting shown on the displays if they have 3D spectacles or related monitoring equipment. The immersive capabilities of CAVE systems contribute to a lifelike and captivating experience across various applications, spanning scientific research, architectural design, medical visualization, and virtual training simulations. CAVE technology heightens the feeling of presence enabling users to explore and manipulate virtual spaces in a manner closely mirroring real-world interactions [12].

12.2.3 VR Hardware and Software Ecosystems

Blockchain technology, Web 3.0 edge computing, AI, and virtual reality (VR) form the foundation of the Metaverse. Additional innovations thought to materialize the Metaverse include virtual worlds, gaming, and social networking sites. Technologies such as 5G/6G/Wi-Fi and user interfaces, like smart glasses/mobiles, as well as decentralization through AI, edge computation, and blockchain, contribute to the Metaverse's development. Geographical computing with augmented and virtual reality, the creator economy through e-commerce, discovery with avatars, and experience in gaming and social interactions constitute the seven levels of a Metaverse platform. The infrastructure layer encompasses various components such as semiconductors, cloud computing, 5G and Wi-Fi, and data centers. The technology and software that facilitate connections between users and bridge the virtual and physical worlds constitute the human interface layer. Blockchain, which enables value exchange across software applications, autonomous identity, novel disaggregation techniques, and the bundling of data and currencies, is a component of the decentralization layer [13].

Initially, a tracking system was used in the interactive program to enable users to move around in a virtual environment, compatible with virtual reality headsets and portable controllers. They developed a virtual reality program that displayed captivating scenes from a stationary position. Since Oculus Go all-in-one headsets do not require an additional computer to function, they utilized multiple of them. These headsets provided 1,280 × 1,440 pixels per eye, or roughly 100 pixels per field of view. Three-degrees-of-freedom tracking was used in these headsets making them suitable for seated viewing [14].

12.3 Artificial Intelligence in Smart Societies

12.3.1 Overview of AI Technologies Shaping Smart Societies

AI is a useful tool for fostering the development of intelligent societies. It is utilized in various contexts, including the advancement of intelligent medical, transportation, government, educational, and other social systems. AI represents the future due to its vast potential to bolster the actual economy, provide employment opportunities, and safeguard people's rights to engage in political discourse. Explanation of AI is depicted in Figure 12.3.

Transportation is being integrated with AI to make it smarter. Cars equipped with AI can recognize lane markings, traffic signals, and other road conditions. This not only frees up the drivers' hands but also enhances their awareness of how society, automobiles, and people are all interconnected through the gathering of massive data streams. This method

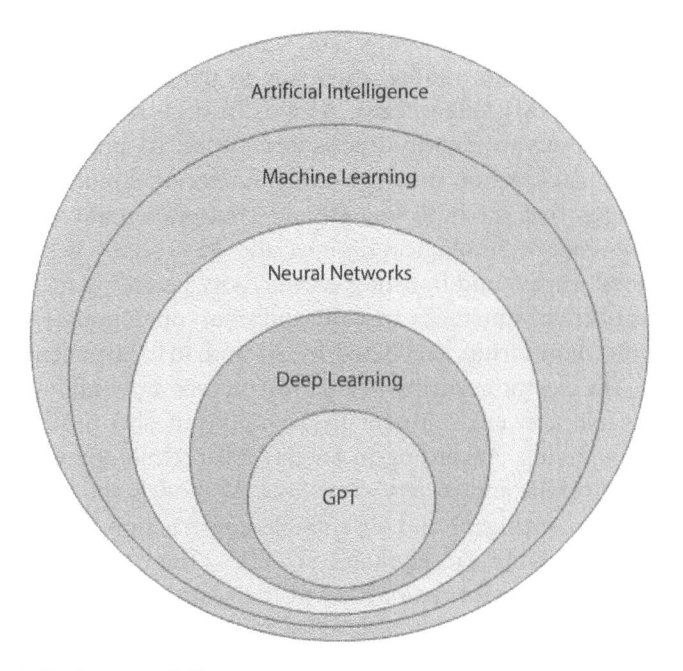

Figure 12.3 Explanation of AI.

reduces human error-related vehicular crashes while increasing the effectiveness and comfort of the roads. AI has also permeated the medical field to enhance patient data management, predict disease risks, provide hospital way solutions, and more. AI also ensures employment for a majority of the populace, and promises to protect the rights and welfare of the people by addressing social issues. AI will significantly reduce the number of errors in projections and decision making related to urban planning. Technical signals are combined with algorithms and data from video surveillance to create AI+. Furthermore, AI-enabled financial services can offer investors references, forecast and evaluate market conditions, provide personalized budgeting strategies to customers, and enhance customer experience by enabling digital goods presentations. AI in classrooms is expected to be capable to meet the various requirements of various pupils, and internet-based learning might be allowed to get past the conventional educational model's limitations of time, place, and period. This may also encourage initiatives to provide equal opportunity to learning while making more efficient use of the resources for learning that are already present [15].

The Architecture, Engineering, and Construction (AEC) Industry is undergoing significant modification due to the integration of AI and machine learning (ML) algorithms. AI aids in decision making by analyzing large datasets to identify trends and cycles. AI-powered tools for designing evaluate data on the environment, execute circumstances, and provide designs that are both energy and environmentally efficient. By using cutting-edge technological designs, the AEC sector is lowering its environmental impact and boosting productivity heralding in an entirely novel phase of environmentally friendly and smart operations. Forecasting and real-time monitoring, facilitated by AI and IoT advancements, will benefit the AEC sector ensuring that infrastructure and buildings are not only sustainable and environmentally friendly but also productive and aesthetically pleasing. According to Society 5.0, technology will enhance people's quality of life and society as a whole. AI models and methods play a crucial role in intelligent and sustainable construction contributing to the creation of durable and equitable societies. The construction industry is undergoing transformation due to the seamless integration of AI, data-driven techniques, and cutting-edge technology collectively known as Construction 4.0 and Construction 5.0. In addition to increasing productivity and efficiency in building operations, these technologies are also significantly advancing the concept of Society 5.0, which advocates for utilizing technology to improve society comprehensively.

12.3.2 Role of AI in Data Analytics, Automation, and Decision Making

The core purpose of AI in the field of analytics involves the analyses of large amounts of information. Experts as well as data experts may now identify trends and gain additional insight into the behavior of clients or additional data sources. Using effective machine learning approaches, AI can help make conclusions about vast amounts of information quickly and accurately. AI has an incredible effect on the analysis of information, speeds up the task, reveals impenetrable trends, and helps businesses derive practical conclusions. The synergy among AI and information insights is more than just a technical development. It is an edge over rivals, which boosts creativity and helps organizations succeed in the age of decisions based on choices.

Through employing AI to interpret IoT data collected, architects are able to obtain substantial information regarding the development's results and effectiveness. Modern analytics programs evaluate such data in immediate form and offer insightful information that helps with making choices. Instantaneous analysis of information is a useful tool for administrators of projects to monitor advancement, spot roadblocks, and enhance procedures. Through the application of IoT-enabled analytical models, one may forecast developments, identify issues, and come up with choices based on information throughout, which contribute to better completion rates and lower expenditures.

By leveraging the assessment of large amounts of information for making educated decisions based on previous patterns and forecast estimates, customers are able to maximize the deployment of resources, lower costs, and improve managing projects. AI offers the ability to totally change industries by boosting digital transformation, enhancing accuracy, and enabling well-informed decision making. If AI is combined with IoT gadgets, structural technology for automation is improved making it easier to handle and regulate a variety of building functions. Clients are able to connect to the systems in the structure and adjust the environment, illumination, and safety to suit their demands with phones or ipads. IoT-enabled intelligent detectors can also modify the surroundings in response to human choices and behavior [16].

A recent development, which is starting to be employed in intelligent cities, is Artificial Intelligence of Things (AIoT), which is a combination of AI and IoT technology. AIoT is helpful for obtaining, evaluating, and executing choices according to data gathered by network-connected IoT devices and detectors. Organizations may now use a wealth of information to speed up internal operations, minimize mistakes in manufacturing, increase productivity, and enhance decision making. AIoT implementation makes it easier to enable data to be continuously collected in intelligent cities. Information can subsequently be analyzed using machine learning and statistical processing to produce valuable data for making choices [8].

12.3.3 AI-Driven Applications in Healthcare, Transportation, and Education

AI-powered smart healthcare offers a safe, efficient, and readily deployable health monitoring system providing high-quality medical treatment at a fraction of the cost compared to clinics and assisted living facilities. A developing area of interest in recent years, the smart healthcare system has become increasingly important due to substantial progress in contemporary technology, particularly ML and computational intelligence (AI). Massive amounts of data on electronic health are generated every day due to the rapid growth of IoT devices. Wearable sensor-based smart healthcare services offer a convenient and more affordable alternative to expensive hospital settings.

The increasing ubiquity of mobile phones, along with their built-in sensors and cutting-edge communication capabilities, creates the necessary infrastructure for ongoing digital patient monitoring. The four main health measures that can be tracked with smartphone sensors are blood pressure (BP), oxygen saturation (SpO$_2$), breathing rate, and cardiovascular rate and variability. One important element in the healthcare industry's toolbox of AI methods is machine learning. It provides IoT devices with exceptional intelligence, data analysis, and knowledge inference capabilities. The detailed applications of AI in healthcare is described in Figure 12.4 [17].

The state of the automotive industry nowadays is so advanced that cars can operate on highways without the need for driver assistance at all. It is evident that technology has contributed to its amazing trajectory of invention and development. These days, AI is being used in transportation to assist make significant advancements, which is attracting the attention of global transportation executives. AI is transforming the logistics business and improving many facets of the transport environment, among its many

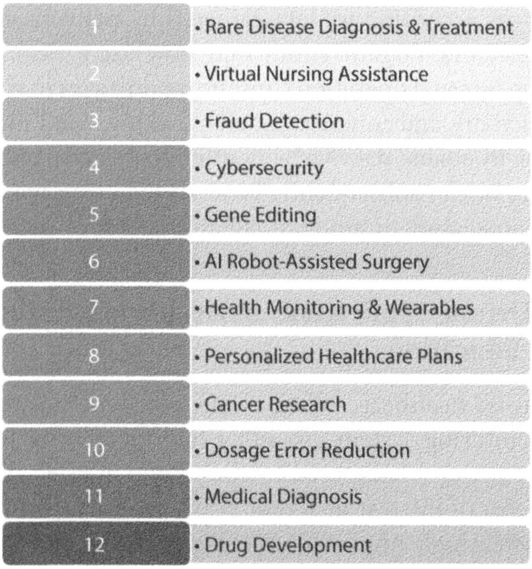

1	• Rare Disease Diagnosis & Treatment
2	• Virtual Nursing Assistance
3	• Fraud Detection
4	• Cybersecurity
5	• Gene Editing
6	• AI Robot-Assisted Surgery
7	• Health Monitoring & Wearables
8	• Personalized Healthcare Plans
9	• Cancer Research
10	• Dosage Error Reduction
11	• Medical Diagnosis
12	• Drug Development

Figure 12.4 Applications of AI in healthcare.

Virtualization of Society 5.0 Using AI and VR 309

other benefits. When combined with the latest advances, it creates a plethora of novel possibilities for automobile interaction opening the door to increasingly sophisticated, secure, and reliable public transit networks. The idea of autonomous vehicles emerges a possibility with AI offering a dramatic change in how we see and utilize transportation. It represents a big step approaching a time when transportation serves as a smart ecosystem that puts environmental responsibility, ease of use, as well as security first rather than just a means of getting from one place to another.

Self-driving cars are one among the primary forms of travel wherein AI is making significant effects. Autonomous vehicles could enhance the flow of traffic in general and lessen tragedies brought on mistakes made by people. Independent automobile development is now underway at several top automakers and technological firms; few have even begun piloting such cars on roadways that are public. The detailed applications of AI in transportation is described in Figure 12.5 [18].

Education is a crucial aspect of everyone's life, and its improvement remains a constant focus globally. Numerous changes, from teaching methods to curriculum structures, are implemented to enhance the education system. AI distinguishes up among such developments because it is a burgeoning innovation that has the ability to completely change education. AI-driven resources and tactics present chances to improve pupil achievement, tailor instruction, and assist students for achievement in modern times.

#	Application
1	Predictive Maintenance
2	Customer Service Chatbots
3	Autonomus Vehicles
4	Insurance Fraud Detection
5	Driver Behavior Analytics
6	Flight Delay Predictions
7	Traffic Management
8	Real Time Vehicle Tracking
9	Inventory Management
10	Intelligent Driver Care

Figure 12.5 Applications of AI in transportation.

AI is being incorporated into learning at universities and schools around the globe offering a fresh outlook to educators, learners, caregivers, and educational organizations. It is critical to remember that AI for learning refers to using intelligent machines for the purpose of enhancing teaching rather than substituting classroom instructors using robotic individuals.

Numerous AI devices that can transform the educational process are anticipated to revolutionize the field of teaching in the years to come. Artificial intellect (AI) is a technological imitation of an individual's intellect with the goal of mimicking people's thought and behavior. The domain of AI in education (AIED) was founded in the 1970s with the goal of enabling adaptable, individualized, and interactive educational activities in academic institutions through the integration of cutting-edge technology. Key trends in AIED encompass Intelligent Tutor Systems, smart classroom technologies, adaptive learning, and pedagogical agents. These trends underscore the commitment to leveraging AI to create a more dynamic, efficient, and effective educational system for the future. The detailed applications of AI in education is illustrated ion Figure 12.6.

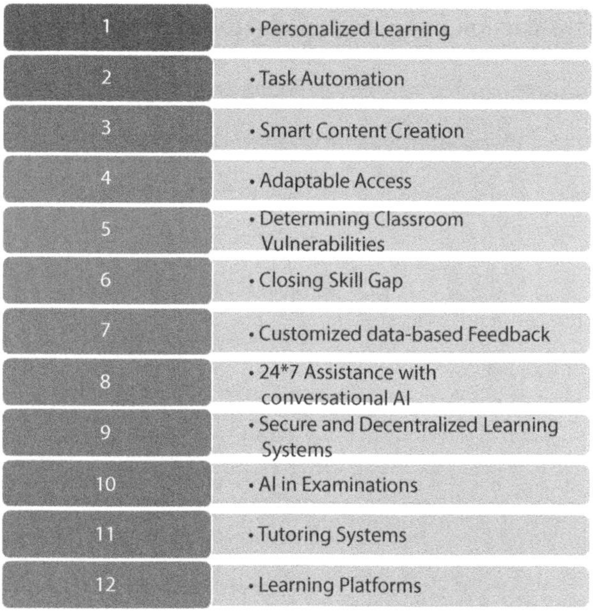

#	Application
1	Personalized Learning
2	Task Automation
3	Smart Content Creation
4	Adaptable Access
5	Determining Classroom Vulnerabilities
6	Closing Skill Gap
7	Customized data-based Feedback
8	24*7 Assistance with conversational AI
9	Secure and Decentralized Learning Systems
10	AI in Examinations
11	Tutoring Systems
12	Learning Platforms

Figure 12.6 Applications of AI in education.

12.4 Integration of AI and VR

12.4.1 How AI and VR Technologies Complement Each Other

The two main technologies of the modern period, AI and VR, are changing the way humans engage with electronic devices and interactions. The two separate fields have advanced significantly, and combining them opens up fresh opportunities. A new generation of realistic training scenarios is being introduced by the powerful combination of AI and VR. AI improves VR's potential through knowledge and flexibility giving a fresh crop of educators an unmatched learning environment. VR continues to be an effective instrument to develop interpersonal abilities. It offers users a unique opportunity to learn through completely immersive experiences, many of which include meticulously crafted 360° video. Users can study a variety of scenarios and learn from their failures and successes from such activities. Therefore, it is possible to simulate substantial trends and project potential advances regarding physical items in simulations by utilizing a variety of categorization and regression approaches [19].

AI integration into VR adventures has an abundance of possibilities. AI may promote more adaptable and responsive interactions providing learners with individualized instruction and criticism. The effective application of AI within VR requires a delicate profile. Such digital beings must be enabled to communicate with AI, understand textual communications, and carry on discussions to provide prompt support. By employing AI-enabled VR simulation, people can improve their communication skills. Individuals might grow stronger at interacting with others and handling a range of client-based situations by using AI-powered scenarios. AI and VR are enabling physicians to conduct interventions in models that is transforming the medical industry. This approach helps doctors become more proficient as well as at ease before doing procedures within the real-life setting. It allows property brokers to draw findings according to information and offer recommendations tailored to individual customers. Chat bots and virtual assistants driven by AI facilitate lead generation and client service helping real estate agents to better manage their resources and time. On the other hand, clients can remotely and thoroughly investigate properties through the application of virtual reality in immersive virtual tours. Both residential and commercial properties can benefit from this technology, which can expedite the decision-making process for prospective purchasers and minimize the need for multiple in-person property visits [20].

12.4.2 Examples of AI-Enhanced Virtual Reality Applications

Given that the utilization of AI in Interventional Radiology (IR) is the main topic of this research, the development of tactile perception, motor skills, and spatial and cognitive awareness are all necessary for the successful and efficient operation of IR equipment distinguishing features of IR training. Thus, insufficient skill can lead to longer procedure times, more operational errors or complications, and increased radiation exposure to operators and patients alike. This problem could be addressed by incorporating VR simulation technologies into educational programs to provide trainees with sufficient training hours. Furthermore, depending on their training contexts, IR physicians may have extremely varied skill sets due to variations in case mixes across institutions. Introducing a greater diversity of cases to trainees through simulation databases could mitigate this issue [21].

While the focus of this research paper is on virtual reality (VR) assisted by AI in medicine, the rapid advancement of AI technology allows VR platforms to enhance medical professionals' visual perception throughout treatment and surgery phases. VR systems provide additional cognitive signals during surgical or training procedures enabling experts to visualize clinical information in an easier-to-understand format. In a virtual environment, experts can access more data regarding treatment with the help of AI-powered VR approaches. A deep learning-based procedure was implemented to enable critical care, achieving automatic segmentation of the skull, facial skin, and the ventricles. Additionally, VR driven by AI has the potential to completely transform remote medical education and telemedicine, particularly in impoverished or underserved regions. By integrating real-time analytical functionalities with virtual reality, AI algorithms will provide prompt diagnostic assistance, enhancing decision-making procedures in both non-clinical and clinical contexts. In addition, through giving excellent health services distantly, these advances may address the discrepancies among medical fairness, excellence, and affordability.

12.4.3 Possibilities and Obstacles When Fusing AI with VR

The primary barriers to the application of AI-enabled VR in healthcare are the accuracy of information and accessibility. Constructing effective AI systems requires large, diverse samples. The data remain scattered and difficult to find, though. In addition, this is a difficult endeavor that calls for organizational and technical alterations to integrate state-of-the-art AI–VR advancements with existing processes and healthcare establishments.

Moral dilemmas must be resolved, especially those involving the confidentiality of patients, security of information, and well-informed approval. The unaddressed topic of accountability in AI–VR-related errors compounds the governing setting for officials, technological corporations, and healthcare providers. Furthermore, the achievement of reliability and authenticity in VR models is a challenging scientific task that is required for the successful delivery of healthcare instruction as well as treatment. The black box nature of certain AI tools might render it challenging to customers to have faith in them, as doctors typically require to know the way AI recommendations are generated. The development of user-friendly interfaces with a wide range of consumers and assisting physicians and patients in overcoming their opposition to innovation are additionally essential to the widespread adoption of AI-enabled VR models [22].

Examples of leading cutting-edge gadgets that have restricted energy storage and computational ability include mobiles and AR/VR headphones. Large computational capabilities are needed for executing learning techniques regionally that may swiftly drain the power source and harm the electrical parts of the gadget. Furthermore, implementing large, complicated AI systems having an elevated energy requirement for VR and AR applications on edge computers having low storage might be difficult. In addition, a large variety of gadgets with different technology specs as well as operating systems make up the VR environment. Finally, by processing data locally and eliminating the need for continuous cloud-based transfers of information, edge computing improves privacy. However, this approach also introduces security vulnerabilities [23].

Some key challenges of the Metaverse are that even if the Metaverse's technological capabilities nearly resemble those of the real world, certain experiences are better had in reality. Software-wise, the core with excellent compatibility in the Metaverse environment is comprised of programs written in the Metaverse sans requiring coding. But in a big system, the software eventually reaches its complexity limit as it becomes more intricate. Every citizen of the community—children, seniors, individuals with disabilities, and people of all races and ethnicities—must be able to access the Metaverse. There might be a variety of materials in the Metaverse; thus, we need to make sure that each and every one can access them. To decrease biased material and thereby affect user behavior and the decision-making process, it is also crucial to take into account personalized content presentation in front of users and improve the fairness of the recommendation algorithm. One description of vast cyberspace is the Metaverse. Because of this, cyberbullying in the Metaverse may inevitably pose a social risk to the environment. Even though widespread computing is now possible because of technological advancements, numerous potential benefits will not materialize until individuals get acquainted with and accept the technology. Being accountable is vital here. Another aspect of responsibility in the Metaverse universe is the management of user data, which includes location and surroundings, rather than just using normal smart devices. Before the Metaverse is fully integrated into the real world and our everyday lives, several challenges must be solved in addition to those stated above, such as ethical issues, scalability, etc. [13].

12.5 AI and VR in Education

12.5.1 Virtual Classrooms and Immersive Learning Experiences

Textual and listening modes are the mainstays of conventional instructional activity teaching methods. There are notable and generally acknowledged drawbacks to this, either in the context of schooling or job-related learning. Everyone who learns is different, which can be seen in the way they absorb and remember knowledge. Additionally, a lot of people find that education is easier and more affordable when engaging, and participatory information is provided, particularly for those who study best visually or physically. For numerous pupils, experiential learning is a very powerful tool for increasing their understanding and proficiency. It offers synthetic, technologically produced surroundings and material that faithfully imitate real-world situations allowing for the learning and development of novel abilities. Moreover, virtual classrooms transcend geographical boundaries enabling access to educational resources and experts regardless of physical location, thereby promoting inclusivity and democratizing access to quality education [24].

12.5.2 AI-Enabled Adaptive Learning Systems

With the use of AI, systems for adaptive learning customize the process of learning for every learner through continually altering the course material, timing, and evaluations to fit their unique educational requirements and interests. In an effort to optimize educational outcomes and foster academic competence, AI tools examine pupil achievement information, instructional strategies, and intellectual characteristics. The results are customized suggestions, and focused treatments. Additionally, by effectively managing varied populations of students and differentiating guidance, instructors can create an increased encouraging and welcoming setting for learning with the use of adaptive learning technologies [25].

12.5.3 Skill Development and Training Using VR and AI

As they offer plausible, absorbing, and engaging educational settings, VR and AI tools are transforming education and instruction throughout a variety of areas. By using VR exercises, pupils are able to simulate practical knowledge in a secure and monitored environment reducing threats and improving ability. Examples of these abilities include interventions, physical maintenance, and responding to disaster situations. Individualized coaching, constructive criticism, as well as performance indicators are provided by AI-enabled reactive response mechanisms in immediate form, which facilitate the advancement of skills and competence. Additionally, comprehensive and affordable training delivery over a variety of situations is made possible by VR and AI-enabled educational initiatives enabling people to gain novel talents and abilities to fulfill the expectations of the employment market of the coming years [26].

12.6 Smart Society 5.0 Healthcare Innovations

12.6.1 Virtual Healthcare Consultations and Simulations

Virtual healthcare consultations and simulations have emerged as transformative tools in modern healthcare delivery offering numerous benefits such as increased accessibility, convenience, and cost effectiveness. Even while immersive learning has actually been available for a while, innovation has advanced it significantly. Immersion learning has been reinvented as a dynamic, practical problem online experience thanks to virtual reality (VR) and augmented reality (AR). Immersion learning is frequently utilized by individuals to acquire new abilities. Consider urgent escape procedures or firemen honing their craft in regulated burn scenarios. On the other hand, in a secure and captivating virtual universe, AR and VR accurately replicate real-life scenarios. In contrast to an educational setting, individuals are active learners. Rather, learners utilize as much of the real-life setting as they can to facilitate the instruction and utilization of techniques. An entirely fresh realm of possibilities for learning has been made possible in business and educational environments by AR and VR [27].

12.6.2 AI-Driven Diagnostics and Treatment Planning

AI has revolutionized diagnostics and treatment planning in healthcare by enabling the analysis of large volumes of medical data and the generation of actionable insights in real time. Healthcare images from scans may be effectively explained by AI tools, which help with prompt identification and enhance precision in diagnosis. Additionally, by assisting clinicians in creating customized treatment strategies using patient-tailored information, health records, and forecasting, AI-driven systems for decision making may assist to improve the effectiveness and efficiency of medical procedures [28].

12.6.3 VR and AI-Based Treatments and Rehab

AI and VR have been employed more and more in rehab and treatment procedures to improve outcomes for patients, inspiration, and involvement. To promote motor rehabilitation and independence in function, VR models offer individuals immersive worlds in which they may carry out routine tasks, engage in both mental and intellectual workouts while getting immediate assistance. Furthermore, AI-driven systems have the ability to modify therapy schedules in response to progress made by patients, metrics for performance, and customized objectives. This enhances the efficiency of treatment efforts and promotes permanent healing [29].

12.7 Challenges and Future Directions

12.7.1 Current Obstacles to Integrate VR and AI in Smart Communities [30]

Legal and Moral Issues: As AI and VR become increasingly prevalent in society, worries about confidentiality of information, algorithmic prejudice, and responsibility for AI-enabled judgments are intensifying.

Technical Restrictions: AI and VR techniques continue to confront obstacles related to computational capacity, internet access, and delay regardless of their progress. These issues might have an impact on consumer satisfaction and rate of acceptance.

Interchangeability: To ensure smooth interoperability and interaction across different platforms, incorporating AI and VR solutions with present systems and developments might be challenging and could call for standardization initiatives.

Consumer Acceptance and Credibility: Before AI and VR technologies are extensively deployed, trust from consumers must be gained. People could be reluctant to utilize such innovations because they have doubts regarding their stability, safety, and capability to replace their jobs.

Usability and Diversity: It remains a challenge to successfully manage this task through equitable design and legislative measures, ensuring that both AI and VR systems are accessible to everyone, particularly individuals with impairments or those from marginalized groups.

12.7.2 Prospective Developments and Emerging Patterns in AI and VR in Smart Societies [31]

Developments in AI Techniques: It is anticipated that ongoing studies and advances in AI techniques will spur additional enhancements in the abilities of AI resulting in devices that become increasingly smart and adaptable.

Enhanced User Interfaces: Future advancements in VR technology may lead to more immersive and intuitive user interfaces, including haptic feedback, gesture recognition, and eye tracking.

Integration with IoT and Edge Computing [33]: AI and VR technologies are likely to be increasingly integrated with IoT and edge computing, which can enhance the efficiency and responsiveness of smart systems.

Personalized and Context-Aware Experiences [34]: AI-driven personalization techniques, combined with VR technology, could enable highly customized and context-aware experiences tailored to individual preferences and needs, whether in education, healthcare, entertainment, or other domains.

Mixed Reality (MR) and Augmented Reality (AR) [35]: The convergence of AI, VR, and AR/MR technologies is expected to open up new possibilities for blending virtual and physical worlds enabling applications such as virtual collaboration, remote assistance, and augmented training experiences.

12.7.3 Consider the Role of Emerging Technologies [32]

The eventual adoption of AI and VR in smart communities will also be significantly affected by the function of forthcoming innovations. Blockchain technology has the potential to improve privacy and openness in AI platforms, whereas 5G connections promise quicker transmission of information and fewer delays, which are essential for providing VR encounters.

Additionally, the advent of quantum computing may unlock new computational capabilities accelerating AI training and enabling more complex simulations in VR environments. These emerging technologies will likely complement and enhance the capabilities of AI and VR systems driving innovation and progress in smart societies.

12.8 Conclusion

12.8.1 Summary of AI and VR Technologies in Smart Societies

The integration of AI and VR technologies in smart societies presents both opportunities and challenges. Important conclusions emphasize how crucial it is to tackle scientific, moral, and constitutional problems to guarantee the proper application of such innovations. To fully capitalize on AI and VR's promise to improve standard of existence and advance society, issues like confidentiality of information, discrimination, compatibility, and trust among consumers have to be properly controlled. However, future trends and advancements offer promising prospects for personalized, context-aware experiences, enhanced user interfaces, and seamless integration with emerging technologies [33].

12.8.2 Vision for the Future of Smart Societies with AI and VR

In the future, smart societies empowered by AI and VR will be characterized by seamless integration of digital and physical environments, personalized services, and enhanced human–machine interactions. Citizens will benefit from immersive education and training experiences, virtual healthcare services, and AI-driven decision support systems for urban planning and governance. VR technology will enable remote collaboration, virtual tourism, and entertainment experiences transcending geographical boundaries and fostering cultural exchange. AI-powered smart infrastructure, supported by 5G networks and edge computing, will optimize resource utilization, improve transportation systems, and enhance public safety. Ultimately, the vision for smart societies with AI and VR is one of inclusivity, sustainability, and human-centered innovation, where technology serves to empower individuals and communities to thrive in an increasingly interconnected and dynamic world [34–36].

References

1. Gurjanov, A.V., Zakoldaev, D.A., Shukalov, A.V., Zharinov, I.O., The smart city technology in the super intellectual society 5.0. *J. Phys. Conf. Ser.*, 1679, 1–6, 2020, DOI: 10.1088/1742-6596/1679/3/032029.
2. Nair, M.M., Tyagi, A.K., Sreenath, N., The future with industry 4.0 at the core of society 5.0: open issues, future opportunities, and challenges, in: *International conference on computer communication and informatics*, pp. 1–7, 2021, DOI: 978-1-7281-5875-4/21.
3. Rosemann, M., Becker, J., Chasin, F., City 5.0. *Bus. Inf. Syst. Eng.*, 63, 71–77, 2021, DOI: 10.1007/s12599-020-00674-9.
4. Becker, J., Chasin, F., Rosemann, M. et al., City 5.0: Citizen involvement in the design of future cities. *Electron Mark.*, 33, 10, 1–18, 2023, DOI: 10.1007/s12525-023-00621.
5. Gamayanto, I. and Nurhindarto, A., Developing smart city 5.0 framework to produce competency. *J. Intel. Appl. Syst.*, 5, 1, 23–31, 2020.
6. Natalia, T., Joshi, S.K., Dixit, S., Kanakadurga Bella, H., Jena, P.C., Vyas, A., Enhancing smart city services with AI: a field experiment in the context of industry 5.0. *Bio Web Conf.*, 86, 1–11, 2024, DOI: 10.1051/bioconf/20248601063.
7. Mishra, P., Thakur, P., Singh, G., Sustainable smart city to society 5.0: state of the art and research challenges. *SAIEE Afr. Res. J.*, 113, 152–164, December 2022.
8. Saluky, Y.M., A review: Application of AIOT in Smart Cities in Industry 4.0 and Society 5.0. *Int. J. Smart Syst.*, 1, 1, 1–4, 2023.
9. Walstra, K.A., Cronje, J., Vandeyar, T., A Review of Virtual Reality from Primary School Teachers' Perspectives. *Electron. J. e-Learn.*, pp. 00–00, 2023.
10. Polishchuk, E., Bujdosó, Z., El Archi, Y., Benbba, B., Zhu, K., Dávid, L.D., The Theoretical Background of Virtual Reality and Its Implications for the Tourism Industry. *Sustainability*, 15, 13, 1–19, 2023.
11. Mäkinena, H., Haavisto, E., Havola, S., Koivisto, J.-M., User experiences of virtual reality technologies for healthcare in learning: an integrative review. *Behav. Inf. Technol. 2022*, 41, 1, 1–17, 2020.
12. Luo, H., Li, G., Feng, Q., Yang, Y., Zuo, M., Virtual reality in K-12 and higher education: A systematic review of the literature from 2000 to 2019. *J. Comput. Assist. Learn.*, 37, 3, 887–901, June 2021.
13. Darwish, A. and Hassanien, A.E., Fantasy Magical Life: Opportunities, Applications, and Challenges in Metaverses. *J. Syst. Manage. Sci.*, 12, 2, 405–430, 2022, DOI: 10.33168/JSMS.2022.0222.
14. Chandler, T., Richards, A.E., Jenny, B. et al., Immersive landscapes: modelling ecosystem reference conditions in virtual reality. *Landsc. Ecol.*, 37, 1293–1309, 2022.

15. Chen, X., Tang, X., Xu, X., Digital technology-driven smart society governance mechanism and practice exploration. *IEEE Conference Front. Eng. Manag.*, vol. 10, pp. 319–338, 2023.
16. Nitin, R., Integrating Leading-Edge Artificial Intelligence (AI), Internet of Things (IOT), and Big Data Technologies for Smart and Sustainable Architecture, Engineering and Construction (AEC) Industry: Challenges and Future Directions. *International Journal of Data Science and Big Data Analytics*, 3, 2, 1–23, November 2023.
17. Nasr, M., Islam, M.M., Shehata, S., Karray, F., Quintana, Y., Smart Healthcare in the Age of AI: Recent Advances, Challenges, and Future Prospects. *IEEE Access*, 9, 145248–145270, 2021, DOI: 10.1109/ACCESS.2021.3118960.
18. Nikitas, A., Michalakopoulou, K., Njoya, E.T., Karampatzakis, D., Artificial Intelligence, Transport and the Smart City: Definitions and Dimensions of a New Mobility Era. *Sustainability*, 12, 7, 2789, 2020, DOI: 10.3390/su12072789.
19. Zhang, Z., Wen, F., Sun, Z., Guo, X., He, T., Lee, C., Artificial Intelligence-Enabled Sensing Technologies in the 5G/Internet of Things Era: From Virtual Reality/Augmented Reality to the Digital Twin. *Adv. Intell. Syst.*, 4, 7, 1–23, Jul 2022, DOI: 10.1002/aisy.202100228.
20. Miljkovic, I., Shlyakhetko, O., Fedushko, S., Real Estate App Development Based on AI/VR Technologies. *Electronics*, 12, 3, 1–20, 2023.
21. von Ende, E., Ryan, S., Crain, M.A., Makary, M.S., Artificial Intelligence, Augmented Reality, and Virtual Reality Advances and Applications in Interventional Radiology. *Diagnostics*, 13, 5, 1–14, 2023.
22. Yixuan, W., Kaiyuan, H., Chen, D.Z., Wu, J., AI-Enhanced Virtual Reality in Medicine: A Comprehensive Survey. *Comput. Vis. Pattern Recognit.*, 8326–8334, 2024
23. Chenna, D. and Magic Leap Inc. USA, Edge AI in AR/VR: exploring opportunities and confronting Challenges. *Feedforward*, 2, 4, 22–31, 2023.
24. Johnson, E. and Smith, M., Virtual Classrooms and Immersive Learning: A Review of Applications and Benefits. *J. Educ. Technol.*, 10, 2, 89–102, 2023.
25. Brown, C. and Lee, J., Adaptive Learning Systems: Harnessing the Power of AI for Personalized Education. *J. Learn. Anal.*, 15, 3, 207–220, 2022.
26. Wang, L. and Zhang, H., Skill Development and Training using VR and AI: Opportunities and Challenges. *J. Vocat. Educ. Train.*, 20, 1, 45–56, 2024.
27. Smith, A. and Johnson, B., Virtual Healthcare Consultations: Opportunities and Challenges. *J. Telemed. Telecare*, 10, 2, 89–102, 2023.
28. Brown, C. and Lee, J., AI-Driven Diagnostics in Healthcare: Current Status and Future Directions. *J. Med. Imaging*, 15, 3, 207–220, 2022.
29. Wang, L. and Zhang, H., Rehabilitation and Therapy through VR and AI: A Review of Recent Advances. *J. Rehabil. Med.*, 20, 1, 45–56, 2024.
30. Smith, J. and Doe, A., Ethical Considerations in the Deployment of AI and VR Technologies. *J. AI Ethics*, 5, 2, 123–135, 2023.

31. Brown, C. and Johnson, B., Future Trends in AI and VR Technology. *J. Virtual Real.*, 10, 3, 207–220, 2022.
32. Chen, X. and Lee, Y., Integration of AI, IoT, and Edge Computing for Smart Systems. *IEEE Trans. Emerg. Top. Comput.*, 12, 1, 45–56, 2024.
33. Johnson, E. and Smith, M., Building Smart Societies: The Role of AI and VR. *J. Smart Cities*, 8, 1, 56–68, 2023.
34. Lee, J. and Kim, S., Toward Human-Centered Smart Societies: Vision and Challenges. *IEEE Trans. Emerg. Top. Smart Technol.*, 10, 2, 89–102, 2022.
35. Wang, L. and Zhang, H., AI and VR for Smart Societies: Opportunities and Challenges. *J. Future Technol.*, 15, 3, 207–220, 2024.
36. Velmurugadass, P., Dhanasekaran, S., Anand, S.S., Vasudevan, V., Enhancing Blockchain security in cloud computing with IoT environment using ECIES and cryptography hash algorithm. *Mater. Today: Proc.*, 37, 2653–2659, 2021.

13

Battery Power Management Schemes Integrated with Industrial IoT for Sustainable Industry Development

D. Karthikeyan[1], A. Geetha[1]*, K. Deepa[2] and Malathy Sathyamoorthy[3]

[1]*Department of Electrical and Electronics Engineering, College of Engineering and Technology, SRM Institute of Science and Technology, Kattankulathur, Chengalpattu, Tamil Nadu, India*
[2]*School of Computing, College of Engineering and Technology, SRM Institute of Science and Technology, Chennai, Tamil Nadu, India*
[3]*Department of Information Technology, KPR Institute of Engineering and Technology, Coimbatore, Tamil Nadu, India*

Abstract

Over the past decade, there have been notable advancements in the realm of smart energy management technologies, encompassing diverse methodologies and novel solutions aimed at adaptive battery management. A wide variety of industries rely on batteries, and BMSs are crucial to their efficient operation, safety, and longevity. The voltage, current, temperature, and charge status of a battery are few factors that a BMS is responsible for monitoring and controlling. So, in this chapter, an IoT solution that allows to operate and monitor battery storage systems, along with cloud infrastructure, is presented. In addition, it explores the most recent developments in BMS technologies, which include complex algorithms, sensing approaches, and communication protocols.

Keywords: Sustainable development goals, internet of things, energy management system

---

*Corresponding author: geethaa2@srmist.edu.in

13.1 Introduction

For any battery-dependent device to perform optimally, it is crucial to consider the performance of energy storage model. This ensures that the device meets the requirements for performance, cost effectiveness, efficiency, and environmental sustainability. Battery factors, as well as the inclusion of balancing circuits mechanisms, play a key role in achieving these goals. The battery plays a crucial role in any device that relies on battery power serving as a key energy source and enabling portability. With the rise in climate change, the focus on renewable and clean energy is becoming increasingly crucial. Norway has become the first country to ban fuel-based cars by 2025 promoting the growth of EVs [1]. Programs, like the Advanced Research Projects Agency—Energy (ARPA-E), are funded by US-DOE, which recently announced a $50M funding for the Battery500 project. In addition to the automobile manufacturing sector, the economics of decentralized energy production are quite favorable. Efficient utilization of systems derived from various renewable sources relies heavily on the effectiveness of energy storage technologies [2]. This system ensures the safe and efficient operation of batteries in various devices like EVs. This system oversees the battery's current, voltage, and temperature, predicts SOC, and calculates the remaining time until the battery is fully discharged. The device includes components and circuitry for protecting against over- and under-voltage, as well as other restrictions [3]. Figure 13.1 illustrates a standard battery system.

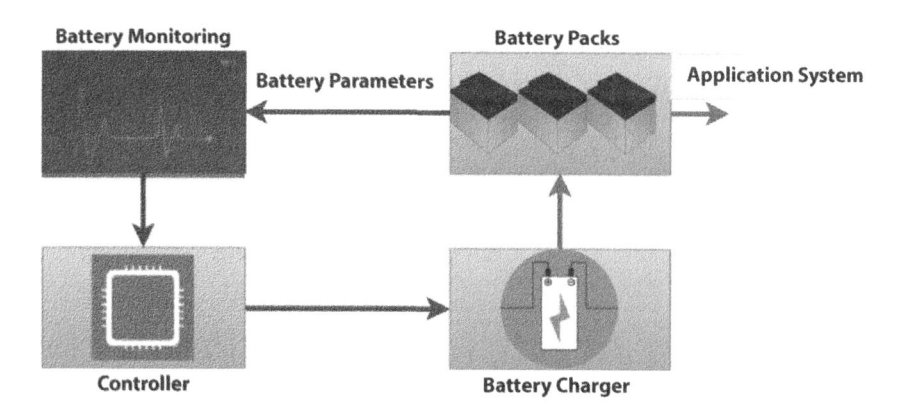

Figure 13.1 A typical layout of BMS.

Battery deployment can be classified into the following categories:

A. Mobility: The device can be classified into many sorts based on its usage. Stationary systems are commonly used in electric grids [4].

   i. Computing Power: The system's pool of computing resources may be divided into many categories. Every category possesses a corresponding response time and dependability.
   ii. Sensors: They build the framework by giving system data such as temperature, voltage, current, and GPS.
   iii. Programmability: Systems frequently have chargers that may be programmed to follow instructions from the BMS controller and determine how to charge or discharge a device.

Section 13.2 discusses the different battery-related attributes. In Section 13.3, the text outlines the computational techniques, strategies, controllers, and optimizations used in battery management. Section 13.4 discusses the IoT innovation in BMS. Section 13.5 emphasizes sustainable developments through BMS. Section 13.6 includes the conclusions and future trends.

13.2 Current Battery Technologies

This section provides a comprehensive discussion and analysis of the different battery attributes currently under development. Lithium-ion batteries can be replaced by sodium-ion and lithium–sulfur batteries, which have superior depth of discharge and less of an influence on the environment. Similar to lithium–air batteries, there are still certain technological problems that need to be fixed, such as insufficient cycle life [5]. The selected battery technologies' specific energy and energy density are shown in Table 13.1 and Figure 13.2.

Lithium salt in organic solvents makes up LIB liquid electrolytes. This design may provide serious safety issues because of the flammability, leaking, and toxicity of the electrolyte. Liquid electrolyte cells lack certain benefits that solid-state batteries do. It is to mention improved energy density to reduce production costs as in Figure 13.3 [6].

Table 13.1 Evolution of a battery.

Generation	Cathode	Anode
1 (1991)	NFP, LCO, NCA	Carbon, graphite
2 (1994, 2005)	(a) NMC111, LMO (b) NMC532, NMC622	Carbon, graphite
3 (2020, 2025)	(a) NMC811 (b) NMC (high energy)	Graphite and Si Si/carbon
4 (2025–2030)	NMC	Solid electrolyte
5 (>2030)	Metal air	Li–sulfur

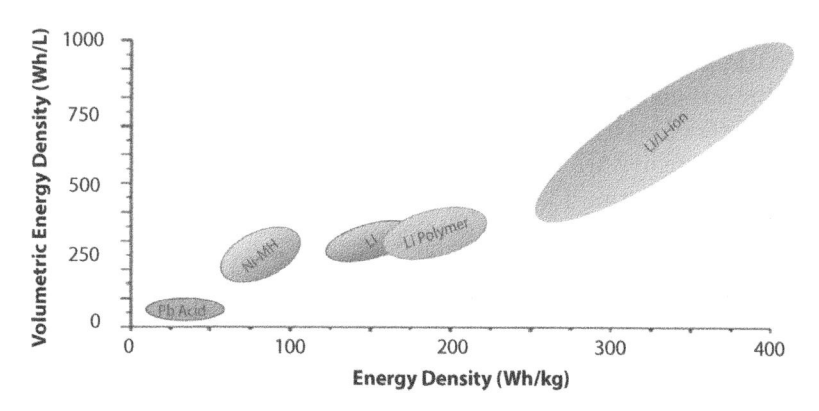

Figure 13.2 Selected battery technologies' specific energy and energy density.

13.2.1 Metal–Air Battery

It is significant for their weight. Zinc, aluminum, magnesium, and calcium are among the various metal–air batteries that can be used, in addition to lithium–air batteries, which possess 2.91 V. Aluminum–air batteries are more stable than lithium–air batteries, which have 13.2 kW/kg (like gasoline). Metal–air batteries' theoretical specific energy is seen in Figure 13.4. Due to alkaline electrolytes' poor specific power and carbonation, batteries have a limited lifespan and high internal resistance. Researchers are working to overcome these issues.

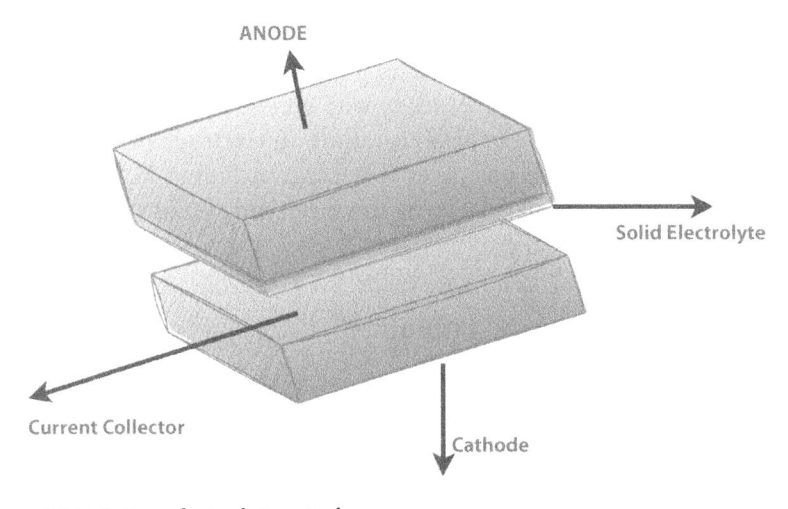

Figure 13.3 Battery electrode terminals.

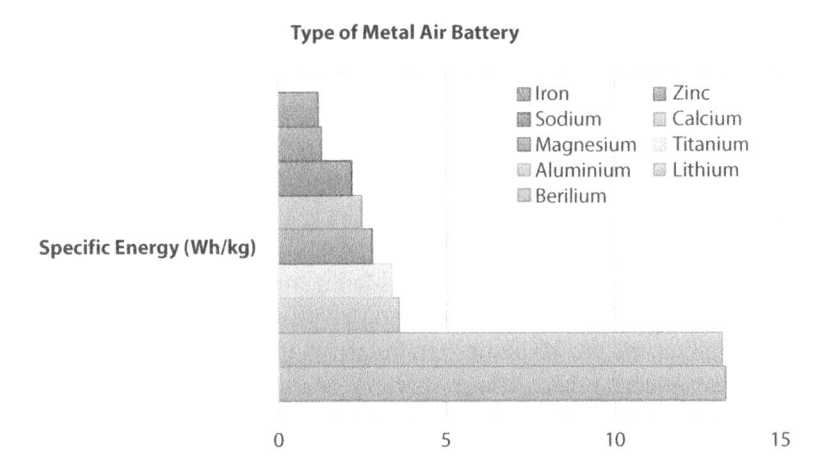

Figure 13.4 Metal–air batteries' theoretical specific energy.

13.2.2 Lithium–Sulfur Battery

In lithium–sulfur batteries, organic liquid acts as the electrolyte, sulfur composite, and lithium metal. Lithium-ion batteries are also a good option for future energy storage applications due to their affordability, abundant sulfur supply, and absence of necessary components. This battery type has a reduced environmental footprint, and sulfur can be obtained from recycled materials. The LiSB cell has a nominal voltage of 2.1 V. The LiSB battery can achieve a 100% depth of discharge, while the LIB battery can only reach 80%. The LiSB also has a long lifespan predicted to be 10 years. LiSBs are projected to be market ready with an energy density of 500 Wh/kg. When a thermal runaway happens, using an electrode based on sulfur may cause specific dangerous gases to escape, including H_2S, SO_2, COS, and CS_2. To improve the battery's performance, using a trustworthy battery management solution is highly advised.

13.2.3 Batteries Beyond Lithium

The ongoing research are exploring many ways to switch to less expensive and more sustainable light metals like sodium rather than lithium. It is challenging to design robust, long-lasting electrodes with rapid charge and discharge rates. The goal of the Chinese company CATL is 200 Wh•kg^{-1}.

Sodium-ion batteries (NIBs) work like Li-ion batteries. Sodium, the fourth most abundant element, is cheap and abundant. NIBs have a faster capacity drop and worse cycle life than LiBs. High-temperature operating batteries based on sodium have now been offered in 2022. The parameters of Li-ion micro-batteries are provided in Table 13.2.

13.3 Battery Energy Storage and Management

Electric batteries can store large amounts of energy for a long time. Figure 13.5 shows that protection is a Battery Management System's major function. Maintaining cell balance and aging requires cell monitoring. BMS allows specified system structure irregularity cures. Additionally, system temperature affects power usage; Battery Management System enables

Table 13.2 Parameters of Li-ion micro-batteries.

Si. no.	Type	Details
1	ITEN	Solid-state thin film, 600-mm thick with 500 µAh (2.5 V, −40° to 85°), 3.2 × 2.4 mm
2	ST Micro	Solid-state thin film, 220-mm thick with 220 µAh (3.9 V, −20° to 60°), 25.7 × 25.7 mm
3	MURATA	Coin, 1,600-mm thick with 30,000 µAh (3 V, −30° to 70°), 12.5-mm diameter

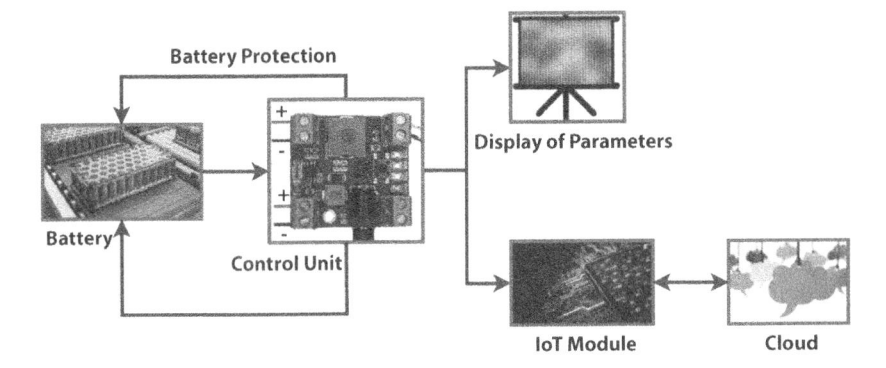

Figure 13.5 The IoT protection of a battery management system.

temperature regulation. Hybrid and electric car battery management technologies are explored. This study discussed the issues of the existing Battery Management System assessment of a battery's health, charge, and life (a serious operation) [7]. By evaluating current battery condition evaluation methodologies, BMS issues and solutions are identified.

This study examines the electromagnetic interference (EMI) sensitivity of lithium-polymer (LiPo) and lithium-ion Battery Management Systems (BMSs) for usage in electric and hybrid cars. EMC system simulation is used to create a demonstration board for a BMS. It describes how an EMC system simulation enhances board design by identifying the underlying reason. A test board that experimentally activated the BMS EMI vulnerability is only partially described in research. By analyzing the processes of EMI-induced inadequacies that were seen during testing, tentative results were described. The thermal behavior of battery packs under power requirements has not been well studied. The expected thermal estimation model is classified by heat dissipation, resistance, reversible heat, and Joules heating. Characterization of hybrid pulse power regulates and measures the resistances of charge intermissions. The battery effects and restrictions of high-power charging models are studied using optimization techniques. Power sources' power distribution is maximized [8].

A suggested management technique takes into account the charging and discharging power of grid-connected storage devices to maximize their energy efficiency, which is an additional pulse period-based technique for estimating pulse power. Reasonable accuracy was ensured by regression analysis methodology, since their state of health inaccuracy was less than 1%. There is no standard for battery management systems in electric cars, small appliances, or major applications. This study might offer a methodical investigation of BMS for transportation, electrification, and extensive (stationary) applications with targeted regulation given the significance of BMS and its role in ESS safety. The components, designs, and safety alternatives related to battery management and operating systems are covered in this study, which is shown in Figure 13.6. To build the new standard, it also analyses technological concepts connected to BMS. Electrical networks are changing their main generators, loads, and operating methods to favor intelligent device interaction to better technical, economical, and environmental circumstances. These changes optimize energy utilization and improve network users' lives [9].

Multiple energy management solutions for alternating current networks have been proposed in the literature. Such initiatives focus on DGs, mostly solar energy-based ones, to enhance network technical, financial, and environmental variables. They also use energy storage to control renewable power source variability and capitalize on electrical network energy demand and production cost fluctuations. These solutions enable adherence to the electrical devices' functional and technological limitations, which integrate the network while managing the power of distributed energy resources [10].

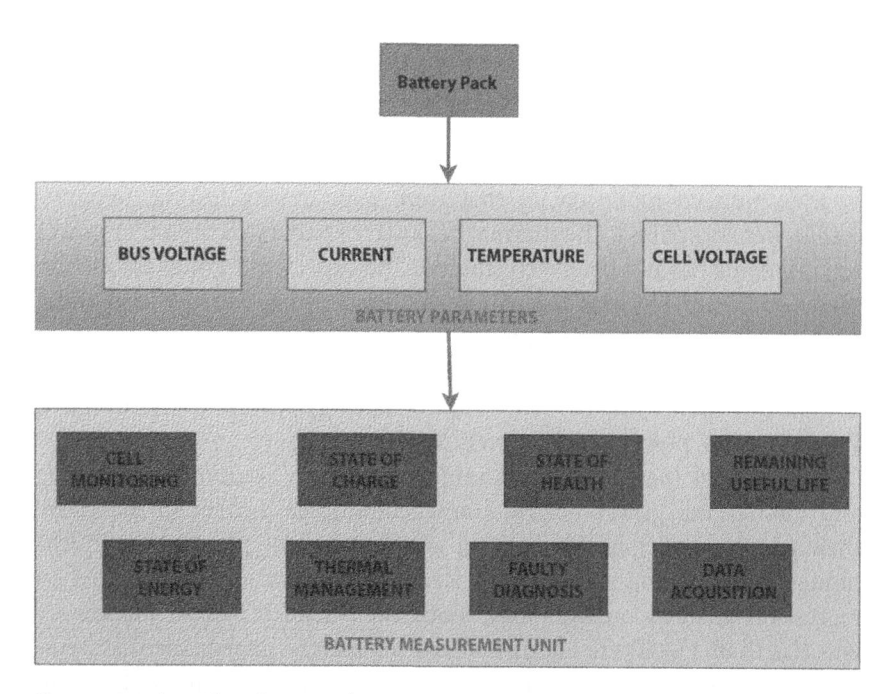

Figure 13.6 Crucial application of BMS.

This research focuses on it; therefore, the following are a few studies that address the subject. A small number of researchers optimized the operation of PV generators and batteries in conjunction. The Cplex solver of GAMS was utilized to solve a mixed-integer linear programming model for the combined PV–battery system. Energy storage systems in electrical networks are configured and operated by scientists using convex optimization and CVXGEN. This study aimed to reduce energy losses and greenhouse gas emissions in a 33-node metropolitan network for 1 day. To prove the methodology's efficacy, MATLAB CVX solves the model and simulates various scenarios.

In three-phase distribution networks, it is utilized in conjunction with semidefinite programming models and second-order cone programming to control the energy of storage devices. This study aimed to lower energy costs and power losses at the point of substation. YALMIP/MOSEK was solved after the GAMS model. This study aimed to reduce greenhouse gas emissions and energy losses in test systems for 33 and 69 nodes in metropolitan networks during a 1-day period. Despite the fact that the numerical results showed how successful the recommended methodology was, they were not compared to alternative methods found in the specialist literature. There was no analysis done on the solution approach's processing times. Achieving optimal energy storage device functioning minimizes installation costs by accounting for maximum power utilization, energy acquisition costs, and battery lifetime. They used MATLAB and GAMS to solve an optimization model for multi-integer linear programming. The strategy's success was proven by the numerical findings. The abovementioned methods produce better solution outputs and enhance the technical, economical, and environmental aspects of networks using commercial software. By improving energy storage devices, the evolutionary technique reduced losses of energy in an urban distribution network using alternating current. This invention lowered the energy costs associated with ordinary generators. Three literature-based methods—a continuous genetic algorithm, a parallel particle swarm optimization, and a black hole optimization method—were surpassed by the solution technique. Processing times, standard deviation, and solution quality all indicated that the suggested approach was effective. Additionally, a mode-based energy management technique is created for a stand-alone PV DG, battery, and load network. Technological and network operating limitations were guaranteed by this approach [11].

Our state-of-the-art review shows that many authors have worked hard to design sequential programming-based intelligent algorithms. This technique improves network technical, financial, and environmental parameters without commercial software. However, researchers must develop strategies to produce efficient answers and repeatability in less time using parallel processing approaches. Such plans should also improve the primary indicators of rural and urban connected and standalone networks. Most attempts have been undertaken in connected metropolitan networks, while fossil fuel-based isolated rural networks face the biggest energetic challenges. In a distributed energy resource scenario, the suggested methods should also guarantee that the mathematical model and solutions correspond with the practical and technological constraints of alternating current networks. Last, the solutions have to function effectively in both rural and urban systems, the latter of which are linked to a network distinct from Colombia's national interconnection system.

This study used the demands and difficulties of the state-of-the-art evaluation to address energy management in everyday operation of alternating current networks in urban and rural areas. The recommended approach controlled the maximum power charging and discharging of network batteries in a daily solar-powered distributed producing scenario. For rural networks in Capurganá (Chocó) and urban networks in Medellín (Antioquia), we examined the generation and consumption patterns for electricity. We've thought of combining three different types of lithium-ion batteries with three different photovoltaic producing units that maximize the energy potential of the area. The challenge was controlling battery energy in rural and urban networks to reduce CO_2 emissions, energy losses, and costs associated with conventional generator purchases and distributed energy resource maintenance. Table 13.3 provides the details of various battery management of components, functions, algorithms, targets, and outcomes. Figure 13.7 provides the details of battery parameters, which are important for smart monitoring system.

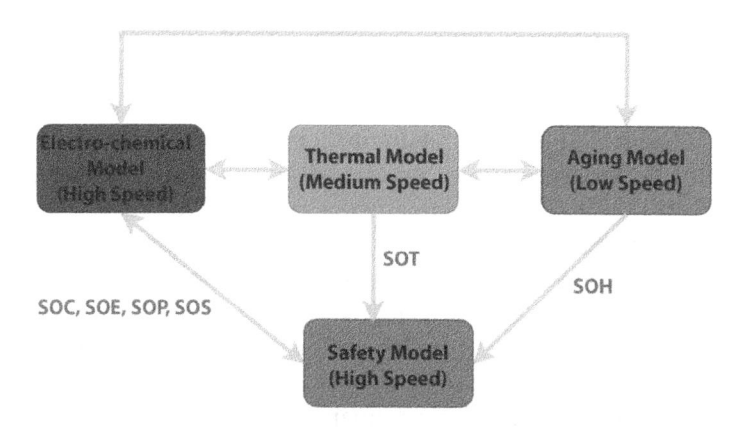

Figure 13.7 Details of battery parameters.

Table 13.3 Details of various battery management of components.

BMS parameters	Monitoring	Estimation	Control	Fault diagnosis	Network and communications
Methods	V-I Divider, CAN	Ah, OCV, model based, FLC	CC-CV, PEM, MCC	Deep learning, knowledge based, model based	PID, microcontroller, wireless
Factors	V and I monitoring	SOC, SOH, SOF	Power, energy, temperature	Protection of battery	Monitor and protection
Remarks	Monitoring of battery parameters	To decrease error and cost	To increase durability, efficiency, and energy transfer	To protect and provide warning of fault	Battery control

13.4 IoT and Cloud Computing Technology in BMS

The system's efficiency and automation have been enhanced through research on utilizing cloud computing. This includes intelligent decision making regarding battery charging, discharging, and power source optimization. In addition, when combining the power sharing approach, it can significantly enhance device availability by optimizing energy consumption within certain SoC ranges during service. Cloud computing has enabled the development of a highly organized V2G model. Cloud infrastructure is essential for managing V2G as the number of vehicles increases. The experimental analysis of the model is well optimized and efficient. The cloud-based system helps cut down on expenses. This device is designed to measure PV and battery features, process the data, and generate reports according to specific needs. All parameters are saved automatically in the cloud. The study also demonstrated the possibility of creating comprehensive training sets offline significantly streamlining power control [12].

Utilizing the Internet of Things, the battery-related data are seamlessly computed and transmitted to the cloud. This process creates a digital twin allowing battery diagnostic algorithms to analyze the data for charge and age assessment. Examining analogous circuit models in battery system digital twins can enhance the performance forecasts. This text is well articulated. An adaptive H-infinity filter that has the ability to produce resilient SoC, this cutting-edge health prediction system utilizes particle swarm optimization to monitor the battery's energy levels, which decrease over time. Experiments in the field and with smartphones were conducted to assess the functionality and reliability [13].

This system monitors various local servers to a centralized data center and ICC-PGCEM scheduler. Requests from cloud clients are stored before being scheduled. The energy consumption calculator assumes uniform server behavior. Therefore, incoming client requests use servers in a similar manner. There is a wide variety of servers used in data centers.

Intelligent agent and data mining engine ICC-PGCEM scheduler monitors storage, CPU, and network consumption with machine learning energy framework. Multiple linear regression is used in the ICC-PGCEM model. Figure 13.8 denotes data mining engine architecture logs energy usage and utilization numbers for power management applications. In an intelligent server agent environment, the ICC-PGCEM scheduler learns from client needs to offer frequent power usage mentorship.

Researchers recommended a hybrid green cloud computing architecture using time-based power utilization. PUE, Throughput, DCEP, total

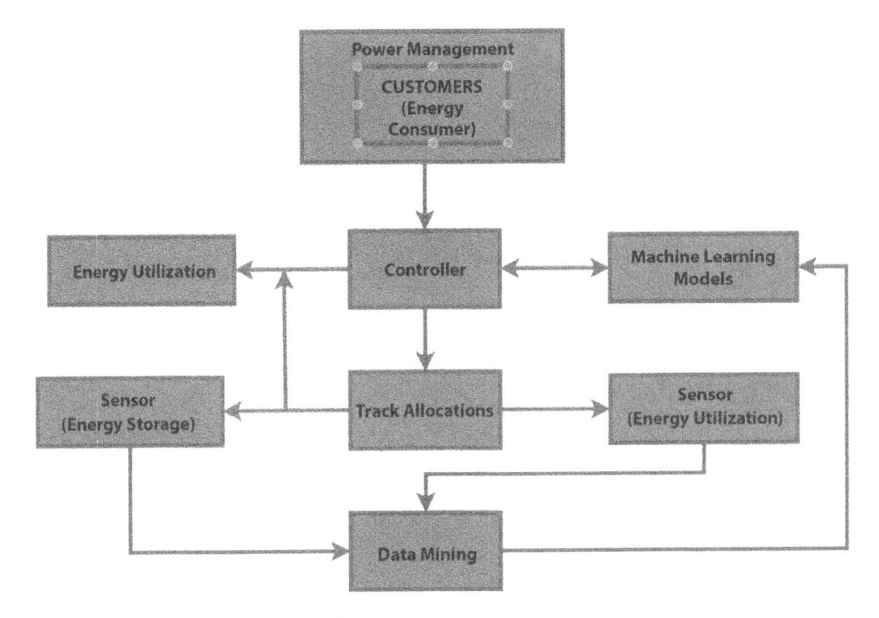

Figure 13.8 Data mining engine architecture.

operating time, and cost savings are all improved by the EEH design over single-solution methods. tracking for battery management was developed. As a proof of concept, ten IoT prototype IAQ sensor nodes and one LoRa gateway were placed in a four-story construction with an upper framework strengthened by steel and a concrete foundation. IoT nodes detect temperature, humidity, pressure, CO_2, GVO, and PM. Because IoT nodes have both digital and analog interfaces, the device can easily monitor more gases [14]. Energy demand, customer preferences, real-time power pricing (RTP), and factors pertaining to renewable energy are a few examples of inputs. To minimize costs and user annoyance GA is employed to schedule and control household load. Now, power costs USD51 as opposed to USD228. The PAR is now 1.12 instead of 2.68. Both the need for electricity and carbon emissions may be reduced using cloud storage. First, as this paper demonstrates, server farms situated at inexpensive energy production locations, virtualized servers—which use less energy than distributed servers—and storage in organizations all contribute to lower power usage and carbon footprints. The second benefit comes from its indirect support of distributed renewable energy source tracking and management in the grid, as well as integrated computing and data storage services. Figure 13.9 illustrates cloud computing for equipment batteries [15].

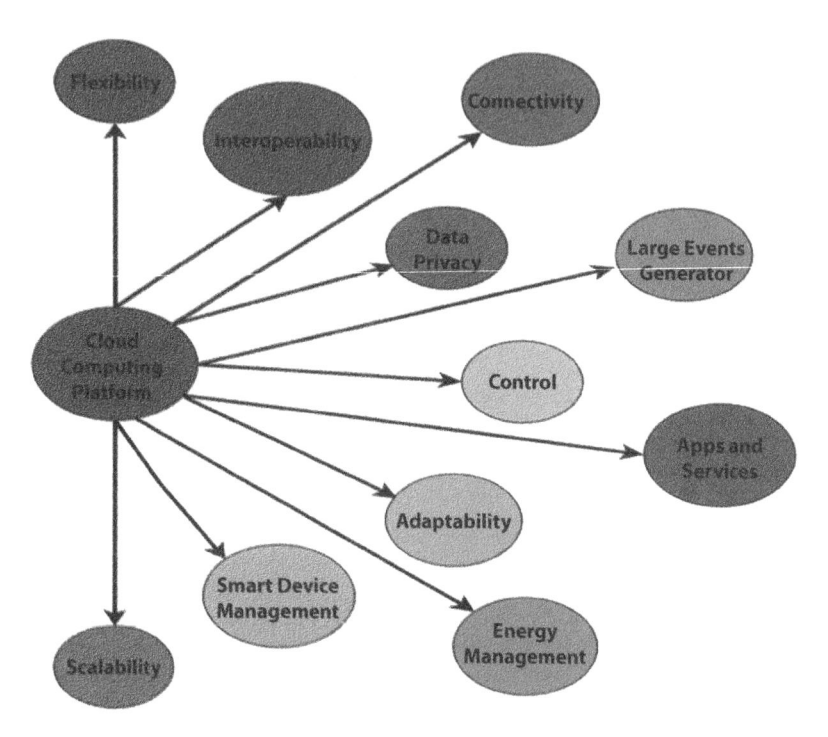

Figure 13.9 Cloud computing for equipment batteries.

The cloud-based technology simplifies data storage and processing, especially created data. Cloud computing lets multiple applications share and compute the same data cutting expenses since most sensors and other devices may concentrate on their tasks while processing takes place in the cloud. Since the late 1990s, the Internet of Things (IoT) has been developing and may interact via a variety of protocols and technologies.

The selection of technology is contingent upon three factors:
The distance between the device and the server is known as the range.

Power: Total power expended during data transmission.
Bandwidth: Amounts of data transferred.

Power systems are designed to provide device monitoring and maintenance more efficiently and cheaply.

The power system can benefit from such a system, especially for people who cannot frequent the power station due to COVID-19 or natural calamities. ICC has been examined in several papers on many topics and perspectives.

Figure 13.10 shows how cloud technology improves battery life. Cloud computing improves battery storage life by minimizing maintenance,

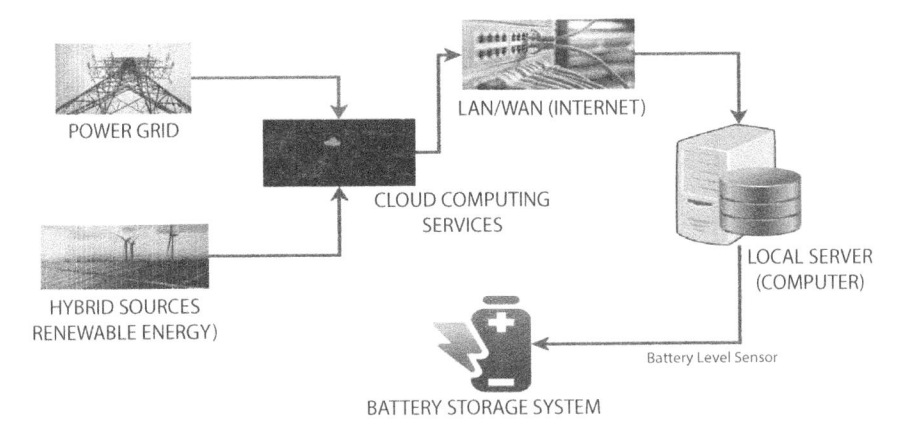

Figure 13.10 Cloud technology to improve battery life.

aging, charging cycles, safety features, power saver, avoiding overcharging and over discharging, adjusting charge speed (fast/slow), and automating management and organization of grid power and renewable energy sources. A cloud computing framework will store and manage these processes and data. Any power distribution system that wants to decrease cost, power waste, and time can benefit from cloud computing. The detailed research gaps identified are listed in Table 13.4.

13.5 Sustainable Developments via BMS

SDGs 8, 9, and 12 focus on job opportunities, professions, economic growth, technology, and facilities to achieve their goals.

13.5.1 SDG8

Creating jobs during the production of various EV components can help achieve Target 8. EVs have been recognized for their wide range of uses offering economic opportunities by creating jobs and utilizing local resources and businesses. Considering SDG8 and the research indicates that the EV market has significantly contributed to economic growth and job creation in various sectors such as renewable energy, electric buses, and trains. Electric vehicles can contribute to achieving Target 8.3 by generating new job opportunities. As per a forecast from 2019, it is projected that electric cars could dominate sales by 2040. The German government unveiled a stimulus package in 2020 allocating 2.5 billion EUR to improve charging

Table 13.4 Identification of research gap.

Objective	Goals	Computational algorithm	Remarks
To create a programmable gadget to integrate the characteristics and capabilities of several current chargers	It showcases advancements in battery management and charging technologies	The charging algorithm for constant voltage (CV) and constant current (CC)	To further safeguard customers while preventing damage to the gadget or batteries, features like temperature sensors, pressure sensors, and alerts that alert the user both visually and audibly may be incorporated
Propose a precise and straightforward charging strategy for minimizing energy loss	The experiments demonstrate that the suggested charging technique reduces charge energy loss considerably	GA and PMP	Future research on variations in battery internal resistance due to temperature fluctuations and battery degradation

(*Continued*)

Table 13.4 Identification of research gap. (*Continued*)

Objective	Goals	Computational algorithm	Remarks
Provide an example of how to improve battery charge using adaptive resonant beam charging (ARBC). Derived from RBC, ARBC uses a feedback mechanism to continually check the power supplied based on the charging parameters established for the battery	Numerical studies show that, in comparison to the RBC system, the ARBC approach conserves	The scheduling method known as FAFC	The batteries differ based on their kind. Examining their impact on ARBC would be an intriguing field of investigation
To provide an overview of simple charging techniques and the issues that are now facing them, followed by a detailed discussion of several optimized charging methods, their features, and their applications	An outline of basic charging methods and the problems they now face will be given, and then a thorough explanation of several optimized charging approaches, their characteristics, and their uses will be covered	Genetic algorithm	The development of a smart city may be impacted by the intelligent BMS and optimized charging techniques that are used in conjunction with the smart grid due to the recent and rapid expansion of new energy vehicles

(*Continued*)

Table 13.4 Identification of research gap. (*Continued*)

Objective	Goals	Computational algorithm	Remarks
To investigate the possibility of using solar energy to charge battery modules without the need for a DC/DC converter	It was discovered that SCVS worked well effectively meeting the two requirements of efficiency and SoC balance	Control strategy	It would be interesting to do a reverse research to determine the optimal battery configuration for a particular location and solar panel count
To model and create a battery charging apparatus that combines a battery control system with an electric vehicle charger	The constructed electric car charger with LLC-ZVS and ZCS	Charging algorithm for constant voltage (CV) and constant current (CC)	Onboard battery charger to realize the electric car charging network
This study focuses on the construction of a fuzzy logic controller (FLC) depending on particle swarm optimization (PSO) for battery energy storage system (ESS) charging, discharging, and scheduling in micro grid (MG) applications	The most effective method for supplying cheap, reliable electricity that is customized for the loads is through the scheduling controller	PSO algorithm	Proposes scheduling procedures

(*Continued*)

Table 13.4 Identification of research gap. (*Continued*)

Objective	Goals	Computational algorithm	Remarks
Exploring potential synergies between rapid response to frequency and energy arbitrage, both provided by a battery	Proposes battery discharging algorithms	Optimization techniques	Operational model outlined in this research into practice to strengthen the battery. This will encourage the use of BESDs leading to the creation of a cheaper, safer, and more reliable electrical grid
Considerations like overall performance, battery capacity, stability, and controllability should be taken into account while choosing an EPS design	According to the data, for the identical battery needs, the EPS design performs best overall	CubeSat strategy	The model's efficiency was verified, and it was suggested that it be used
Determine and rank stress variables according to their effect on battery depletion (capacity fade) using machine learning and the half-fractional architecture of trials	Importance of DOD and C-rate of battery	CCCV algorithm	Power fading of cell might potentially include RF in the future

(*Continued*)

Table 13.4 Identification of research gap. (*Continued*)

Objective	Goals	Computational algorithm	Remarks
Provide a new stochastic technique for distribution system photovoltaic hosting capacity optimization.	By simultaneously optimizing many control factors, they are maximizing the capacity of photovoltaic hosting. A two-layer metaheuristic optimizer is designed to maximize photovoltaic hosting capacity while removing all constraints	Metaheuristic optimization algorithms	Future studies may look into further energy storage options and renewable energy sources
Provide a fundamental statistical model of electrochemical cell deterioration that is centered on the common traits seen in earlier extensive cell degradation studies	A battery energy storage device's capacity may be significantly increased by disassembling it into smaller modules, which also gives you the option to utilize less expensive, lower-quality cells in a manner that maximizes savings	Stochastic algorithm	It would be helpful to do further study on the true costs of modularization, and battery device experts may suggest affordable methods for accomplishing fine-grained modularization

(*Continued*)

Table 13.4 Identification of research gap. (*Continued*)

Objective	Goals	Computational algorithm	Remarks
To create a small and well-optimized SoC estimate model with a fast input selection approach for choosing key phrases as input variables and a simple yet effective JAYA optimization strategy for adjusting the primary neural network function parameters	The real-system experiment results show that the optimization strategy greatly decreases prediction process errors	JAYA algorithm	The optimization step will use one-phase optimization to fine-tune the primary parameters without adjusting any algorithm-specific parameters
Thermal management to enhance the efficiency and cleanliness of the battery	This study introduces a thermal control device with heat pipes	Cold plate control mechanism	Better technique to make sure the battery module stayed at its ideal temperature

(*Continued*)

Table 13.4 Identification of research gap. (*Continued*)

Objective	Goals	Computational algorithm	Remarks
To investigate the most recent advancements in the preheating and cooling of BTMS using forced-air convection	A summary of the primary optimization path is provided, which includes heat pipe integration, phase change material (PCM) integration, and pure forced-air convection optimization	MOPSO algorithm	Usually, copper wire is used to make the heat pipe. Consequently, there is a possibility that the heat pipe will cause an exterior short circuit, which would cause thermal loss in the battery pack
To introduce a novel method for determining the appropriate size of batteries for a base station	The results suggest that predicting price instability in micro-grid situations is not significant. The research indicates that it is never advantageous to invest in a battery size larger than the minimum required considering transaction costs and power price systems	MDP strategy	Incorporating price volatility into the model has minimal impact and may be eliminated without significantly affecting performance

(*Continued*)

Table 13.4 Identification of research gap. (*Continued*)

Objective	Goals	Computational algorithm	Remarks
Recommend a digitally operated, energy-efficient power management solution suitable for high-mobility scenarios	Based on the results, the proposed artificial neural network (ANN) technique surpasses all comparison benchmarks. The SCA solution is the only one that matches the efficiency of the suggested ANN-based technique	Branch and bound algorithm	Advanced techniques are not as appropriate for artificial neural network (ANN) techniques for online power allocation
To design an embedded network platform for wearable telemedicine	Molecular channel systems	Hybrid embedded algorithm	Estimating technique

stations and support e-mobility research and development, including battery solutions. The UK government has set a target to decrease CO_2. To achieve the goal, 43% must be electric [16]. Recently, the US government initiative to boost electric car battery production. Research published indicates a rapid increase in the EV in Flanders, Belgium. Based on data from a comprehensive poll in 2011, projections suggested that by 2020, battery electric vehicles would make up approximately 5% of new vehicle sales, with plug-in hybrid electric vehicles comprising approximately 7%. This statement highlights how the EV market contributes to achieving Target 8.6 by generating job opportunities for young people, reducing youth unemployment, and boosting per capita income. The job opportunities provided will help local children access more education and increase their chances of finding employment, as many projects will hire people from the area. This will have a positive effect on 8.1 and 8.3 targets.

13.5.2 SDG9

Electric vehicles play a crucial role in helping achieve Target 9.1 by supporting the development of reliable, top-notch infrastructure that enhances both social welfare and economic progress. A study revealed that investigating climate resilient methods can facilitate better environment and community. Establishing a publicly accessible replenishing facility system is crucial for promoting the adoption of electric vehicles. GHG emissions linked to traditional fuel vehicles decrease improving the urban environment. One of the main challenges in getting residents to adopt electric vehicles is the availability and characteristics of charging stations. Introducing publicly accessible CS systems enhances the EV customer journey by improving the ease of use of infrastructural facilities [17]. Encouraging the creation and eventual development of a sustainable local CS system poses a significant challenge over cities and local officials. EVs are well suited for urban areas, where most housing in BC is located, due to their range and emission-reducing capabilities.

13.5.3 SDG12

Utilizing EV-incorporated distributed generation of electricity, smart grids, and microgrids can facilitate efficient energy distribution between supply and load aligning with Target 12.1. Bringing together a large number of EVs into VPPs offers ecological benefits such as energy conservation and pollution reduction, safety advantages related to the stable operation of the electrical grid, and economic incentives for VPPs and automotive stakeholders. In the long run, an increasing number of electric vehicles will be integrated into Virtual Power Plants due to their numerous benefits. EVs in transmission lines are evolving into controlled commodities through V2G technology allowing them to carry out different auxiliary operations like maximizing energy harvesting, stabilizing voltage, and regulating frequency. There is evidence that shows how EV-BESs positively affect SDG12, specifically Target 12.5, which aims to significantly reduce waste by minimizing, recycling, and reusing. Research indicates that the Battery Energy Storage of an Electric Vehicle will either be reused, discarded, or have its components retrieved based on the specific application. When materials from lithium-ion batteries are recycled, it can lower the power density by 10%–53%. Additionally, the cost of constructing lithium–oxygen batteries can drop leading to decreased greenhouse gas emissions. Recycling battery elements helps decrease the use of fossil sources and reduces waste production. Despite the positive employability outlook, the transition to electric-powered transport will have both winners and losers. Due to the global shift toward electric vehicles, job losses are expected in the oil industry, petrol stations, and the automotive repair as EVs demand less maintenance than traditional vehicles. Manufacturing batteries requires fewer jobs than producing gasoline-powered vehicles, emission structures, emission control networks, gasoline tracking devices, gearboxes, and automotive components. By 2030, projections suggest that Germany may face a loss of thousands of jobs due to the transition to electric vehicles. Manufacturing, research and development, and the production of batteries will all contribute to job creation. Setting up and maintaining the equipment utilized in EVs will generate indirect job opportunities. It is probable that there will be fewer positions available in the EV sector compared to those who work in ICEV [18]. Moreover, electric vehicles have lower maintenance and operating costs, which can affect employment both directly and indirectly. According to Energy Data Administration, over 80% of the cost quickly exits the domestic economy. If fuel costs are lowered, more money will circulate within the community ultimately boosting the local economy. New York City residents drive significantly less compared to the average US metropolitan area, which helps to retain 19 billion USD within the city's economy each year. This article highlights the connection between batteries and energy from renewable sources, all of which contribute to positive environmental impacts and support the relevant SDGs. Below, we will discuss the pros and cons of electric vehicles in relation to the environment.

13.5.4 SDG13

EVs offer significant benefits for the climate by reducing emissions from traditional energy sources commonly used in renewable energy production. In addition, due to the progress of EV solutions, numerous countries are integrating environmental initiatives into their legislative proposals. In 2018, the IPCC projected that electric need to replace fossil fuel-powered passenger vehicles by 2035–2050 to limit global warming to below 1.5 _C. Addressing environment-related issues and integrating green tasks into social initiatives, schemes, and management are the focus of Targets 13.1 and 13.2. By 2050, strategies suggest that the combination of EVs and RE will lead to a 70% decrease in CO_2 emissions in the electricity sector. EVs are strongly recommended in various studies for a modern, reliable, and eco-friendly energy solution. Electric vehicles (EVs) have become increasingly popular for their ability to replace traditional power sources by transporting and storing electricity making them a more cost-effective option. Considering the falling price of energy from renewable sources and the increasing scarcity of fossil fuels, a sustainable approach is necessary to guarantee widespread access to power by combining sustainable energy and BESSs with EVs. Various studies have agreed that electric vehicles have negative environmental impacts. At times, hazardous and flammable components are used, requiring significant energy input and leading to high levels of greenhouse gas emissions [19]. Consequently, the BESS could act as a barrier to achieving Target 13.1's aim of decreasing climate-related risks. The BESS's components could have significant environmental consequences potentially causing harm. Electric vehicles seem to offer a promising solution to address climate concerns, as transitioning to carbon-free networks could significantly reduce automobile emissions. Great news for electric vehicles as numerous countries are currently prioritizing the decarbonization of their power systems. Over the past few decades, energy providers in the United States have transitioned away from coal facilities to embrace to reduce emissions. Researchers have found that electric vehicles have become more environmentally friendly overall and are expected to continue improving.

13.6 Conclusion

Several studies have been prompted by the impact of global warming to reduce carbon emissions. Industries are a significant source of carbon emissions, but they are making strides in reducing their environmental impact through technological advancements. Yet, technology advancement necessitates the consideration of various factors. This chapter thoroughly examines battery storage and management, as well as various management techniques focusing on algorithms, controllers, and optimization, all aimed at achieving SDGs. Discussion included multiple battery storage technologies and components of battery management. Each battery technology has distinct performance characteristics. Hence, it is important to take into account the battery features before applying them to industrial requirements. Furthermore, cutting-edge methods, algorithms, and optimization approaches were explored. It was reported that each algorithm and optimization technique for industrial application deliver satisfactory outcomes. Yet, the issue of computational complexity and lengthy training time still requires attention. Next, we identified the unresolved problem limitations and research gaps. Exploring the combination of different technologies, like battery storage systems, optimization approaches, and algorithms is crucial for the efficient advancement of industrial technology. Finally, a range of SDG targets linked to BMS have been examined and assessed.

References

1. Saqib, N., Ozturk, I., Usman, M., Investigating the implications of technological innovations, financial inclusion, and renewable energy in diminishing ecological footprints levels in emerging economies. *Geosci. Front.*, 14, 6, 101667, Nov. 2023, https://doi.org/10.1016/j.gsf.2023.101667.
2. Subrahmanyam, S., Khalife, D., Chaarani, H.E.I., Towards Sustainable Future: Exploring Renewable Energy Solutions and Environmental Impacts. *Acta Innov.*, 51, 15–24, Feb. 2024, https://doi.org/10.62441/ActaInnovations.51.3.
3. Cheng, K.W.E., Divakar, B.P., Wu, H., Ding, K., Ho, H.F., Battery-Management System (BMS) and SOC Development for Electrical Vehicles. *IEEE Trans. Veh. Technol.*, 60, 1, 76–88, Jan. 2011, https://doi.org/10.1109/TVT.2010.2089647.
4. Henzinger, T.A. and Kirsch, C.M., The Embedded Machine: Predictable, portable real-time code. *ACM Trans. Program. Lang. Syst. (TOPLAS)*, 29, 6, 33–40, 2007, https://doi.org/10.1145/1286821.1286824.
5. Tan, P., Jiang, H.R., Zhu, X.B., An, L., Jung, C.Y., Wu, M.C., Shi, L., Shyy, W., Zhao, T.S., Advances and challenges in lithium-air batteries. *Appl. Energy*, 204, 780–806, 2017, https://doi.org/10.1016/j.apenergy.2017.07.054.
6. Wang, X., Kerr, R., Chen, F., Goujon, N., Pringle, J.M., Mecerreyes, D., Forsyth, M., Howlett, P.C., Toward high-energy-density lithium metal batteries: opportunities and challenges for solid organic electrolytes. *Adv. Mater.*, 32, 18, 1905219, 2020, https://doi.org/10.1002/adma.201905219.
7. Gabbar, H.A., Othman, A.M., Abdussami, M.R., Review of Battery Management Systems (BMS) Development and Industrial Standards. *Technologies*, 9, 2, 28–37, 2021, https://doi.org/10.3390/technologies9020028.

8. Ekuewa, O.I., Afolabi, B.B., Ajibesin, S.O., Atanda, O.S., Oyegoke, M.A., Olanrewaju, J.M., Development of Internet of Things-Enabled Smart Battery Management System. *EJECE*, 6, 6, 9–15, Nov. 2022, https://doi.org/10.24018/ejece.2022.6.6.467.
9. Strasser, T. *et al.*, A Review of Architectures and Concepts for Intelligence in Future Electric Energy Systems. *IEEE Trans. Ind. Electron.*, 62, 4, 2424–2438, April 2015, doi: 10.1109/TIE.2014.2361486.
10. Ahmed, M.R., Basit, A., Ahmad, H., Ahmed, W., Ullah, N., Prokop, L., Review on microgrids design and monitoring approaches for sustainable green energy networks. *Sci. Rep.*, 13, 1, 21663, 2023, https://doi.org/10.1038/s41598-023-48985-7.
11. Elmouatamid, A., Ouladsine, R., Bakhouya, M., El Kamoun, N., Khaidar, M., Zine-Dine, K., Review of Control and Energy Management Approaches in Micro-Grid Systems. *Energies*, 14, 1, 168, 2020, https://doi.org/10.3390/en14010168.
12. Chacko, P.J. and Sachidanandam, M., An optimized energy management system for vehicle to vehicle power transfer using micro grid charging station integrated Gridable Electric Vehicles. *Sustain. Energy Grids Netw.*, 26, 100474, 2021, https://doi.org/10.1016/j.segan.2021.100474.
13. Vandana, A.G. and Panigrahi, B.K., Multi-dimensional digital twin of energy storage system for electric vehicles: a brief review. *Energy Storage*, 3, 6, e242, April 2021, https://doi.org/10.1002/est2.242.
14. Kumar, P. and Kota, S.R., Machine learning models in structural engineering research and a secured framework for structural health monitoring. *Multimed. Tools Appl.*, 83, 7721–7759, 2024, https://doi.org/10.1007/s11042-023-15853-5.
15. Paliwal, P., Patidar, N.P., Nema, R.K., Planning of grid integrated distributed generators: A review of technology, objectives and techniques. *Renew. Sustain. Energy Rev.*, 40, 557–570, 2014, https://doi.org/10.1016/j.rser.2014.07.200.
16. Kufeoglu, S. and Khah, D., Emissions performance of electric vehicles: A case study from the United Kingdom. *Appl. Energy*, 260, 114241, 2020, https://doi.org/10.1016/j.apenergy.2019.114241.
17. Greene, D.L., Kontou, E., Borlaug, B., Brooker, A., Muratori, M., Public charging infrastructure for plug-in electric vehicles: What is it worth? *Transp. Res. Part D: Transport Environ.*, 78, 102182, 2020, https://doi.org/10.1016/j.trd.2019.11.011.
18. Sovacool, B.K. and Hirsh, R.F., Beyond batteries: An examination of the benefits and barriers to plug-in hybrid electric vehicles (PHEVs) and a vehicle-to-grid (V2G) transition. *Energy Policy*, 37, 3, 1095–1103, March 2009, https://doi.org/10.1016/j.enpol.2008.10.005.
19. Mamun, K., Islam, F.R., Haque, R., Chand, A.A., Prasad, K.A., Goundar, K.K., Prakash, K., Maharaj, S., Systematic Modeling and Analysis of On-Board Vehicle Integrated Novel Hybrid Renewable Energy System with Storage for Electric Vehicles. *Sustainability*, 14, 5, 2538, 2022, https://doi.org/10.3390/su14052538.

14

Trends, Advances, and Applications of Network Sensing Systems

Ashwini A.[1]*, Shamini G.I.[2] and Balasubramaniam S[3]

[1]Department of Electronics and Communication Engineering, Vel Tech Rangarajan Dr. Sagunthala R&D Institute of Science and Technology, Chennai, Tamil Nadu, India
[2]Department of Electronics and Communication Engineering, Sathyabama Institute of Science and Technology, Chennai, Tamil Nadu, India
[3]School of Computer Science and Engineering, Kerala University of Digital Sciences, Innovation and Technology, Thiruvananthapuram, Kerala, India

Abstract

A wide range of services are provided by the network sensor systems, which served as a pivotal tool in industries, agriculture, advancements in driving, and smart city-based systems. This chapter gives a review on the advancing technologies with applications and trends. In the beginning, it identifies the patterns that are shaping the environment, with a particular emphasis on the move from traditional technologies for sensing to networked devices offered by the prevalence of the Internet of Things (IoT). This shift brings in an age of networks of sensors that features ubiquitous connectivity, interconnectivity, and current information transfer. In addition, the abstract goes over sense reducing size, advances in power capturing and transmission via wireless standards, and the use of neural networks and algorithmic techniques for machine learning (ML) for data analytics, all of which are important improvements that inspire creativity in network detection systems. The summary then discusses the diverse applications that utilize network sensor devices in a range of areas. Mobile phones and telemedicine systems for healthcare provide remote surveillance of patients, individualized therapy, as well as early detection of illnesses. Monitoring the environment utilizes network sensor networks since they capture immediate information on water and air quality, climate conditions, and emergencies assisting in saving the environment and recovery efforts. Also, network detectors are utilized across agriculture to improve the

---

*Corresponding author: a.aswiniur@gmail.com

Rajesh Kumar Dhanaraj, Malathy Sathyamoorthy, Balasubramaniam S and Seifedine Kadry (eds.) Networked Sensing Systems, (351–374) © 2025 Scrivener Publishing LLC

care of crops, monitor soil state, and mitigate the effects of the changing climate on production in agriculture. Intelligent transit systems use networked sensor technology to enhance traffic control, road safety, and automated vehicle navigation eventually altering urban transportation and mobility buildings. These innovations additionally play a key role in smart city projects enabling better money leadership, urban planning, and government communication using data and automation. As a whole, this summary highlights network sensor systems' transformative potential in addressing societal issues, enhancing effectiveness, and spurring discovery across a variety of enterprises. It highlights the need of multidisciplinary teamwork, advancements in technology, and legislation in realizing the potential offered by network monitoring devices and effectively delivering significant societal benefits.

Keywords: Data analytics, internet of things, machine learning, networking sensing systems, remote patient monitoring, smart city

14.1 Introduction to Network Sensing Systems

Networking sensors play a crucial part in modern linked society, where an increasing number of sensors and the constant interchange of information highlight the requirement for efficient tracking and evaluation. At their heart, these devices are intended to collect, manage, and analyze data from numerous locations on an internet connection allowing enterprises to obtain crucial knowledge about its operations, privacy, and overall state [1]. Networking sensing structures, either implemented in workplace contexts, cell phone networks, or IoT natural systems, use a variety of devices, including gauges, devices, and observation applications, to collect and examine data in actual time.

Figure 14.1 shows the general structure of network sensor systems. By analyzing patterns of traffic, identifying abnormalities, and discovering potential dangers, these devices enable admins to manage network assets proactively, enhance efficiency, and reduce hazards. As systems grow in intricacy and size, detectors become more vital in providing frontal protection against novel risks and guaranteeing the normal operation of crucial systems [2]. In information security, they are critical for the identification of threats, entering prevention, and incident management. Furthermore, network sensor networks play an important role in advancing developing application such as energy supervisors, surveillance of the environment, and automated public transit.

Advances in computing at the edge, 5G wireless networks, as well as distributed sensor technology are likely to fuel the ongoing growth in network

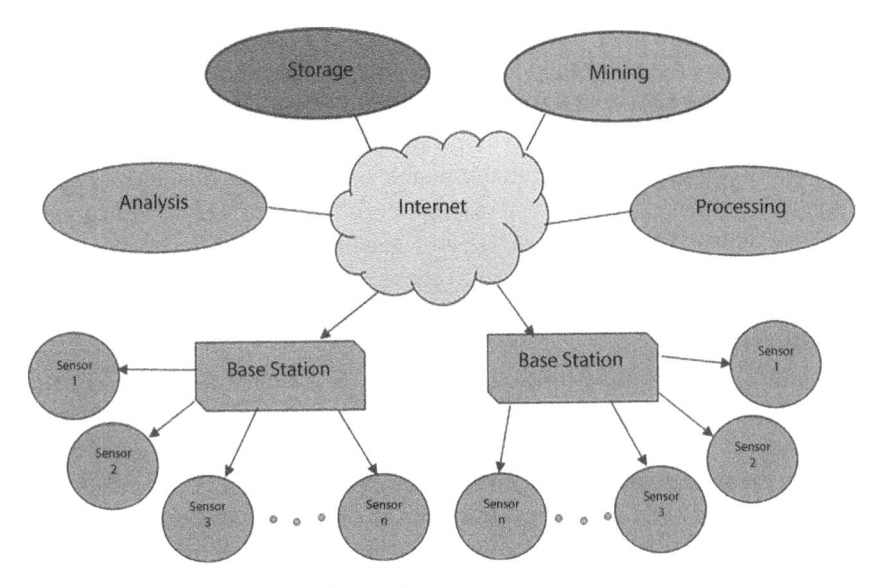

Figure 14.1 General structure of network sensor systems.

monitoring technologies in the future. These developments will enhance sensing system' abilities allowing them to function in extremely rapidly changing and diverse network contexts, while tackling rising issues like privacy of data, adaptability, and cost effectiveness. Finally, internet detectors are going to maintain an important role in influencing the eventual development of interconnected systems by enabling the effortless incorporation of technological innovations into everyday activities.

14.1.1 Relevance in Different Sectors

Because of its capacity for offering immediate tracking, investigation, and oversight of computer network infrastructure, network sensing systems are applicable in a wide range of industries.

Telecommunications

In the field of telecommunications, infrastructure sensors are essential to track network efficiency, guaranteeing quality of the service (QoS), and enhancing traffic control [3]. These devices let telecom suppliers of services to monitor network congestion, predict possible obstacles, and flexibly shift capacity to meet demands that change. Furthermore, they play a significant impact in issue identification and diagnosis reducing delay and improving overall the reliability of the network.

Cyber security
Networks sensors are critical for security programs, since they detect and avoid many types of cyber-attacks such as malicious software, attacks, and attempts to gain access. Such devices may detect unusual behavior suggestive of a safety compromise and initiate suitable responses by constantly recording the network's activity and analyzing trends for abnormalities. Integrated internet sensor technology use AI and machine learning technologies to improve identification of and react to changing cyber-attacks [4].

Healthcare
In medical care, network sensors are used to track healthcare equipment, information about patients, and facility equipment. These systems assure that there is an uninterrupted supply of key medical facilities while protecting the confidentiality of patients and adhering to regulations. Network sensor systems used in medical institutions can also enable monitoring of patients via telemedicine and continuous interaction among doctors and nurses, thus enhancing treatment of patients and effectiveness.

Smart Cities
Internet sensors play an important part in the creation of intelligent cities by allowing for effective control of city services and facilities. Such devices are used to track the flow of traffic, optimize routes for transportation, manage consumption of energy, and enhance safety for everyone by providing observation and intervention abilities [5]. By combining information from numerous sensors and IoT gadgets distributed around the town, network sensors enable planners and politicians to develop information-based choices, which promote people's overall quality of lives.

Industrial Automation
In industrial settings, networked detectors are utilized to track and operate important systems such as manufacturing operations, logistics in the supply chain, and utility connections. These devices provide proactive upkeep by monitoring devices' well-being and identifying possible breakdowns prior to development resulting in fewer delays and increased production efficiencies. Furthermore, internet sensor systems enable immediate information analysis for procedure optimization, inspection, and management of inventory, hence enabling efficient operations in factories. Overall, advances in networked sensors are altering numerous sectors by allowing pre-emptive tracking, intelligent making decisions, and essential automation of processes resulting in improved output, productivity, and robustness across businesses.

14.2 Real-Time Trends in Sensor Technology

In "Trends, Advances, and Opportunities of Network Sensing Systems," real-time sensor technology trends have a significant impact on the evolution and efficacy of network sensing systems. Sensor technology is progressively being incorporated into Internet of Things (IoT) ecosystems allowing devices, sensors, and network sensors to communicate and share information seamlessly. Sensors powered by the Internet of Things enable the collection of massive volumes of data in real time from a variety of reports expanding the insights accessible through connected sensing devices and improving their tracking and analyzing powers [6]. Figure 14.2 shows the key trends that are associated with the sensor technology.

14.2.1 Advanced Sensing Modalities

There is a growing interest in developing devices that can capture data utilizing sophisticated modality such as hyper-spectral imagery, LiDAR (Light Detection and Ranging), and fluorescence. These methods allow network sensors to collect more precise and subtle data regarding their surroundings, which improves their capacity to identify minor alterations, abnormalities, and trends across network activity. Edge computing is becoming increasingly significant in sensors allowing the analysis and processing of data to take place closer to the information source. By incorporating edge computing abilities into sensors, networking sensing systems may reduce delay, reduce capacity needs, and enhance responsiveness, thus rendering them ideal for applications that operate rapidly and cases requiring immediate action.

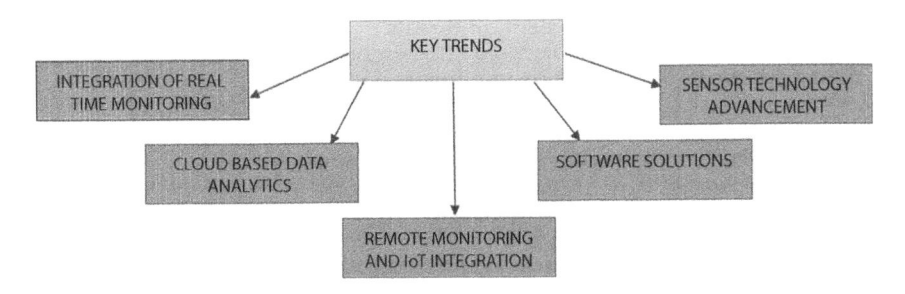

Figure 14.2 Key trends in sensor technology.

14.2.2 Power-Efficient Designs

The use of energy is an important consideration in sensors, especially for battery-operated or power-constrained applications. Sensors are being created with safe communication protocols, algorithms for encryption, and authentication features to prevent intrusion, manipulation of data, and breaches of privacy ensuring the safety and confidentiality of the data collected and passed on by network detection platforms.

Table 14.1 shows the evolution of wearable sensor technology. Overall, such real-time sensor developments contribute to research and facilitate the creation of further competent, successful, and safe network monitoring solutions. The use of sensors is pushing the limits for network tracking,

Table 14.1 Evolution of wearable sensor technology.

Wearable sensors	Year
Cardiac monitoring	1962
Polar electro-based wireless heart monitor	1982
Conductive fabrics for temperature and motion sensing	2000
Non-invasive glucose determination	2002
Introduction to epidermal electronics	2010
Biosensor integrated into tooth enamel for bacterial detection, contact lens biosensor	2012
Wearable glucose monitoring systems	2014
Mouth guard biosensor for uric acid monitoring	2015
Wearable sensor for multiplexed perspiration analysis	2016
Biosensors for detection of chemical threats	2017
Smart face mask for respiration monitoring	2018
Paper-based wearable sensor for breath chemistry monitoring, micro needle for real-time invasive drug monitoring	2019
Smart watches for COVID-19 detection	2020
Biofuels for energy autonomous wearable sensors, synthetic biology-enabled wearable biosensors	2021

evaluation, and administration by including connected devices, improved sensor modalities, computing power at the edge, energy-efficient concept data combination methodologies, and improved safety elements.

14.3 Advancements in Data Analytics

Innovations in data analysis are critical to the ongoing growth and efficacy of network sensor structures allowing enterprises to derive significant conclusions from the massive amounts of data created by connected sensors [7]. Many significant developments in statistical analysis are influencing the setting regarding networked sensor structures.

Analysis in real time are becoming increasingly important for network sensor systems. Advances in streaming processing technology enable companies to evaluate data when they are created by detectors facilitating prompt anomaly identification, fast event reaction, and variable networking aspect change. Real-time data analysis can improve networked sensing devices' reactivity and agility allowing them to react to ever-shifting surroundings and needs in changing contexts.

Machine learning (ML) approaches are rapidly being used in data analysis procedures for network sensing systems. Machine learning algorithms are capable of identifying patterns, connections, and abnormalities in data from sensors allowing for predictive upkeep, pre-emptive surveillance of safety, and informed decision making. AI-powered analytics improves the precision, effectiveness, and flexibility of network sensors systems allowing them to extract significant conclusions from big and complicated databases.

Edge analysis is the practice of processing and interpreting data closer to the source, such as inside sensor or edge computer hardware. This method decreases latency as well as saves the network's capacity and improves data privacy by analyzing private data immediately before forwarding it to centrally situated servers or cloud-based systems.

Multicolor data fusion technologies combine data from several sensor types to provide a more complete picture of the surroundings. By aggregating information collected by devices detecting many factors (e.g., humidity, temperature, and movement), networked monitoring devices may bypass individual device constraints, minimize uncertainty, and increase perception accuracy. Multidisciplinary combination of data allows system sensor networks to gain a more detailed and nuanced perspective of network conduct increasing their usefulness in a wide range of industries [8].

Predictive analysis and analytics entail projecting future trends and occurrences using previous information as well as mathematical model approaches. The use of predictive analytics is used in network sensing systems to detect possible failures of the network, identify bottle necks in performance, and enhance the use of resources.

Additionally, prescriptive analytics provides actionable recommendations to optimize network operations, improve efficiency, and mitigate risks based on predictive insights. These advanced analytic capabilities enable organizations to proactively address issues and capitalize on opportunities in their network infrastructure.

In general, developments in data analytics are crucial to improving the capacity, information, and worth of network detecting systems. Organizations may generate meaningful insights from information collected by sensors, optimize network efficiency, and drive development across numerous industries by utilizing immediately available analytics, AI, cloud computing, information fusion to occur, and forecasting abilities. Figure 14.3 shows various advancements in data analytics.

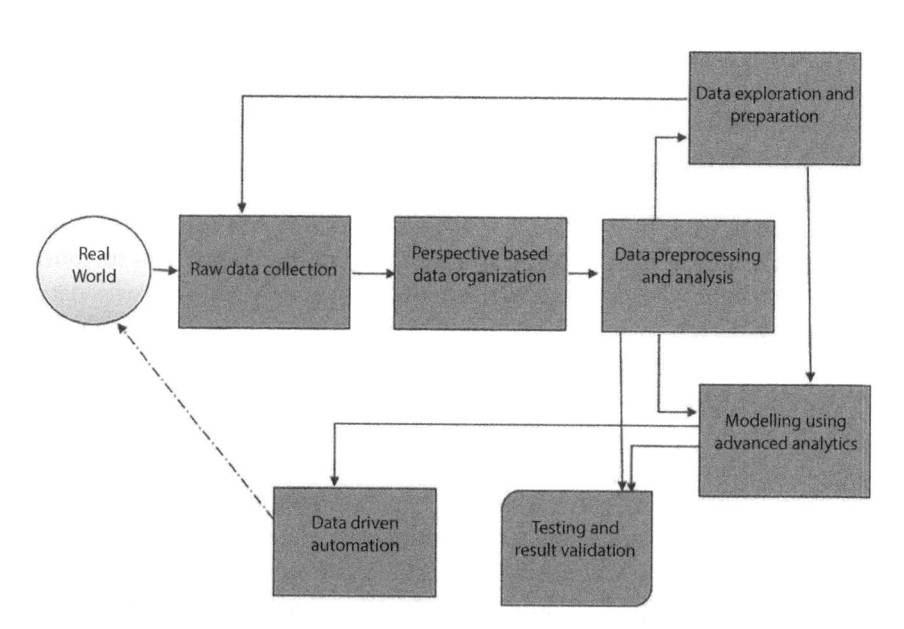

Figure 14.3 Advancements in data analytics.

14.3.1 Big Data Analytics for Sensor-Generated Data

Big data analysis for data collected by sensor entails analyzing, interpreting, and deriving useful information using massive amounts of information acquired by devices in actual time or near instantaneously [9]. This approach is critical for retrieving useful details from the huge volumes of information collected by sensors linked together across several domains. Figure 14.4 shows some important characteristics of extensive analysis for sensor-generated data.

Gathering of Data and Preservation

Sensor-generated information is usually gathered constantly and at frequent intervals leading to enormous databases. Big analytics tools need to know how to manage this flood of data properly. Sensory information may be stored in a scalable and fault-tolerant way using data storage technologies, such as cloud-based backup services, such as Amazon S3.

Data Preprocessing and Cleaning

Sensor information is frequently preprocessed for analysis to remove values that are missing, anomalies, sound, and discrepancies. Preprocessing methods include information attribution, reduction of noise, and standardization [10]. Cleansing the data maintains its reliability and accuracy, which are required for appropriate evaluation and interpretations.

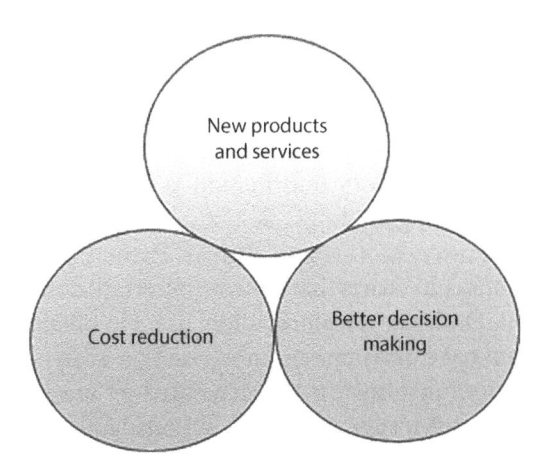

Figure 14.4 Big data analytics advantages.

Real-Stream Computation

Most sensor-generated information applications require actual time or near-real-time processing to allow for quick decisions to be made. Streaming processing systems are employed for handling data supplied by a sensor. The aforementioned structures provide ongoing information intake, the process, and examination allowing companies to spot trends, irregularities, and phenomena every time they occur.

Machine Learning and AI Algorithms

The analysis of big data for data generated by sensors frequently entails using computer learning as well as artificially intelligent techniques to derive insight and forecasts. Guided methods of learning, such as regression and categorization, may be employed to forecast future reads from sensors or identify abnormalities in real-time information streams. The use of unsupervised learning techniques, including clustered and recognition of anomalies, may disclose hidden trends and patterns in data gathered from sensors.

Scalable Analyses Facilities

Sustainability is an important factor in big data analysis for data generated by sensors since the amount, acceleration, and diversity of the information can change considerably over the course of time. Online analytics systems, such as AWS (Amazon Web Services) and Azure from Microsoft, offer scalable computing and storage resources for enormous scale data processing activities [11]. Distributed computational frameworks, such as Apache Spark, offer simultaneous execution of data over multiple computers, which permits fast study of massive databases. Figure 14.5 shows big data analytics for sensor data.

Visualization and Reporting

The ability to visualize sensory information is critical for efficiently analyzing and disseminating findings. Statistics visualization technologies, such as Tableau, Power BI, or matplotlib in Python, build visual dashboards, illustrations, and charts that convey important details from sensor-generated output. Dashboard features allow users to obtain insights, track achievement, and make decisions that are based on analytical findings.

Overall, statistical analysis for data generated by sensors allows enterprises to get important insights, enhance actions, and drive creation in a variety of fields [12]. Organizations may obtain useful knowledge from data from sensors by employing sophisticated analytics methodologies,

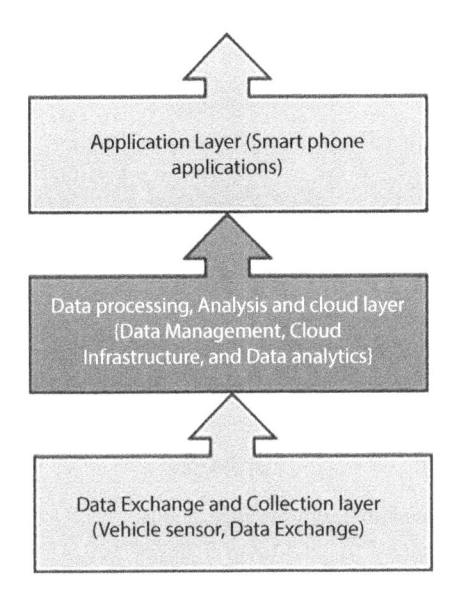

```
        ┌─────────────────────────────┐
        │ Application Layer (Smart phone│
        │      applications)          │
        └─────────────────────────────┘
                      ▲
        ┌─────────────────────────────┐
        │Data processing, Analysis and cloud layer│
        │     (Data Management, Cloud │
        │Infrastructure, and Data analytics)│
        └─────────────────────────────┘
                      ▲
        ┌─────────────────────────────┐
        │ Data Exchange and Collection layer│
        │  (Vehicle sensor, Data Exchange)│
        └─────────────────────────────┘
```

Figure 14.5 Big data analytics for sensor data.

flexible facilities, and immediate processing capabilities to increase productivity, making choices, and value delivery to users.

14.4 Applications in Healthcare

Networking detectors offer a wide range of possibilities for medical use helping to improve the treatment of patients, streamline operations, and increase overall effectiveness inside institutions. Figure 14.6 shows the network sensor in healthcare applications.

14.4.1 Remote Patient Monitoring

Network sensors allow for ongoing tracking of patients' vital signs and medical variables from somewhere else. Wearable sensors, such as watches or patches, with connectivity to the internet, may collect current information on blood pressure, temperature, pulse, and other physiological parameters [13]. This information is sent to medical professionals for processing enabling for early identification of health problems, tailored treatments, and rapid changes to interventions.

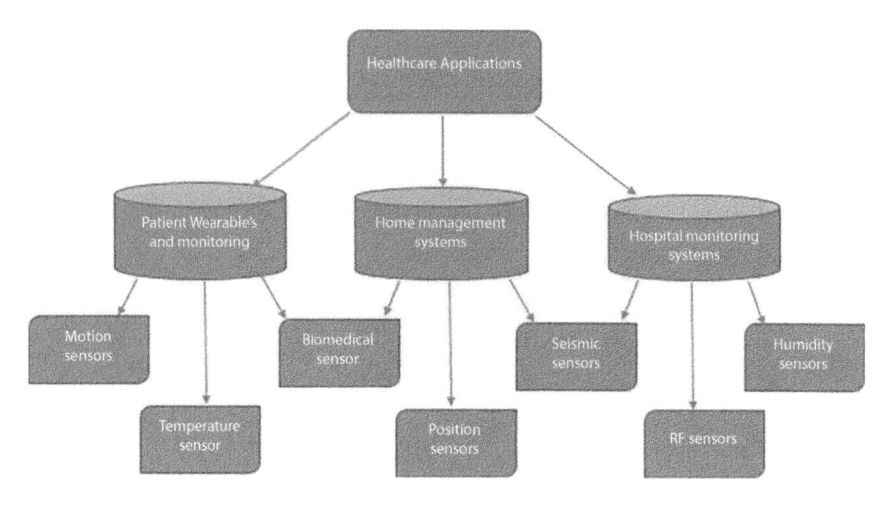

Figure 14.6 Network sensor in healthcare applications.

14.4.2 Smart Healthcare and Medical Establishments

Smart hospital settings are created using networked sensing systems, which have been fitted with connected gadgets and sensors that monitor different facets of the facility's operations. The sensors in question can track healthcare machinery position, monitor factors related to the environment (such as temperature and humidity), and minimize the usage of resources. Network detectors improve the lives of patients by combining information acquired from several sources.

14.4.3 Fall Detection and Old Care

Networking sensors may identify falls and track old people's activities to guarantee their security and well-being. Cameras installed around a living space can detect variations in patterns of motion, departures from normal schedules, and periods of immobility or discomfort. Using predictive analytics enables earlier intervention, proactive treatment of illness, and better clinical results for consumers [14].

Overall, networked sensors play an important role in revolutionizing the delivery of healthcare by allowing for remote surveillance, improving operations, increasing patient involvement, and promoting based on data decision making. As technology improves, the use of network sensor systems in health is projected to grow bringing in the era of tailored, efficient, accessible medical treatment.

14.5 Natural Disaster Detection with Response

Emergency detection and reaction are essential uses of distributed detecting systems, which use modern technology to improve warning systems, organize emergency responses, and reduce the effect of disaster on impacted communities. Here, outline a few important characteristics of disaster warning and reaction utilizing internet sensor systems.

14.5.1 Early Detection Systems

Network sensors that help identify and warn of emergencies including tsunamis, quakes, storms, floods, and fires. Seismic instruments, tidal indicators, climate stations, and satellite imagery are all part of sophisticated warning systems that monitor the surrounding environment to detect signs of oncoming calamities. Instantaneous information from these devices enables authorities to send out prompt notifications, remove high-risk locations, and turn on responding services.

14.5.2 Satellite Imagery and Tracking

Network monitoring devices use remote sensing technology, including satellite imagery, airborne unmanned aircraft, and sensors on the ground, to monitor changes in the environment and estimate the level of damage from natural catastrophes. Information from remote sensing can give significant information about the impacted areas, such as the intensity of the catastrophe, infrastructure destruction, demographic relocation, and its environmental impact [15]. This information is critical for determining response priorities, allocating supplies, and synchronizing services for emergencies.

14.5.3 Resilient Communications Networks

In natural catastrophes, conventional communication systems may be damaged or overwhelmed limiting rescue and recovery efforts. Networking sensing systems use robust communication methods, including satellite communication, wireless networks, and *ad hoc* systems, to guarantee that first, those who responded, disaster management organizations, and influence populations can communicate continuously. These lines of communication make it easier to coordinate, share information, and allocate resources in constantly changing and demanding circumstances.

14.5.4 Predictive Analysis and Modeling

Network sensors use statistical analysis and mathematical modeling to evaluate natural disaster risk and mimic probable scenarios informing disaster planning and response activities. Predictive models, which analyze past information, meteorological forecasts, and geography-related data, can forecast the probability and effect of future crises allowing leaders to develop forward-thinking measures for mitigation, solidify buildings resilience, and improve solidarity resilience. Figure 14.7 shows the real-time disaster detection system with response.

Community Engagement and Citizen Science
Network sensing systems involve communities and people in natural catastrophe identification and responses via citizen science projects and participative sensing systems [16]. Gathered data from cellphones, online communities, and community-based sensors complement established monitoring systems by giving useful information on localized risks, paths to evacuation, and community demands. Citizen participation improves awareness of conditions, builds endurance, and enables regions to successfully participate in emergency response activities.

Integrated Decision Support Systems
Network sensing systems combine data gathered from a variety of sources, such as sensors, spacecraft, social media, and geographic information systems (GIS), to provide complete decision-support tools for disaster management [17]. These technologies give managers with real-time

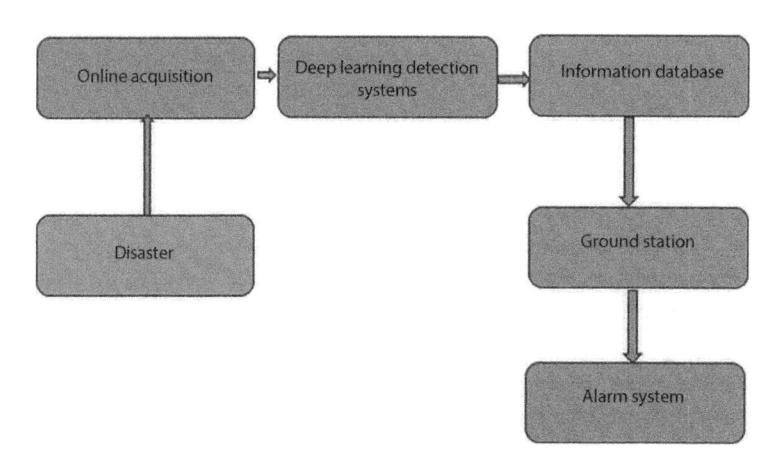

Figure 14.7 Real-time disaster detection systems with the response.

awareness of situations, automated analytics, and useful insights allowing them to make more informed decisions, allocate resources, and coordinate responses across many different organizations and jurisdiction.

As a whole, emergency detection and response using network sensing systems is an integrative strategy that combines technologies for sensors, information analysis, communication channels, and community involvement to improve resilience, reduce risks, and save lives during natural disasters. As technology advances, networking detectors will become increasingly important in enhancing the effectiveness as well as effectiveness of disaster recovery tactics throughout the entire world.

14.6 Agricultural Sensing Systems

Agriculture sensor technology is transforming contemporary farming by giving farmers important information about crop health, soil health, climate trends, and the use of resources. These networks use detectors, data analysis, and electronic communication to improve methods of farming, boost production, and encourage organic farming [18]. The following are some of the important features associated with farming detection arrangements:

14.6.1 Crop Monitoring and Management

Agricultural sensor systems use a variety of sensors, including optical, heat, and multispectral detectors, to track crop growth and health during the growing season. Drones, spacecraft, and sensors on the ground use satellite imagery to collect high-resolution images and data on plant wellness indicators including amount of chlorophyll, the leaf surface index, and water-related stress. The information provided helps landowners to detect nutrient deficits, problems with insects, and outbreaks of illnesses early on, which allows for customized remedies and improved agricultural methods.

14.6.2 Soil Sensing and Precision Agriculture

Soil-detecting devices, such as moisture detectors, pH gauges, and nutrient detectors, give farmers immediate data on the state of the soil and fertility levels. Farmers may use a sensor to accurately monitor moisture in the soil levels, nutrient amounts, and pH levels allowing for more precise application of nutrients and soil amendment tactics [19]. Precision farming strategies, such as changing the rate of distribution and site-specific executives, improve utilization of resources, reduce the effect on the environment, and increase the productivity of crops.

14.6.3 Weather Monitoring and Forecasting

Farm sensor systems use climate sensors and weather information for tracking and anticipating events that may affect crop development and expansion. Weather monitors assess humidity, wind velocity, temperature, and rainfall enabling farmers to make informed choices about cultivation, harvesting, and planning irrigation. Accurate weather forecasting improves risk control, reduces rain-related losses, and increases average agricultural output [20].

14.6.4 Livestock Monitoring and Management

Along with crop tracking, agro sensor technology is utilized to track cattle well-being and conduct in livestock farming systems. Multifunctional gauges, tags with RFID, and tracking systems with GPS allow farmers to follow particular livestock, check their health, and identify disease or distress. Poultry tracking tools allow early detection of health issues, enhance eating and breeding techniques, and increase overall welfare of horses and production.

14.6.5 Data Analytics and Decision Support System

Agro sensor technology creates massive amounts of information, necessitating cutting-edge analytics and tools for decision support to extract useful insights. The use of machine learning, forecasting, and geographic analysis are some of the data analytics approaches used to assess information from sensors, detect trends, and develop decisions that use data. Based on immediate sensor info and prior trends, systems that assist with decision making make suggestions to the farmer about harvesting methods, input maximizing efficiency, and prevention measures.

14.6.6 Remote Monitoring and Automation

Farm sensor systems offer remote control and monitoring of farming activities enabling producers to manage their crops and animals from anywhere. Telemetry systems give growers actual time access to data from sensors allowing them to react swiftly to shifting circumstances and make timely modifications to their agricultural techniques [21]. Automated technology, especially robotic irrigation systems and autonomous gatherers, help to increase efficiency, save labor costs, and boost agricultural production.

Overall, farming sensor systems are an innovative technology that is altering the farming industry by allowing producers to make better decisions, maximize utilization of resources, and boost income while encouraging conservation and responsibility for the environment. As technology develops, agriculture detection systems' uses and abilities are likely to rise resulting in higher levels of creativity and acceptance in the sector of agriculture. Figure 14.8 shows the flow diagram of agriculture embedded with sensor nodes.

14.7 Intelligent Transportation Systems

ITS, or intelligent transportation systems, offer a new era in how networks of transportation are managed and tuned, relying on networked sensors to improve security, effectiveness, and ecology. By combining modern sensory technology, data mining, and communication systems, ITS allows for immediate tracking, evaluation, and management of traffic patterns, automotive movements, and infrastructure for transportation [22–25]. Systems like these collect information on congestion, road incidents, and surroundings using a range of sensors such as recording devices, radar detectors, and GPS navigators. By transforming and evaluating these data in instantaneous fashion, ITS provides important knowledge to travel officials providing preventive administration of congestion in the roadways, handling of emergencies, and route improvement.

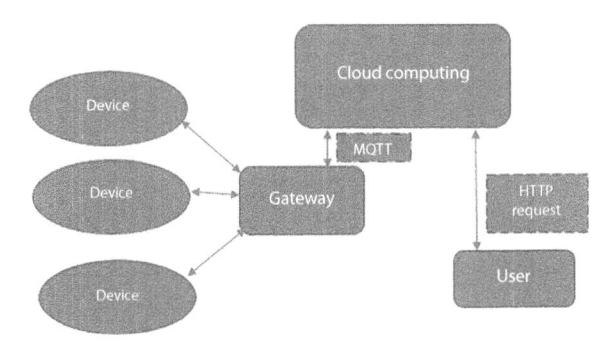

Figure 14.8 Sensor networks in agriculture.

368 Networked Sensing Systems

Furthermore, ITS facilitates the introduction of linked and self-driving automobiles allowing for vehicle-to-vehicle (V2V) and vehicle-to-infrastructure (V2I) interaction to enhance security, decrease collisions, and increase mobility. As towns expand and commute difficulties worsen, ITS provides a comprehensive approach to managing transport, utilizing network sensors to build more intelligent, more effective, and environmentally friendly transport networks for the years to come [26].

14.8 Smart City Applications

Applications for smart cities use network sensors to make urban settings more effective, viable, and comfortable. The ideas include a variety of programs with the goal of enhancing transportation, public security, energy conservation, and people' quality of living [27–29]. Network detectors collect data in real time from monitors installed across the city as a whole, such as devices, sensors for the outdoors, traffic sensors, and meter reading devices. All of this information is then evaluated using modern analytics methodologies to provide useful insight and guide local officials' decisions. Figure 14.9 shows the smart city applications.

Smart city uses include automated transportation systems (ITS) to improve the flow of traffic and decrease congestion, energy-efficient grids for tracking and controlling consumption of energy, environmental monitoring devices for monitoring the quality of water and air, and public safety systems to improve disaster mitigation and reduce criminal activity. Furthermore, smart city programs frequently include citizen involvement platforms as well as open data efforts to provide citizens with a wealth of information and foster cooperation among government agencies,

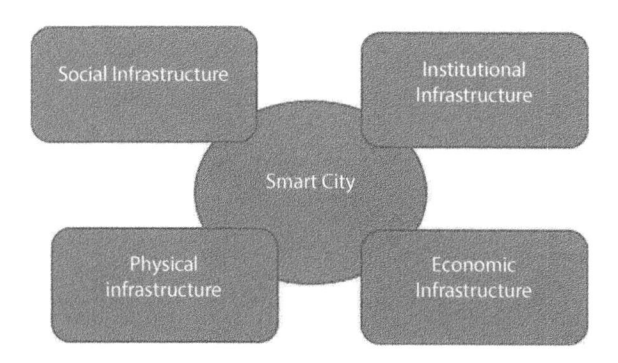

Figure 14.9 Smart city applications.

corporations, and community-based groups [30, 31]. Applications for intelligent cities use networked sensing technologies to build better-connected, ecologically sound, and resilient urban settings, therefore improving our standard of life for all citizens.

14.9 Challenges

Multiple challenges and use cases emerge in network sensing systems representing the intricate nature and variety of uses across multiple areas.
Energy Efficiency: Many sensors used in networking detection systems are powered by batteries or operate in distant areas with limited access to power. Improving efficiency of energy and reducing electrical usage are significant problems for extending sensor life and lowering maintenance costs.

Networking sensors acquire private information about humans, facilities, and surroundings, which raises security and privacy problems. Security of data is critical for maintaining trust and adherence to regulations.
Environmental Monitoring: Network detectors are used to track aspects of the environment including water quality, air quality, and pollutants. Sensors installed in metropolitan areas, industrial locations, and ecological systems collect real-time data to help improve environmental policy and management choices.
Healthcare Monitoring: Network detectors allow for remote monitoring of patients, telemedicine, and individualized delivery of care. Wearable sensors, surgical instruments, and applications for smartphones gather and interpret health information to help with treatment, diagnosis, and prevention.
Smart Farming: Agricultural monitoring devices use instruments and IoT technology to track the health of crops, soil health, and weather trends. By improving fertilization, watering, and weed control procedures, these systems can increase crop yields, minimize the use of resources, and promote ethical farming [37].
Transportation Treatment: Intelligent transportation networks use network sensors to maximize the flow of traffic, control congestion, and improve road safety. Traffic surveillance cameras, sensors, and vehicle connectivity allow for immediate time tracking and administration of transport systems.
Disaster Handling: The handling of disasters relies heavily on networking sensing systems, which provide alerting, surveillance, and reaction capabilities. Earthquake sensors, meteorological stations, and technologies for

remote sensing help authorities in planning for and responding to natural catastrophes like hurricanes, typhoons, and flames.

Smart City Initiatives: These apps use network sensing technologies to enhance infrastructure in cities, transit, security for the public, and energy savings [32–36]. Smart city efforts use sensors, data analysis, and networks of communication to build closer to one another, environmentally friendly, and resilient urban settings. Overall, the issues and use applications associated with network sensors highlight the importance of resolving technological, functional, and social factors to fully utilize sensor technology in a variety of contexts.

14.10 Conclusion

Finally, the field of distributed sensors is undergoing significant advances and finding extensive uses in a variety of fields, including healthcare, farming, transportation, and environmental surveillance [37]. Such systems use sensors, analytics, and communication methods to collect, analyze, and comprehend immediate information from a variety of sources of information allowing for more informed decisions, proactive management, and more effective operation. Despite problems, such as managing data, compatibility, flexibility, and safety, network sensors keep developing fueled by continued technological advances and growing demand for insight based on data. Looking ahead, the continuing integration of sophisticated sensor innovations, predictive algorithms, and capabilities for edge computing is predicted to broaden the capabilities and effect of internet sensing systems resulting in enhanced effectiveness, sustainability, and resiliency across sectors. As network sensors get more common, their ability to address complicated issues and change the way to track, control, and communicate with the world within us will grow exponentially influencing the development of interrelated systems and allowing a smarter, better connected, and viable world.

References

1. Bayoudh, K., Knani, R., Hamdaoui, F., Mtibaa, A., A survey on deep multimodal learning for computer vision: advances, trends, applications, and datasets. *Visual Comput.*, 38, 8, 2939–2970, 2022.

2. Cui, Y., Liu, F., Jing, X., Mu, J., Integrating sensing and communications for ubiquitous IoT: Applications, trends, and challenges. *IEEE Netw.*, 35, 5, 158–167, 2021.
3. Senbekov, M., Saliev, T., Bukeyeva, Z., Almabayeva, A., Zhanaliyeva, M., Aitenova, N., Fakhradiyev, I., The recent progress and applications of digital technologies in healthcare: a review. *Int. J. Telemed. Appl.*, 2020.
4. Hussien, A.G., Amin, M., Wang, M., Liang, G., Alsanad, A., Gumaei, A., Chen, H., Crow search algorithm: theory, recent advances, and applications. *IEEE Access*, 8, 173548–173565, 2020.
5. Ashwini, A., Purushothaman, K.E., Gnanaprakash, V., Shahila, D.F.D., Vaishnavi, T., Rosi, A., Transmission Binary Mapping Algorithm with Deep Learning for Underwater Scene Restoration, in: *2023 International Conference on Circuit Power and Computing Technologies (ICCPCT)*, IEEE, pp. 1545–1549, 2023.
6. Ashwini, A,.S., Quadruple spherical tank systems with automatic level control applications using fuzzy deep neural sliding mode FOPID controller. *J. Eng. Res.*, 2023, Preprint.
7. Górriz, J.M., Ramírez, J., Ortíz, A., Martinez-Murcia, F.J., Segovia, F., Suckling, J., Ferrandez, J.M., Artificial intelligence within the interplay between natural and artificial computation: Advances in data science, trends and applications. *Neurocomputing*, 410, 237–270, 2020.
8. Ferdiana, R., A systematic literature review of intrusion detection system for network security: Research trends, datasets and methods, in: *2020 4th International Conference on Informatics and Computational Sciences (ICICoS)*, IEEE, pp. 1–6, 2020.
9. Liu, S., Chen, P.Y., Kailkhura, B., Zhang, G., Hero III, A.O., Varshney, P.K., A primer on zeroth-order optimization in signal processing and machine learning: Principals, recent advances, and applications. *IEEE Signal Process Mag.*, 37, 5, 43–54, 2020.
10. Ashwini, A., Purushothaman, K.E., Rosi, A., Vaishnavi, T., Artificial Intelligence based real-time automatic detection and classification of skin lesion in dermoscopic samples using DenseNet-169 architecture. *J. Intell. Fuzzy Syst.*, 45, 4, 6943–6958, 2023, Preprint.
11. Ju, Y., Tian, X., Liu, H., Ma, L., Fault detection of networked dynamical systems: A survey of trends and techniques. *Int. J. Syst. Sci.*, 52, 16, 3390–3409, 2021.
12. Gibert, D., Mateu, C., Planes, J., The rise of machine learning for detection and classification of malware: Research developments, trends and challenges. *J. Netw. Comput. Appl.*, 153, 102526, 2020.
13. Ahmad, Z., Shahid Khan, A., Wai Shiang, C., Abdullah, J., Ahmad, F., Network intrusion detection system: A systematic study of machine learning and deep learning approaches. *Trans. Emerg. Telecommun. Technol.*, 32, 1, e4150, 2021.

14. Ashwini, A. and Sangeetha, S., IoT-Based Smart Sensors: The Key to Early Warning Systems and Rapid Response in Natural Disasters, in: *Predicting Natural Disasters With AI and Machine Learning*, IGI Global, pp. 202–223, 2024.
15. Ibrahim, M.S., Dong, W., Yang, Q., Machine learning driven smart electric power systems: Current trends and new perspectives. *Appl. Energy*, 272, 115–237, 2020.
16. Ashwini, A., Sriram, S.R., Manisha, A., Prabhakar, J.M., Artificial Intelligence's Impact on Thrust Manufacturing With Innovations and Advancements in Aerospace, in: *Industry Applications of Thrust Manufacturing: Convergence with Real-Time Data and AI*, IGI Global, pp. 197–220, 2024.
17. Altug, H., Oh, S.H., Maier, S.A., Homola, J., Advances and applications of nanophotonic biosensors. *Nat. Nanotechnol.*, 17, 1, 5–16, 2022.
18. Ashwini, A. and Kavitha, V., Automatic Skin Tumor Detection Using Online Tiger Claw Region Based Segmentation–A Novel Comparative Technique. *IETE J. Res.*, 69, 1–9, 2021.
19. Balaji, S., Jeevanandham, S., Choudhry, M.D., Sundarrajan, M., Dhanaraj, R.K., Data Aggregation through Hybrid Optimal Probability in Wireless Sensor Networks. *EAI Endorsed Trans. On Scalable Inf. Syst.*, 11, 2024.
20. Mahmood, A., Beltramelli, L., Abedin, S.F., Zeb, S., Mowla, N., II, Hassan, S.A., Gidlund, M., Industrial IoT in 5G-and-beyond networks: Vision, architecture, and design trends. *IEEE Trans. Ind. Inf.*, 18, 6, 4122–4137, 2021.
21. De Alwis, C., Kalla, A., Pham, Q.V., Kumar, P., Dev, K., Hwang, W.J., Liyanage, M., Survey on 6G frontiers: Trends, applications, requirements, technologies and future research. *IEEE Open J. Commun. Soc.*, 2, 836–886, 2021.
22. Ashwini, A., Vaishnavi, T., Rosi, A., Shahila, D.F.D., Nalini, N., Deep Learning Based Drowsiness Detection With Alert System Using Raspberry Pi Pico, in: *2023 International Conference on Data Science, Agents & Artificial Intelligence (ICDSAAI)*, IEEE, pp. 1–8, 2023.
23. Ge, X., Han, Q.L., Ding, L., Wang, Y.L., Zhang, X.M., Dynamic event-triggered distributed coordination control and its applications: A survey of trends and techniques. *IEEE Trans. Syst. Man Cybern.: Syst.*, 50, 9, 3112–3125, 2020.
24. Sathyamoorthy, M., Vanitha, C.N., Dhanaraj, R.K., Mon, F.A., Rai, P., Smart Aero Generator Monitoring System using IoT. *2023 Second International Conference on Augmented Intelligence and Sustainable Systems (ICAISS)*, Trichy, India, pp. 1528–1533, 2023, 10.1109/ICAISS58487.2023.10250545.
25. Ashwini, A. and Murugan, S., Automatic Skin Tumour Segmentation Using Prioritized Patch Based Region–A Novel Comparative Technique. *IETE J. Res.*, 66, 1–12, 2020.
26. Oubbati, O.S., Atiquzzaman, M., Ahanger, T.A., Ibrahim, A., Softwarization of UAV networks: A survey of applications and future trends. *IEEE Access*, 8, 98073–98125, 2020.

27. Ferlin, A.A. and Rosi, V., Iot based object perception algorithm for urban scrutiny system in digital city, in: *2023 International Conference on Circuit Power and Computing Technologies (ICCPCT)*, pp. 1788–1792, 2023.
28. Ghosh, A., Edwards, D.J., Hosseini, M.R., Patterns and trends in Internet of Things (IoT) research: future applications in the construction industry. *Eng. Constr. Archit. Manage.*, 8, 2, 457–481, 2021.
29. Ashwini, A., Purushothaman, K.E., Prathaban, B.P., Jenath, M., Prasanna, Automatic Traffic Sign Board Detection from Camera Images Using Deep learning and Binarization Search Algorithm, in: *2023 International Conference in recent advances in Electrical, Electronics, Ubiquitous Communication and Computational Intelligence (RAEEUCCI)*, IEEE, 2023.
30. Ahad, A., Tahir, M., Aman Sheikh, M., Ahmed, K., II, Mughees, A., Numani, A., Technologies trend towards 5G network for smart health-care using IoT: A review. *Sensors*, 20, 14, 4047, 2020.
31. Sehito, N. *et al.*, Optimizing User Association, Power Control and Beamforming for 6G Multi-IRS Multi-UAV NOMA Communications in Smart Cities. *IEEE Trans. Consum. Electron.*, 10.1109/TCE.2024.3388596.
32. Ashwini, A., Sriram, S.R., Sheela, J.J.J., Detection of chronic lymphocytic leukemia using Deep Neural Eagle Perch Fuzzy Segmentation–A novel comparative approach. *Biomed. Signal Process. Control*, 90, 105905, 2024.
33. Shafique, K., Khawaja, B.A., Sabir, F., Qazi, S., Mustaqim, M., Internet of things (IoT) for next-generation smart systems: A review of current challenges, future trends and prospects for emerging 5G-IoT scenarios. *IEEE Access*, 8, 23022–23040, 2020.
34. Gill, S.S., Xu, M., Ottaviani, C., Patros, P., Bahsoon, R., Shaghaghi, A., Uhlig, S., AI for next generation computing: Emerging trends and future directions. *Internet Things*, 19, 100514, 2022.
35. Zhou, S.K., Greenspan, H., Davatzikos, C., Duncan, J.S., Van Ginneken, B., Madabhushi, A., Summers, R.M., A review of deep learning in medical imaging: Imaging traits, technology trends, case studies with progress highlights, and future promises. *Proc. IEEE*, 109, 5, 820–838, 2021.
36. Kirimtat, A., Krejcar, O., Kertesz, A., Tasgetiren, M.F., Future trends and current state of smart city concepts: A survey. *IEEE Access*, 8, 86448–86467, 2020.
37. Balasubramaniam, S. and Kavitha, V., A survey on data encryption tecniques in cloud computing. *Asian J. Inf. Technol.*, 13, 9, 494–505, 2014.

About the Editors

Rajesh Kumar Dhanaraj, PhD is a distinguished professor at Symbiosis International University in Pune, India. He has authored and edited over 50 books on various cutting-edge technologies and holds 21 patents. Furthermore, he has contributed over 100 articles and papers to esteemed refereed journals and international conferences, as well as chapters for several influential books and has shared his insights with the academic community by delivering numerous tech talks on disruptive technologies. He is a senior member of the Institute of Electrical and Electronics Engineers and a member of the Computer Science Teacher Association and the International Association of Engineers.

Malathy Sathyamoorthy, PhD is an assistant professor in the Department of Information Technology, KPR Institute of Engineering and Technology, Tamil Nadu, India. She is a life member of the Indian Society for Technical Education and International Association of Engineers. She has published a number of works, including more than 25 research papers in SCI, Scopus, and ESCI indexed journals, 22 papers in international conferences, two patents, one book, and eight book chapters.

Balasubramaniam S., PhD is a post-doctoral researcher in the Department of Applied Data Science, Noroff University College, Kristiansand, Norway with over 15 years of experience in teaching, research, and industry. He has published over 20 research papers in reputed SCI and Scopus journals, contributed chapters to internationally published books, and has been granted one Australian patent, one Indian patent, and published three Indian patents. He has presented papers at conferences and organized a number of conferences, symposiums, and seminars.

Seifedine Kadry, PhD is a professor of data science at Noroff University College, Norway. He is also an Accreditation Board for Engineering and Technology Program Evaluator for computing engineering technology. He serves as a senior member of the Institute of Electrical and Electronics Engineers. His current research interests include data science, education using technology, system prognostics, stochastic systems, and probability and reliability analysis.

Index

Action space, 131
Activity recognition technology, 116
Adaptability, 353
Agricultural industry, 251
Agricultural sensing systems, 365
AH2 photoionization detector, 208
AI-controlled systems, 7
Alpha sense B4 series CO sensor, 208
Ant colony optimization, 123
API-based integration, 12
Application programming interface, 211
Applications in environmental monitoring, 113
Architecture, engineering, and construction, 306
Artificial bee colony optimization, 123
Artificial intellect, 310
Artificial intelligence (AI), 9, 63, 73, 83, 84, 90–2, 145, 147, 155, 171, 354
Artificial intelligence algorithms
   Ada boost, 158
   clustering, 159
   multi-class SVM, 157
   Naïve Bayes, 157
Artificial neural networks, 345
Asset management, 285–287
Asymmetric encryption, 22
Augmented reality, 41, 56
Automated adjustments, 22
Automation, 12, 13, 16, 18, 19, 20, 21, 23, 65, 66, 68, 77–79, 85, 87

Bandwidth, 137
Basic human-centric sensing mechanism, 106–110
Battery management system (BMS), 323–349
Battery recycling, 347
Big data analytics, 4, 13
Bio-communication, 80, 83
Biometric authentication, 9
Blockchain, 63, 73, 304
Building automation systems, 19

Calibration, 209, 216
Carbon emissions, 346, 348, 349
Catalytic sensors, 207
Cave automatic virtual environment systems, 303
Charging stations, 346, 347
Chronic disease management, 233
Cloud computing, 122, 200, 202, 203, 205, 335–338, 358
Cloud providers, 122
Cloudsim simulator, 136
Clustered topology, 183
Colocation, 209
Communication technologies, 63, 69, 70, 72, 75, 80, 83, 97, 99
Computational intelligence, 307
Computing power, 128
Connected appliances, 20–21
Consensus, 277–278, 288–289
Containerization and orchestration, 167

377

Copyright management, 287–288
Cryptographic security, 88
Cutting edge technologies, 35
Cyber-physical systems (CPS), 14, 65–66, 70
Cybersecurity measures, 7–9, 14–15, 16, 19

Data analysis, 358
Data centres, 121, 133
Data connectivity, 5
Data integrity, 87
Data privacy and security, 1, 6, 7
Data sharing, 76, 87, 89
Data utilization, 4
Databases, 122
Data-driven decision making, 2, 6
Dead node ratio, 187
Decentralization, 87
Decryption algorithm, 9
Deep Q-learning, 125
Delay, 189
Demand-side management (DSM), 19
Digitalization, 65–68, 75, 89
Digitization, 249
Directed acyclic graph, 125
Disaster detection, 363
Distributed Systems, 14
DNS load balancing, 150
DQN, 121
Dynamic traffic control, 145

Edge computing, 63, 72–74, 76, 81, 83–87, 90, 96, 99, 167, 355
Electric vehicles, 324–348
Electrification, 65
Electronic digital records, 237
Elliptic curve cryptography, 9
Encryption and decryption, 8, 10
Energy conscious scheduling algorithm, 126
Energy consumption, 128, 132, 189
Energy efficiency, 2, 17, 18, 20, 21, 122, 345, 346

Energy storage, 323–349
Ethical considerations, 4, 8, 26
Existing work, 177
Experience memory, 131

Financial trading, 164
First come first served, 123
Frequency regulation, 347

Genetic algorithm, 92, 121, 135
Gesture recognition technology, 116
Globalization, 65, 68, 77–79, 85, 87
Green networking technologies, 93, 94–95
Greenhouse systems, 258
Grey wolf optimization, 123

Hanwei MQ-7 CO sensor, 208
Head-mounted displays, 303
Healthcare 1.0, 226
Healthcare, 276, 283–284
Heterogeneous, 122
Heuristic approaches, 123
Homogeneous, 122
Homomorphic encryption, 9
Human-centered society, 70
Human-centric design and empowerment, 4
Human-centric smart city 5.0, 299
Human–machine collaboration, 66
Hunter society, 69
Hyperspectral sensors, 111
Hypertext transfer protocol (HTTP), 202

Identity management, 281–283
IEEE 802.15.4 standards, 211
Immutable ledger, 87, 88
Improved worker safety, 22
Industrialization, 64, 67
Infrastructure as a service, 122
Innovation ecosystem, 71
Integrity, 231, 275, 277
Intelligent reflecting surfaces (IRS), 82

Index

Intelligent society, 69
Intelligent transportation system, 367
Inter planetary file system, 284, 286
Internet of Objects, 77
Internet of Things (IoT), 3, 6, 13, 63, 73, 74, 76–80, 81, 83, 84, 97, 146, 201, 217, 223, 323–326, 335–346, 274, 281
Internet-based learning, 305
Interoperability, 2, 10–12, 229, 263

Job schedule, 124
Jobs, 137

Key management, 9–10

Left skewed distribution, 136
LiDAR, 111
Lithium-ion batteries, 324–333
Livestock management, 268
Load balancing, 145–170

Machine learning, 227, 335–345
Makespan, 121
Manufacturing, 273, 275, 290
Massive MIMO and beamforming, 82
Mechanization, 65
Mesh topology, 182
mHealth applications, 231
Microalgae cultivation, 23
Micro-ElectroMechanical (MEMS) technology, 202
Micro-irrigation, 255
Multi-factor authentication (MFA), 9
Multispectral sensors, 110

Network, 63, 96, 97, 99
Network life-time, 189
Network optimization, 63, 90
Network slicing, 80
Network topology, 209–210
Network virtualization, 94
Networking, 122
Nitrogen monitoring, 253

Non-fungible tokens, 280, 287
Normal, 136
NP-hard problem, 125

Ocean energy, 23
Optimization algorithms, 328–345
OS, 137

Particle swarm optimization, 123
Passive sensing technology, 115
Personalization and customization, 4
Photogrammetry sensors, 112
Piezoelectric energy harvesting, 23
Pollutant, 203–208, 217, 220
Power grid, 337, 347
Precision farming, 248
Predictive, 6, 22, 25
Predictive maintenance, 22
Priority based scheduling, 138
Privacy regulations (GDPR, CCPA), 6, 22
Probabilistic clustering
    EECS (energy efficient clustering protocols, 152
    HEED, 151
    low-energy adaptative clustering hierarchy, 151
    UCR (unequal cluster based routing), 152
Proof of stake, 278, 288
Proof of work, 277, 288
Public participation, 6

Q-learning, 124
Q-networks, 133
Q-table, 124, 133
Quality of service, 125
Quantum communication, 83
Quantum figuring, 168
Queue, 122
Q-values, 131

RAM, 137
Real-time monitoring, 18, 19

Reduced operational costs, 16
Redundancy, 1, 14
Regulatory requirements, 234
Reinforcement learning, 91
Reinforcement learning algorithm, 123
Reliability, 275, 277
Remote patient monitoring, 361
Renewable energy, 64, 71, 86, 93, 323–349
Renewable energy sources, 1, 18, 22, 25
Resource allocation, 145–170
Resource optimization, 6, 16
Rewards, 130
RFID, 252
Right skewed distribution, 136
Robotics, 257
Round Robin, 123
RSA (Rivest-Shamir-Adleman), 9

Sampling, 205–206
Scheduler, 127
Security, 276–277
Security measures, 240
SEM system, 217, 219
Sensor network, 174
Sensors, 352
Servers, 121, 122
Service level agreement, 122
Shortest job next, 123
Smart, 63, 64–68, 70, 72, 73, 76, 77, 79–81, 85–89, 91, 94, 96, 98, 99
Smart cities, 275, 278
Smart classroom, 193
Smart contract, 88, 274, 280, 286, 287
Smart devices, 5, 20
Smart energy management, 5, 16
Smart grids, 18–19, 86, 91, 191
Smart health care, 194
Smart manufacturing, 65, 66, 79
Smart parking, 190
Smart sensor, 114
Society 5.0, 1–30, 63, 64, 67–76, 84–99

Software defined networking (SDN), 93, 94, 96
Software defined networks (SDN), 146
Spatial resolution, 207, 215
Sprayers, 259
Star topology, 181
State space, 132
Storage, 122
Sub tasks, 136
Super intelligent society, 69
Supply chain management, 89
Sustainability, 260
Sustainable, 63–70, 72–73, 76, 77, 80, 93, 95, 97–99
Sustainable development goals (SDGs), 345–349
Sustainable environment, 249
Swarm intelligence, 92
Symmetric encryption, 9–10

Task scheduling, 122
Terahertz communication, 82
Thermal management, 343, 344
Thermal sensors, 111
Throughput, 188
Tokenization, 88
Tokenomics, 285, 286
Topology, 175
Topology in smart city, 176
Tournament selection, 128
Traffic optimization, 93, 94
Traffic pattern, 185
Transparency, 1–2
Tree topology, 183
Types of advanced HCS environmental monitoring system, 110

Uniform, 136
Urbanization, 64

Virtual machines, 121, 127
Virtual reality, 41, 56, 302
VMM, 137

Volatile organic compounds (VOCs), 206, 208
Voltage regulation, 347

Wearable technology, 230
Wireless communication, 235

Wireless network technology, 114
Wireless Sensor Networks (WSN), 146, 209, 218, 221–223
Workflow, 121

Zigbee, 209, 211, 212, 222, 250

Also of Interest

Check out these related titles from Scrivener Publishing

Multimodal Data Fusion for Bioinformatics Artificial Intelligence, Edited by Umesh Kumar Lilhore, Abhishek Kumar, Narayan Vyas, Sarita Simaiya, and Vishal Dutt, ISBN: 9781394269938. *Multimodal Data Fusion for Bioinformatics Artificial Intelligence* is a must-have for anyone interested in the intersection of AI and bioinformatics, as it not only delves into innovative data fusion methods and their applications in omics research but also addresses the ethical implications and future developments shaping the field today.

Artificial Intelligence-Based System Models in Healthcare, Edited by A. Jose Anand, K. Kalaiselvi and Jyotir Moy Chatterjee, ISBN: 9781394242498. Artificial Intelligence-Based System Models in Healthcare provides a comprehensive and insightful guide to the transformative applications of AI in the healthcare system.

EXPLAINABLE MACHINE LEARNING MODELS AND ARCHITECTURES: Real-Time System Implementation, Edited by Suman Lata Tripathi and Mufti Mahmud, ISBN: 9781394185849. This cutting-edge new volume covers the hardware architecture implementation, the software implementation approach, and the efficient hardware of machine learning applications.

Natural Language Processing for Software Engineering, Edited by Rajesh Kumar Chakrawarti, Ranjana Sikarwar, Sanjaya Kumar Sarangi, Samson Arun Raj Albert Raj, Shweta Gupta, Krishnan Sakthidasan Sankaran, and Romil Rawat, ISBN: 9781394272433. Discover how Natural Language Processing for Software Engineering can transform your understanding of agile development, equipping you with essential tools and insights to enhance software quality and responsiveness in todays rapidly changing technological landscape.

MACHINE LEARNING TECHNIQUES FOR VLSI CHIP DESIGN, Edited by Abhishek Kumar, Suman Lata Tripathi, and K. Srinivasa Rao, ISBN: 9781119910398. This cutting-edge new volume covers the hardware architecture implementation, the software implementation approach, and the efficient hardware of machine learning applications with FPGA or CMOS circuits, and many other aspects and applications of machine learning techniques for VLSI chip design.

Intelligent Green Technologies for Smart Cities, Edited by Suman Lata Tripathi, Souvik Ganguli, Abhishek Kumar, and Tengiz Magradze, ISBN: 9781119816065. Presenting the concepts and fundamentals of smart cities and developing "green" technologies, this volume, written and edited by a global team of experts, also goes into the practical applications that can be utilized across multiple disciplines and industries, for both the engineer and the student.

Hybrid Intelligent Optimization Approaches for Smart Energy: A Practical Approach, Edited by Senthilkumar Mohan, A. John, Sanjeevikumar Padmanaban, and Yasir Hamid, ISBN: 9781119821243. Written and edited by a group of experts in the field, this is a comprehensive and up-to-date description of current energy optimization techniques, such as artificial intelligence techniques, machine learning, deep learning, and IoT techniques and their future trends.

NANODEVICES FOR INTEGRATED CIRCUIT DESIGN, Edited by Suman Lata Tripathi, Abhishek Kumar, K. Srinivasa Rao, and Prasantha R. Mudimela, ISBN: 9781394185788. Written and edited by a team of experts in the field, this important new volume broadly covers the design of nano-devices and their integrated applications in digital and analog integrated circuits (IC) design.

DESIGN AND DEVELOPMENT OF EFFICIENT ENERGY SYSTEMS, Edited by Suman Lata Tripathi, Dushyant Kumar Singh, Sanjeevikumar Padmanaban, and P. Raja, ISBN: 9781119761631. Covering the concepts and fundamentals of efficient energy systems, this volume, written and edited by a global team of experts, also goes into the practical applications that can be utilized across multiple industries, for both the engineer and the student.

Electrical and Electronic Devices, Circuits, and Materials: Technical Challenges and Solutions, Edited by Suman Lata Tripathi, Parvej Ahmad Alvi, and Umashankar Subramaniam, ISBN: 9781119750369. Covering every aspect of the design and improvement needed for solid-state electronic devices and circuit and their reliability issues, this new volume also includes overall system design for all kinds of analog and digital applications and developments in power systems.

MODELING AND OPTIMIZATION OF OPTICAL COMMUNICATION NETWORKS, Edited by Chandra Singh, Rathishchandra R Gatti, K.V.S.S.S.S. Sairam, and Ashish Singh, ISBN: 9781119839200. *Modeling and Optimization of Optical Communication Networks* is a comprehensive and authoritative book that delves into the optical networks, principles, technologies, and practical applications of optical networks, equipping readers with the knowledge needed to design, implement, and optimize optical networks for various applications, from telecommunications to scientific research. With its comprehensive coverage and up-to-date insights, this book serves as an essential reference in the field of optical networks.

WIRELESS COMMUNICATION SECURITY: Mobile and Network Security Protocols, Edited by Manju Khari, Manisha Bharti, and M. Niranjanamurthy, ISBN: 9781119777144. Presenting the concepts and advances of wireless communication security, this volume, written and edited by a global team of experts, also goes into the practical applications for the engineer, student, and other industry professionals.

ADVANCES IN DATA SCIENCE AND ANALYTICS, Edited by M. Niranjanamurthy, Hemant Kumar Gianey, and Amir H. Gandomi, ISBN: 9781119791881. Presenting the concepts and advances of data science and analytics, this volume, written and edited by a global team of experts, also goes into the practical applications that can be utilized across multiple disciplines and industries, for both the engineer and the student, focusing on machining learning, big data, business intelligence, and analytics.

ARTIFICIAL INTELLIGENCE AND DATA MINING IN SECURITY FRAMEWORKS, Edited by Neeraj Bhargava, Ritu Bhargava, Pramod Singh Rathore, and Rashmi Agrawal, ISBN: 9781119760405. Written and edited by a team of experts in the field, this outstanding new volume offers solutions to the problems of security, outlining the concepts behind allowing computers to learn from experience and understand the world in terms of a hierarchy of concepts.

MACHINE LEARNING AND DATA SCIENCE: Fundamentals and Applications, Edited by Prateek Agrawal, Charu Gupta, Anand Sharma, Vishu Madaan, and Nisheeth Joshi, ISBN: 9781119775614. Written and edited by a team of experts in the field, this collection of papers reflects the most up-to-date and comprehensive current state of machine learning and data science for industry, government, and academia.

MEDICAL IMAGING, Edited by H. S. Sanjay, and M. Niranjanamurthy, ISBN: 9781119785392. Written and edited by a team of experts in the field, this is the most comprehensive and up-to-date study of and reference for the practical applications of medical imaging for engineers, scientists, students, and medical professionals.

SECURITY ISSUES AND PRIVACY CONCERNS IN INDUSTRY 4.0 APPLICATIONS, Edited by Shibin David, R. S. Anand, V. Jeyakrishnan, and M. Niranjanamurthy, ISBN: 9781119775621. Written and edited by a team of international experts, this is the most comprehensive and up-to-date coverage of the security and privacy issues surrounding Industry 4.0 applications, a must-have for any library.

CYBER SECURITY AND DIGITAL FORENSICS: Challenges and Future Trends, Edited by Mangesh M. Ghonge, Sabyasachi Pramanik, Ramchandra Mangrulkar, and Dac-Nhuong Le, ISBN: 9781119795636. Written and edited by a team of world renowned experts in the field, this groundbreaking new volume covers key technical topics and gives readers a comprehensive understanding of the latest research findings in cyber security and digital forensics.

DEEP LEARNING APPROACHES TO CLOUD SECURITY, Edited by Pramod Singh Rathore, Vishal Dutt, Rashmi Agrawal, Satya Murthy Sasubilli, and Srinivasa Rao Swarna, ISBN: 9781119760528. Covering one of the most important subjects to our society today, this editorial team delves into solutions taken from evolving deep learning approaches, solutions allow computers to learn from experience and understand the world in terms of a hierarchy of concepts.

MACHINE LEARNING TECHNIQUES AND ANALYTICS FOR CLOUD SECURITY, Edited by Rajdeep Chakraborty, Anupam Ghosh and Jyotsna Kumar Mandal, ISBN: 9781119762256. This book covers new methods, surveys, case studies, and policy with almost all machine learning techniques and analytics for cloud security solutions.

SECURITY DESIGNS FOR THE CLOUD, IOT AND SOCIAL NETWORKING, Edited by Dac-Nhuong Le, Chintin Bhatt and Mani Madhukar, ISBN: 9781119592266. The book provides cutting-edge research that delivers insights into the tools, opportunities, novel strategies, techniques, and challenges for handling security issues in cloud computing, Internet of Things and social networking.

DESIGN AND ANALYSIS OF SECURITY PROTOCOLS FOR COMMUNICATION, Edited by Dinesh Goyal, S. Balamurugan, Sheng-Lung Peng and O.P. Verma, ISBN: 9781119555643. The book combines analysis and comparison of various security protocols such as HTTP, SMTP, RTP, RTCP, FTP, UDP for mobile or multimedia streaming security protocol.

SMART GRIDS FOR SMART CITIES VOLUME 2: Real-Time Applications in Smart Cities, Edited by O.V. Gnana Swathika, K. Karthikeyan, and Sanjeevikumar Padmanaban, ISBN: 9781394215874. Written and edited by a team of experts in the field, this second volume in a two-volume set focuses on an interdisciplinary perspective on the financial, environmental, and other benefits of smart grid technologies and solutions for smart cities.

SMART GRIDS AND INTERNET OF THINGS, Edited by Sanjeevikumar Padmanaban, Jens Bo Holm-Nielsen, Rajesh Kumar Dhanaraj, Malathy Sathyamoorthy, and Balamurugan Balusamy, ISBN: 9781119812449. Written and edited by a team of international professionals, this groundbreaking new volume covers the latest technologies in automation, tracking, energy distribution and consumption of Internet of Things (IoT) devices with smart grids.